CRITICAL TECHNOLOGIES AND COMPETITIVENESS

CRITICAL TECHNOLOGIES AND COMPETITIVENESS

D. MALANA (EDITOR)

Nova Science Publishers, Inc.
Huntington, New York

Editorial Production: Susan Boriotti
Office Manager: Annette Hellinger
Graphics: Frank Grucci and Jennifer Lucas
Information Editor: Tatiana Shohov
Book Production: Patrick Davin, Cathy DeGregory, Donna Dennis, Jennifer Kuenzig, Christine Mathosian, Tammy Sauter and Lynette Van Helden
Circulation: John Bakewell, Lisa DeGangi and Michael Pazy Mino

Library of Congress Cataloging-in-Publication Data
available upon request

227 Main Street, Suite 100
Huntington, New York 11743
Tele. 631-424-6682 Fax 631-424-4666
e-mail: Novascience@earthlink.net
e-mail: Novascil@aol.com
Web Site: http://www.nexusworld.com/nova

Printed in the United States of America

TABLE OF CONTENTS

PREFACE

This new book examines some of the crucial issues and technologies, which are apt to play a significant role in the first part of the 21st century. Technology is charging forward at an unprecedented rate shaking the ground we stand on just as we become comfortable. This volume presents articles on technologies, which are likely to help revolutionize society. It also looks competitiveness and its role in serving as the engine of economic progress.

D. Malana

FLAT PANEL DISPLAY TECHNOLOGY: WHAT IS THE FEDERAL ROLE?

Glenn J. McLoughlin and Richard M. Nunno

INTRODUCTION

Flat Panel Display technologies likely will help revolutionize the way information is transmitted, received and used. What often is referred to as FPD technology is really an array of technologies currently involving a variety of public and private, United States and foreign, research and development (R&D). Regardless of the type of R&D, FPD technology generally represents the application of very thin screens displaying alphanumeric information and images generated by computers and other electronic devices. FPD technology already is having a significant use in U.S. military weapon systems and its impact on commercial applications--especially in computer products--may be enormous.

The importance of FPD technology rests upon several considerations. It is likely that the office, home and school of the very near future will use combinations of information and telecommunications technologies to meet a variety of needs. High-powered yet lightweight FPD technologies may play an important part in these new applications and uses of information and telecommunications goods and services. But while current display technology (cathode ray tube, or CRT) is reliable, has excellent color and contrast and is reasonably inexpensive to produce, it also is big, bulky, and consumes a great deal of electrical power compared to FPD technology.[1] Therefore, greater demand for display technologies which can provide clear resolution for either miniaturized or large screen images, for both military and commercial users, may push commercial FPD developments. However, high-volume, high-quality manufacturing capability is considered a prerequisite for ensuring technology development and commercialization.

The current public policy debate regarding FPD technology revolves around a report released by the Department of Defense (DOD) on dual-use technology.[2] This report includes an appendix on FPD technologies and provides the DOD strategy--called by some the FPD Initiative--for meeting its future needs for FPD technology. DOD's core argument is that continued Federal support for FPD technology is needed because of national security. FPD technology is important for modern military weapons, and was prominent in the United States effort in

---

[1] Savitz, Eric J. Dangerous Display: Flat Panel Floundering Holds Risks for U.S. Industry. Barron's, June 28, 1993. p. 8-9.

[2] U.S. Dept. of Defense. Industrial Capabilities for Defense. Washington: Office of the Assistant Secretary of Defense for Economic Security, Sept. 1994. [pages unnumbered]

the 1991 Persian Gulf war. U.S. manufacturers of FPD products, either commercially or for the military, represent a very small percentage of the global production capacity--currently dominated by the Japanese. The largest supplier of FPD technology, the Japanese company Sharp, has given some indication that it would prefer to sell to commercial buyers and not to the U.S. military. DOD argues that early, assured, and affordable access to domestic FPD technology sources is necessary in the face of heavy reliance on foreign suppliers for most FPD technology.

While there is likely to be support for the DOD position for the FPD Initiative, policymakers may look beyond the core argument and address other issues. These questions may include:

- What are the details of the FPD Initiative, and is it a continuation of current Federal policy or does it represent a substantial change in Federal funding policy and program direction?

- What is the U.S. Display Consortium, and what role does it play in private and public sector support of FPD technology?

- What are other policy issues and options which may be addressed within the context of the FPD debate?

- What are the national and economic security implications for the United States if the FPD Initiative is--or is not--adopted by Federal policymakers?

- Do other recent Federal technology programs and initiatives offer lessons for Federal policymakers?

Central to all of these questions is the Federal role (particularly that of the Department of Defense) in supporting FPD technology and whether the DOD role should go beyond providing direct support for current military needs of FPD technology to broader support of a commercial FPD industry. Congress will likely be asked to strike a balance between two national policy goals. The first goal is to address the concern that the United States be militarily prepared going into the 21st Century. Supporters of the FPD Initiative contend that FPD technology has, and will, play an important role of this readiness. The second goal is the need to extend DOD support of FPD technology, which may be challenged by policymakers committed to reducing the size of the Federal deficit. They would do this by cutting those programs where either the domestic market or global trade could provide answers in place of the Government. Some may seek to modify the FPD Initiative, or place it within the context of a reduced DOD role in technology development. Some observers argue that failure to achieve a balance between these goals may create unwanted policy outcomes difficult to rectify in the future: the inability for the U.S. military to gain access to FPD technology when it is needed, or concerns that the role and scope of the Federal Government have not been properly curtailed.

FLAT PANEL DISPLAY (FPD) TECHNOLOGY, MARKET, AND TRADE

FPD Technologies[3]

Generally, FPD technology, regardless of its type or application, is a thin screen upon which are displayed alphanumeric information and images generated by computers and other electronic devices. There are many types of FPD technologies which are being researched and developed worldwide. The predominant FPD technology today is the liquid crystal display (LCD) technology, which has been used in small electronic devices (e.g., calculators) for over 20 years. LCD technology incorporates liquid materials (complex chemical compounds called polymers) whose crystals become re-oriented in response to an applied voltage, producing light or dark images by changing the polarization of the light impinging onto the display screen. A matrix of electronic cells called *pixels* is arranged on the screen and controlled to produce an image, whose sharpness, or resolution, depends on the pixel density.[4] LCD technology operates as an on/off switch by letting light through the liquid material when in the "off" state, and making the viewing surface dark when in the "on" state. By allowing varying amounts of light through the display, different shades of gray can be created.

There are two forms of LCD technology--passive and active matrix. Passive matrix LCD (PMLCD) technology, the most common form of FPD application, applies horizontal and vertical electrodes that combine voltages at intersection points to turn on pixels at those points. Since the electric signal must travel through the entire PMLCD each time it reaches individual pixels, each pixel receives the voltage for only a small fraction of the signal.[5] This results in a decreased contrast ratio (the ratio of brightness between on and off states), a narrower viewing angle, and decreased resolution of the overall display.[6] PMLCD technology primarily has applications in laptop computers, as well as watches and calculators. While PMLCD products are easy to fabricate, the

---

[3] Sources include: U.S. Dept. of Defense. Building U.S. Capabilities in Flat Panel Displays. Washington, Sept. 1994. 108 p.; Debb, Stephen and Webster Howard. Flat Panel Displays. Scientific American, Mar. 1993. p. 90-97; Werner, Kenneth. The Flat Panel's Future. IEEE Spectrum, Nov. 1993. p. 18-26; William Omara. New York, NY: Van Nostrand Rheinhold, 1993; and Flat Panel Displays. Stanford, CT. International Resource Development, Inc., Aug. 1986. 204 p.

[4] This is measured by the number of pixels per unit area on the display screen.

[5] The typical signal period is 1/60 second, and a typical voltage is applied to each pixel 1/240 of the period.

[6] Another problem with PMLCD is "crosstalk," or the varying intensities of pixels that are adjacent to pixels in the opposite state--pixels in the "on" state next to those in the "off" state. This also results in a decreased clarity of image.

circuitry needed to address each pixel is complex, which slows down the screen's response time, a particular disadvantage for video programs. The slowness of response time makes PMLCD technology impractical for screens larger than 10 inches in diagonal, a distinct disadvantage for military use.

Active Matrix LCD (AMLCD) technology may actually be one of two technologies. The first type of AMLCD technology uses diodes made from a three layer structure of metal-insulator-metal (like that now used in manufacturing semiconductor chips). A second AMLCD technology uses thin film transistors applied at each pixel. The result of applying thin film transistors at each pixel is a much higher image resolution than from PMLCD technology, therefore resulting in higher resolution in commercial products.

Another advantage of AMLCD, regardless of either technology used, is a much wider viewing angle than PMLCD technology (about a 160-degree viewing range compared to about a 30-degree viewing range for PMLCD technology). The user can tilt the display panel at a very wide angle and still get a clear picture. A PMLCD device affords its clearest picture only when viewed at a 30-degree angle. Thin film transistor AMLCD technology is more prevalent than the metal-insulator-metal structure, but both types are making progress, and commercial applications are already available. However, AMLCD products are still very expensive, and it would take several years for a company to recover the capital costs of entering the AMLCD manufacturing business. In addition to cost, high-volume manufacturing of AMLCD products also raises issues of the reliability of pixels. No U.S. company is now producing AMLCD devices in high volume and only one now produces any 8- to 10-inch diagonal size AMLCDs. The leading AMLCD producers are Japanese firms, some of whom have demonstrated displays larger than 10 inches, but have yet to produce screens larger than 10.5 inches in any volume. For commercial applications, particularly for large-screen, high resolution television systems, the inability to manufacture large display screens has inhibited the further development of AMLCD technology. Researchers are studying new processes to construct thin film transistors which make up the pixels by varying the temperatures and materials used in the process.

Still, fabricating an AMLCD product requires equipment, materials, and techniques similar to that used for manufacturing integrated circuits. Therefore, engineers have drawn from the large amount of knowledge accumulated from decades of manufacturing integrated circuits and are applying this to AMLCD production. Some industry leaders have expressed hope that the current revival of the U.S. semiconductor industry may lead to infrastructure development that might also support some of the technologies needed for AMLCD manufacturing.

Another class of display technology is called emissive, because it generates its own light, unlike LCD technology, which can only reflect or absorb light from an outside source (its on/off switch function). Emissive displays can generally consume less power, and achieve much higher brightness, improved contrast, and a wider viewing angle compared to LCD technology. The electroluminescence display (ELD) technology is an emissive technology, in which a light-generating phosphor material is placed between electrode plates. Images are generated using an "addressing scheme" similar to LCD technology.[7]

ELD has been used mostly in niche commercial markets, such as financial and ATM banking machines, industrial and scientific test equipment, and in transportation and medical applications. They are used in military applications primarily where ruggedness and wide viewing angle are important.

The military has used ELD technology in eyeglass mounted screens (or goggles) called helmet-mounted displays. Currently, this has many military applications for both ground and airborne troops. There is, however, much work still to be done in commercializing ELD technology, particularly in producing phosphors with greater brightness and range of color, as well as improving the techniques used to activate the phosphor.

Another important technology is plasma panel technology, which can be made larger (up to 60 inches diagonally) than other types of FPD technology and potentially can be manufactured at a lower cost. Plasma displays are made of two glass face-plates and a thin phosphor material coating on one of the plates. Light is generated when an inert gas is discharged between the face-plates causing the phosphor to glow. Different phosphors in the matrix of cells produce different colors in each pixel. Plasma displays are produced in relatively low volumes for lower information content applications, such as stock exchange trading tote boards, transportation applications, and certain military applications on surface ships and submarines.

Field emission display (FED) technology is a solid state vacuum display that can project electrons onto a glass screen.[8] Research on FED technology is still in its early stages. Recent prototypes use a cold emission of electrons from an array of thousands of small semiconductor cones called microtips that serve as cathodes (the electrodes that emit the electron beam). Experiments are being conducted to develop different techniques and processes. For example, one area targeted is the improvement of microtip uniformity across a pixel. Once developed, commercial FED technology is anticipated to be brighter and sharper than LCD technology, and consume 1/3 to 1/2 the power.

Another class of display technology, projection display, performs as the name suggests--it projects the image from a distance. One military application is the "head up display," now used in aircraft cockpits to display information onto the windshield so that pilots do not have to look down to see it. Another projection display technology, called digital micro-mirror devices (DMD), is made from an array of tiny moveable mirrors (16-square microns), each corresponding

---

[7] Each pixel has an "address" or coordinate in which information can be directed to specific pixels in varying combinations.

[8] This is a similar technology to CRT, the most common form of video projection and used in most televisions.

to a pixel of a display. Each mirror can be controlled electronically to move independently up to 100,000 times per second and reflect pulses of light to produce moving color images. The DMD projects an image onto a screen, which may make it best suited for large-screen displays. Since this technology appears to work best in large screen format, its military application would best be suited for a war room where troop, aircraft and other movements could be observed.

Two other FPD technologies are light emitting diodes (LED) and vacuum florescent displays, both used in large-sized displays where relatively simple information is shown (although theses technologies were originally used for small applications). These technologies have been used for many years in billboards and large road signs. Due to their large features and the difficulty with which they are integrated into defense systems, they will probably not be used for advanced information applications. Other FPD technologies--such as electrophoresis, electrochromics, and micro-curl--bear mentioning primarily because R&D has been undertaken in these fields by FPD companies. However, to date, the scientific and engineering communities have mostly lost interest in these technologies beyond existing R&D.

Beyond the immediate future, many scientists believe that future generation FPD technology will integrate peripheral microelectronic circuitry, so that much of the computer system will be incorporated within the glass panel of the FPD technology. Should that occur, if U.S. computer systems manufacturers have little expertise in manufacturing FPD technology, they may not be able to compete globally with companies which have a more thorough and intimate understanding of FPD technology.

FPD Market Growth and the Global Competition

Estimates of the potential size of the total--commercial and military--FPD global market forecast rapid growth over the next decade. The Stanford Resources Inc. Study of the Flat Panel Display Market, 1993-1994, often cited by both industry and Government experts, forecasts rapid growth.[9] It estimates that, in 1993, the total worldwide market for FPD technologies was $4.6 billion. By 2000, Stanford Resources, Inc. expects the global market to grow to $14.2 billion.[10] (See figure 1.) Among the technologies most likely to

---

[9] The U.S. display consortium, the U.S. industry consortium, and the DOD Dual-Use Office use the SRI numbers in their estimates of future FPD markets. The SRI data seem consistent with summaries of other, proprietary reports done on the potential FPD market size.

[10] In 1993 dollars, not adjusted for inflation.

FIGURE 1.

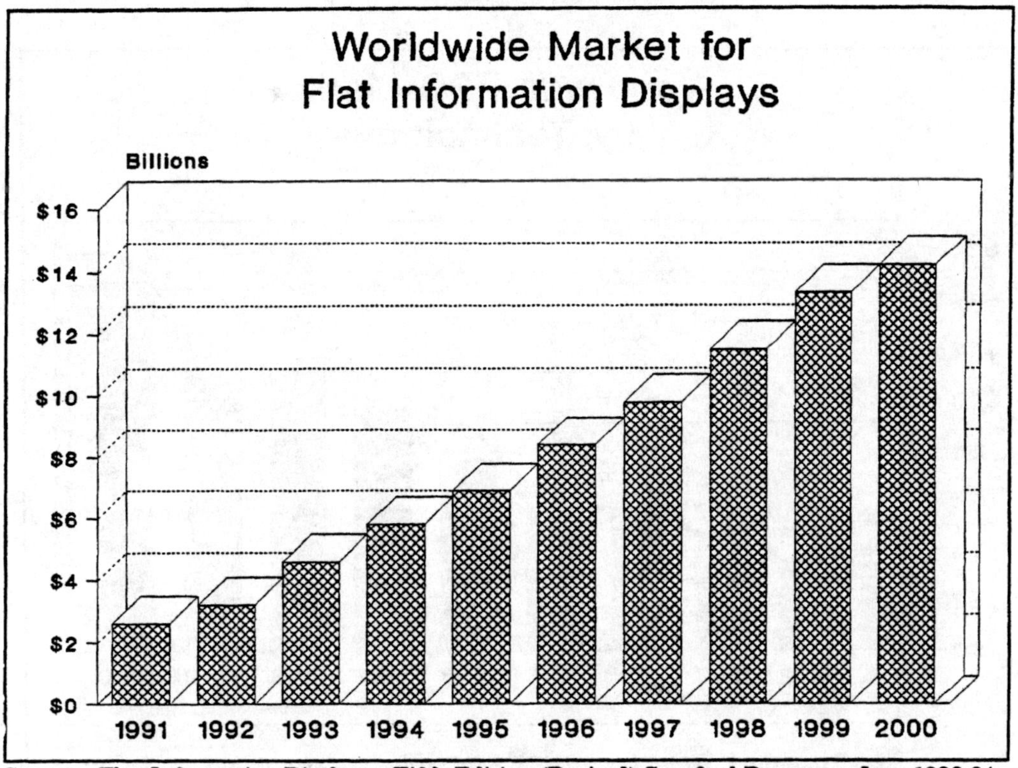

Source: Flat Information Displays: Fifth Edition (Revised) Stanford Resources, Inc., 1993-94.

continue to pay off in application (and described further below), the most prominent is liquid crystal display (LCD) technology, with electroluminescence, plasma and other technologies remaining much smaller. (See figure 2.)

FIGURE 2.

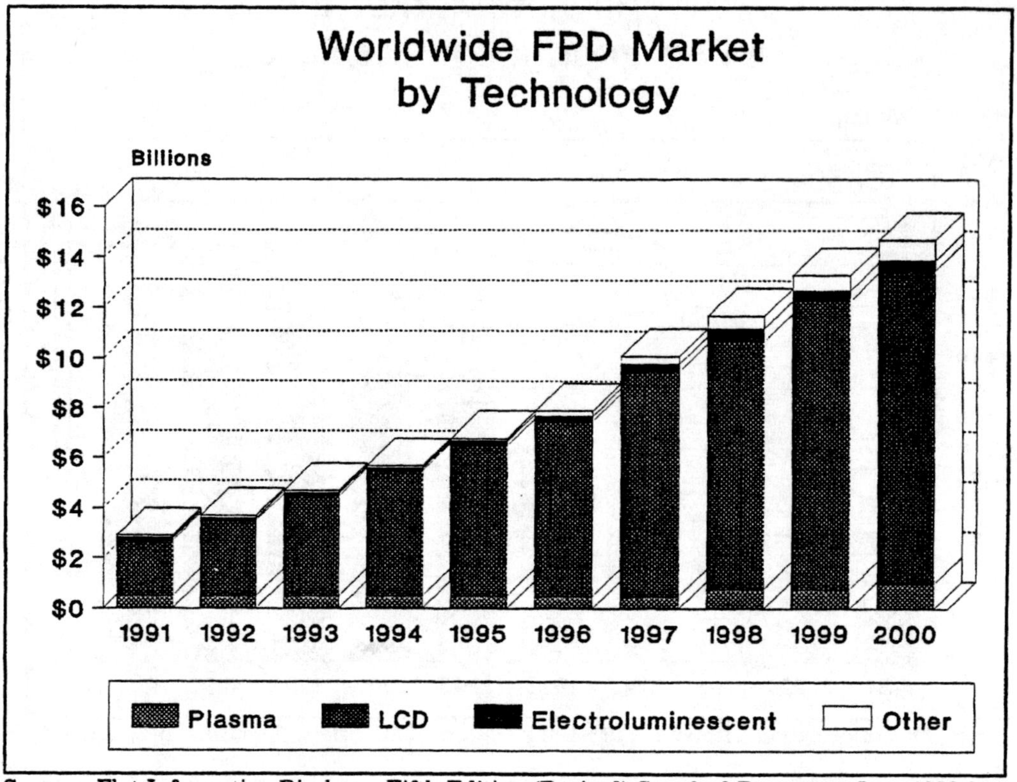

Source: Flat Information Displays: Fifth Edition (Revised) Stanford Resources, Inc., 1993-94.

Computer applications--use of FPD technology in laptop, personal and other computers--also are expected to grow substantially, with consumer use growing at a rapid rate. (See figure 3.)

FIGURE 3.

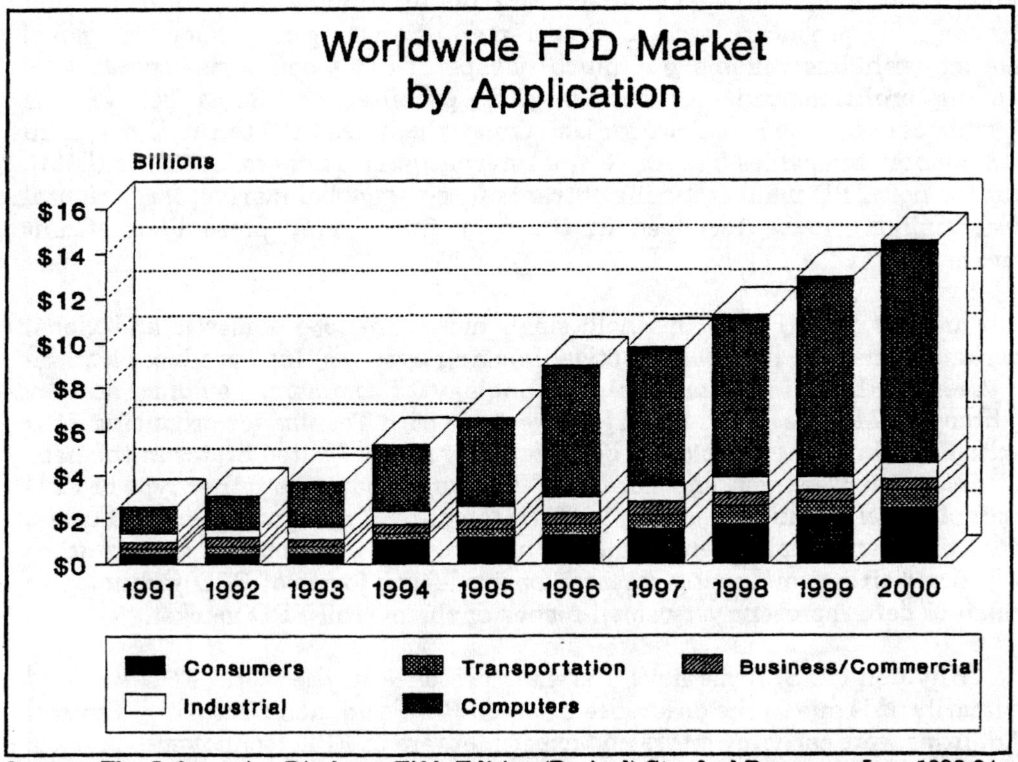

Source: Flat Information Displays: Fifth Edition (Revised) Stanford Resources, Inc., 1993-94.

In the global market, the primary suppliers of FPD technologies today are Japanese firms. Sharp Corporation is the world's leading producer of FPD technology, primarily both passive and active matrix liquid crystal display technologies (described previously). Sharp has moved from a firm which assembled electronic devices in the 1950s to a semiconductor firm in the 1960s, to a firm which produced electronic calculators with integrated circuits in 1970s. In the 1980s, Sharp diversified into producing a variety of electronic devices and products, an important step for developing FPD technologies and incorporating them into a range of commercial products. By 1994, Sharp held just over 43 percent of the global market for FPD production.[11] In 1991, it became the first company to assemble liquid crystal display technology in the United States, at its plant in Camas, Washington.

Other Japanese firms producing FPD products are NEC (second to Sharp in market share), Toshiba, Hosiden, and Hitachi. Together, Japanese firms control approximately 90 percent of the world FPD production. Almost all of the Japanese FPD production is for commercial use; Japanese military use of FPD technology is negligible when compared to commercial applications.

---

[11] Hess, Pamela. Sharp Will Open Domestic Production Facility for AMLCDs in October. Inside the Pentagon, v. 10, no. 38, Sept. 22, 1994. p. 4.

The Koreans also are entering the FPD market, producing basic FPD products at high volume manufacturing plants. Some are concerned that Korean FPD producers such as Goldstar and Samsung may flood the global market with less reliable but much cheaper FPD technologies, potentially eroding profit margins for current FPD producers.[12] Some believe this development may be beneficial for U.S. firms which use FPD technologies, such as computer companies like Apple and International Business Machines (IBM). But for new FPD manufacturing entrants into the global market, large capital costs unlikely to be recouped in the near future may present significant barriers.[13]

In 1994, U.S. FPD firms hold small niches in both domestic and global markets. One U.S. firm, OIS (Optical Imaging Systems) Inc., produces a state-of-the-art FPD liquid crystal display technology for commercial avionics and the military. IBM has a Japan-based joint venture with Toshiba for producing FPD technology, but does not plan to build capacity in the United States in the near future. Planar Systems, another U.S. FPD firm, produces another type of FPD technology, electroluminescence, for U.S. computer firms and other FPD users. Motorola, Texas Instruments, Micron, and InFocus Systems are among other U.S. firms either conducting research or producing forms of FPD technologies which to date represent very small niches of the overall FPD market.

Why don't U.S. firms have a greater presence in the world FPD market? Primarily, it is due to the difference between R&D and manufacturing. Several U.S. firms were early inventors and experimenters in FPD technologies and are superb at developing new FPD technologies. However, the U.S. commercial manufacturing base for FPD products is not as developed. A survey of U.S. firms which either closed or sold FPD production facilities prior to 1990 found several common reasons why these firms were no longer in the industry: the belief that advanced displays were not central to the firm's business strategy; the cost of capital for establishing an FPD manufacturing line; the fear that Japanese competition is too strong to overcome; and the belief that the global economy allows purchase of FPD technologies from any source, domestic or foreign.[14]

According to the Department of Defense's estimations, the DOD share of the total U.S. market is small. Reportedly, DOD purchases less than one percent of all units sold in the United States, comprising less than three percent

---

[12] Crystal Diplomacy. The Economist, May 14, 1994. p. 70.

[13] Capital costs for an FPD fabrication plant are similar to that of a semiconductor fabrication line; approximately $400-600 million.

[14] As cited in: Hart, Jefferey. The Anti-Dumping Petition of the Advanced Display Manufacturers of America: Origins and Consequences. World Economy, v. 16, Jan. 1993. p. 99; and also see: U.S. Congress. Office of Technology Assessment. The Big Picture: HDTV and High-Resolution Systems. Washington, U.S. Govt. Print. Off., June 1990. p. 71.

of dollar sales.[15] Most estimates of the DOD market share by 2000 place it below five percent of sales in the total U.S. market.[16] Even if DOD procurement of FPD technology remains constant at current levels, the commercial market is likely to grow so fast that DOD's part will remain small.

Still, the DOD share of the market, though small, is important. Some contend that DOD's funding and spending programs can provide the necessary nurturing of technology commercialization which may not occur in the global marketplace.[17] Should DOD seek to establish a broad dual-use strategy which would include the goal of increasing the U.S. FPD manufacturers' market share and establishing a technology base which has military but also commercial payoffs? Some contend that more than just military-captive suppliers of FPD technology are needed in a global economy. Suppliers of FPD technology, dependent on the DOD for business, could address the short-term technology needs of the U.S. military. However, DOD contends that without a strategy that addresses the current state of the global market--and a very small U.S. FPD technology base--the cost of FPD technology developed just for military use might be exorbitant and that there would likely be the necessity of continuous reinvestment to avoid obsolescence.[18] A U.S. FPD industry which primarily serves DOD may not have a negative impact on commercial users of FPD products, if the users can get the FPD technologies when they want them at relatively low cost, regardless of origin. But many in the U.S. FPD industry also contend that it is important to expand all aspects of technology commercialization and manufacturing, and include the needs of DOD as a part of this expanded technology base and market. The DOD position on FPD technology and other policy issues are presented later in this report.

---

[15] Samuelson, Robert J. Flat Screens and Subsidies. The Washington Post, May 18, 1994. p. A23.

[16] Carey, John. Why Washington Is Anointing Flat Panels. Business Week, May 16, 1994. p. 36-37.

[17] The role of the Federal Government, and especially DOD, in fostering high technology development is discussed in: Flamm, Kenneth. Targeting the Computer: Government Support and International Competition. Washington: The Brookings Institution, 1987. 266 p.; and Creating the Computer: Government, Industry and High Technology. Washington: The Brookings Institution, 1988. 282 p.

However, these two books examine the Federal role mostly during the late 1950s and early 1960s, when there was little international competition for new commercial technologies, when Federal support could nurture new companies in high-risk, high-payoff fields over years, if not decades, and when the cost of capital was very favorable to new entrants in nascent fields such as semiconductor chip production and computers.

[18] Memorandum from Dr. Richard Van Atta, Special Assistant for Dual-Use Technology Policy, Department of Defense, Jan. 27, 1995. p. 9.

Trade

The Japanese dominance of the FPD world market is closely intertwined with U.S.-Japan trade disputes over the last decade. By the late 1980s, U.S. electronics firms had lost significant market shares (or the markets themselves) in videocassette recorders (VCRs), televisions and other consumer electronics products. The Japanese had similar successes in dominating markets of those technologies which enable or drive consumer electronics products--most notably semiconductor chips, and then FPD technology. By the early 1990s, U.S. industry leaders were claiming that Japanese companies were "dumping" FPD products into the United States to capture market share and drive out U.S. competitors.[19] The U.S. Government responded in 1991 by applying tariffs on foreign FPD technology imported into the United States.[20] However, over a two-year period, U.S. firms which *use* FPD technologies (including International Business Machines (IBM) Inc., Apple Computer Inc., Sun Microsystems, and others) protested that the tariffs made Japanese FPD technologies too expensive to use in U.S. products. In essence, this problem pitted small and barely established U.S. FPD companies against larger firms which were their potential customers. In 1992, one of the U.S. FPD firms which had supported the petition changed its position and joined U.S. computer companies to call for a removal of the tariff.[21] In 1993, the Department of Commerce lifted the tariff.

The imposition of the tariff for the two-year period apparently did little to reduce low-cost Japanese FPD imports into the United States and to help the U.S. FPD industry. The U.S. FPD industry did not grow substantially during that two-year period and U.S. computer manufacturers were faced with higher

---

[19] The claim of "dumping" actually was in the form of a petition alleging predatory pricing, in which Japanese FPD manufacturers were setting prices so far below the average cost of production as to eliminate competitors and to secure a monopolistic or oligopolistic advantage. The Advanced Display Manufacturers of America represented U.S. FPD companies. The FPD technology cited by these firms was the Active Matrix Liquid Crystal Display (AMLCD) technology, more fully explained in the previous section. Hart, Jefferey. The Anti-Dumping Petition of the Advanced Display Manufacturers of America: Origins and Consequences. World Economy, v. 16, Jan. 1993. p. 85.

[20] The largest tariff, as implemented by the Department of Commerce on foreign AMLCD, was 62.67 percent. Japan Economic Institute. Notes-Flat Panel Displays. Washington, JEI, Jan. 29, 1993. p. 8-9.

[21] The company in question, OIS, Inc., reportedly gave several reasons for its change in position: it had won several contracts to manufacture FPD products and compete with Japanese suppliers, it was concerned that the FPD industry position was causing a deep wedge in the U.S. electronics industry, and top management at the firm had been replaced after a corporate buy-out. Japan Economic Institute. Notes-Flat Panel Displays. Washington, JEI, Dec. 18, 1992. p. 9.

prices for FPD components in their products. Japanese FPD manufacturers continued to develop new technologies and gain an overwhelming market share over U.S. FPD manufacturers.

For some, the failure of the 62 percent tariff to reduce imports of Japanese FPD products meant that Federal policymakers needed to take a more direct and activist approach. It seemed apparent to some policymakers that trade policy alone would not address both market needs and industry capabilities in this field. In 1994, the Office of the Assistant Secretary of Defense for Economic Security began preparing a strategy and programs to address DOD's technology needs in the post-Cold War era. As a part of this assessment, DOD examined its FPD technology needs and how a changing industrial base may serve the United States. As the first of several industrial and technology assessments, DOD examined FPD technology within the context of a technology critical to national security needs.[22] Others, however, contend that in a global economy, the availability of FPD technology is primarily a trade issue. Attempting to create a DOD technology strategy should not be an attempt to compensate for what they view as a failed trade policy. Absent a national security need, some experts believe that market forces should continue to completely determine suppliers.

The other significant trade issue affecting U.S. FPD technology development is the General Agreement on Tariffs and Trade (GATT), a multilateral trade agreement promoting open trade among nations. Several nations are currently considering final ratification.[23] The new GATT agreement is called the Uruguay Round (named after the site where the multilateral negotiations began). A major provision of the Uruguay Round Agreement is the Subsidies Code.[24] The Subsidies Code intends to limit any signatories' subsidies targeted towards a specific industry or group of industries within the subsidizing country, which in turn might result in harm to the

---

[22] It was reported that Laura Tyson, chair of the Council of Economic Advisers, suggested FPD technology as a good technology/trade case study for the Clinton Administration. Carey, John. Why Washington Is Anointing Flat Panels. Business Week, May 16, 1994. p. 36. This story is disputed by DOD officials, who contend that the DOD FPD study reflects a long-standing concern regarding the availability and access of FPD technology for military applications. Memorandum from Dr. Richard van Atta, Special Assistant for Dual-Use Technology Policy, Jan. 2, 1995. p. 10.

[23] Legislation approving and implementing the new GATT agreement was signed by President Clinton on Dec. 8, 1994.

[24] The formal title of this provision is The Agreement on Subsidies and Countervailing Measures. U.S. Library of Congress. Congressional Research Service. Research and Development Subsidies and the GATT, by Lennard G. Kruger. CRS Report 94-416. Washington, May 6, 1994. p. 1.

trading interests of other member countries.[25] An important concern are those actions by governments which would assist *applied* industrial research for commercial use. Several U.S. policymakers have stated that the FPD Initiative, as allowed under the GATT Subsidies Code, would essentially create a U.S. Industrial Policy in the form of an FPD production subsidy.[26] Others contend that the Subsidies Code is consistent with existing U.S. Government-industry cooperation, such as the SEMATECH consortium. (See appendix A.)

FEDERAL FPD PROGRAMS

Federal Funding for FPD R&D is carried out or supported by several agencies (table 1). Since FY 1991, the Department of Defense (DOD), mainly

TABLE 1.
U.S. Government R&D Programs in FPD Technology
($ millions)[a]

Agency	FY1991	FY1992	FY1993	FY1994	FY 1995
DOD:	**76.0**	**76.8**	**155.2**	**121.6**	**95.0**
ARPA	74.3	75.0	152.2	84.8	83.0
TRP (ARPA)	0	0	0	25.0	0
Army	0.9	1.0	2.2	1.0	1.2
Navy	0.4	0.4	0.4	0.4	0.4
Air Force	0.4	0.4	0.4	0.4	0.4
other DOD	0	0	0	10.0	10.0
Dept. of Commerce:	**1.8**	**5.3**	**6.0**	**5.2**	**4.3**
NIST[b]	---	0.4	0.4	0.4	0.4
ATP	1.8	4.9	5.6	4.8	5.5
DOE	**0.3**	**1.0**	**3.5**	**6.0**	**7.0**
NASA	**1.3**	**1.3**	**1.3**	**1.3**	**1.3**
NSF	---	**0.3**	**0.4**	**0.4**	**0.4**
TOTAL	**79.4**	**84.7**	**166.4**	**134.5**	**108.0**

[a] Agencies' estimates of their total flat panel display funding
[b] NIST funding in addition to the ATP program.

---

[25] Ibid., p. 1.

[26] Jacobson, Ken. Danforth Brands Flat-Panel Plan "Production Subsidy." New Technology Week, v. 8, no. 19, May 6, 1994. p. 1 and 7.

through its Advanced Research Project Agency (ARPA), has been funding FPD R&D at higher levels than in other agencies. While FPD programs in ARPA and DOD (mainly through the four armed services) focus on DOD's needs, programs at other agencies emphasize commercial needs and the development of the industry's manufacturing capability. The Department of Energy (DOE) FPD program focuses on technologies for long-term commercial market applications. The National Institute for Standards and Technology's (NIST) Advanced Technology Program (ATP) supports competitive industry R&D programs in FPD technology on a cost-shared basis.

The current debate on the Federal role for supporting technology development with industry may affect current and future FPD programs. On February 22, 1995, the House of Representatives approved a total of $502 million in rescissions for DOD's Technology Reinvestment Program (TRP) and $107 million in rescissions for the Department of Commerce's Advanced Technology Program (ATP) for FY 1995. At this time, the Senate is considering these rescissions. Both programs provide substantial support for FPD technology development with industry. Other agencies' FPD programs may be affected by rescissions for FY 1995 as well. In addition, Congress will likely closely scrutinize the Clinton Administration's FY 1996 request for support of FPD technology across agencies, and may consider future reductions in these programs.

DOD's FPD Programs

DOD has established clear requirements for specialized FPD technology. DOD has placed a high priority on securing *early*, *assured* and *affordable* access to leading-edge, customized FPD products at affordable prices, for applications in current and future weapon systems. ARPA's FPD program, called High Definition Systems (HDS), is managed in ARPA's Electronic Systems Technology Office. HDS, part of ARPA's Integrated Command and Control program element, focuses on developing the technology and manufacturing capability for FPDs to be used in all DOD applications involving visual and graphic information. This includes projection equipment, head mounted displays, three-dimensional imagery, display architectures and processors, compression algorithms, and high speed data transmission. These efforts are aimed at demonstrating a manufacturing capability of components necessary for military systems that collect, process, store and display high resolution images. DOD intends to use the technology and skills developed in this program to produce FPD devices in the near term, primarily within the next five years. A limited FPD production capability is already available for DOD's custom use.

Officials at ARPA consider High Definition Systems an ideal example of a "dual use" technology program because it embraces the vision of an integrated commercial and defense industrial base. The underlying reason for its support of High Definition Systems is that the only way DOD can hope to ensure access to state-of-the-art FPD technology in the future is by fostering the development of a commercial FPD industry that can keep up with rapid technological advances overseas. Although ARPA focuses mostly on DOD's FPD military technology requirements, it is also pursuing alternative technologies, which show signs of both military and commercial long-term payoff.[27]

ARPA provides funding for several FPD technology facilities, to study the economics and technology of FPD manufacturing. One project is an AMLCD manufacturing demonstration facility jointly funded with Optical Imaging Systems, Inc., (OIS) focusing on avionic displays for cockpit and tank applications, awarded in August 1993, before the FPD Initiative was announced. This is a five-year, $100 million joint effort, with OIS contributing more than half the funding. In July 1994, ARPA awarded another AMLCD manufacturing test bed to the team of AT&T, Xerox, and Standish, to develop production methods for advanced AMLCD devices. This award is also cost shared 50/50 between Government and the industrial partners. A third facility is the National Center for Advanced Information Components Manufacturing (NCAICM), which supports the development of equipment and materials used to build FPD devices. Finally, ARPA, together with several universities and research centers, has established the Phosphor Technology Center of Excellence, to focus on engineering problems associated with phosphors, a critical material for emissive displays and AMLCD products.

Overall, more than 50 percent of ARPA's FPD funding is cost-shared with industry. Projects that are at the earlier stages of research are funded more heavily by ARPA and less by industry than those that are closer to a financial pay-off to companies. The more a project is funded by industry, the more industry takes the lead in directing and monitoring the project, easing some of the burden from ARPA.

In addition, ARPA's Technology Reinvestment Program (TRP) funds some FPD R&D in specific projects that are cost-shared with industry. While no FY 1993 TRP funding supported FPD technology programs, several FPD-related projects were awarded using FY 1994 funds, totalling $25 million. These projects include three teams of companies developing ELD and FED technologies military applications in cockpits and helmet-mounted displays. However, the rescissions of the TRP for FY 1994 and FY 1995 may dramatically alter current programs. There is planned competition for FPD technology for FY 1996, depending on the DOD R&D budget level, and the future of the TRP.

DOD's Future Years Defense Plan (FYDP) projects funding for FPD R&D over five years to be about $587 million. As shown in table 2, core R&D programs support ongoing product technology, manufacturing processes, industry consortia, and manufacturing test-beds. The R&D Incentives program is comprised of a series of awards for product development and improvement of manufacturing processes R&D to firms demonstrating a commitment to current generation and high volume manufacturing. R&D incentives are contingent on U.S. industry providing at least 50 percent matching funding to ensure that new R&D undertaken is considered commercially viable by industry. Core R&D programs are more heavily supported by the Government. The manufacturing

---

[27] Personal conversation with D. Slobodin, ARPA, Sept. 8, 1994.

test bed was a one-time FY 1994 appropriation to build an AMLCD facility with Xerox/AT&T/Standish (mentioned above). "Purchase incentives" in table 2 are funds to be managed by the Office of the Secretary of Defense (OSD).

TABLE 2.
DOD's FPD Funding Plan
($ millions)

Fiscal Year	1994	1995	1996	1997	1998
Core R&D	46.6	85[a]	68	68	68
Manufacturing Test Bed	50	0	0	0	0
R&D Incentives	25	0	50	50	74
Purchase Incentives	0	10	10	TBD	TBD
DOD Total[b]	121.6	95	128	118	142

[a] Increased by Congress (request was $68 million) for earmarked programs.
[b] Totals subject to final rescissions for FY 1994 and FY 1995, as well as congressional appropriations for FY 1996 and beyond.

Source: Funding data obtained from ARPA Display Technology Program Manager. Sept. 8, 1994, updated Feb. 10, 1995.

DOD also supports FPD R&D through the armed services' programs. These programs support technology development for particular weapons systems, and their components, of interest to the services (such as aircraft cockpits, tanks, and ships). The armed services are making modest investments in retrofitting FPD technology for existing weapons systems--that is, adapting existing and available FPD technology for their specific military needs. However, total armed services R&D in FPD technology is very small compared to ARPA's funding of FPD R&D (table 1). The Army's FPD R&D consists of testing FPD technology for shock resistance and other environmental characteristics at its Communications Electronics Command, part of Army Research Labs, at Ft. Monmouth, New Jersey. The Navy's FPD office is at the Naval Command, Control and Ocean Surveillance Center RDT&E, San Diego, California. Air Force FPD is supported at Griffiss Air Force Base, New York, Wright Paterson Air Force Base, Dayton, Ohio, and other locations. According to DOD, the services expect to continue funding FPD R&D at the approximately the same levels as in previous years.

The primary emphasis of DOD's armed services' efforts is to mainly stay abreast of the technology, evaluate products for use in service platforms, and consult with their program offices on new construction work in order to advise contractors on service-specific requirements. Very little original research is conducted by the services, except when it is funded by ARPA. One possible exception to this is the National Information Display Laboratory at David Sarnoff Research Center, funded by the Air Force, which is conducting basic research on image processing and user support.

In addition to ARPA and the services' funding for FPD technology, in 1994, an additional $40 million was provided to DOD as Defense Production Act funds under Title III of Defense Appropriations. Of these, $30 million were non-RDT&E funds used to ensure FPD production capabilities at the Xerox/AT&T/Standish manufacturing test bed, and $10 million was for R&D activities at that test bed. In FY 1995, another $10 million was appropriated for DOD's FPD purchase incentive program.

Other Federal FPD Programs

DOE's FPD program is part of its Technology Transfer program in the office of Defense Programs. FPD work is conducted at four of DOE's National Labs: Sandia, Los Alamos, Lawrence Livermore, and some in Oak Ridge. DOE's FPD program is pursuing breakthrough technologies and new manufacturing processes that can potentially "leap-frog" the foreign lead. DOE's laboratories utilize their experience in materials, precision engineering, agile manufacturing, plasma physics, computer modeling, and electronics packaging, to develop FPDs using new technologies. DOE's largest FPD program is in electroluminescence displays. Several DOE labs conduct ELD research, such as a joint project between Sandia National Laboratory and Motorola Inc., to investigate the electroluminescent properties of new organic polymers for FPD applications. DOE is also developing field emission displays (FED) in a program run by Lawrence Livermore National Laboratory together with an industry partner, and in another program at Sandia. Los Alamos is conducting research on plasma displays and is developing models to help understand plasmas, as well as new types of phosphors for plasma displays.

Since FY 1992, DOE has funded FPD technology at about $6 million per year distributed among 11 two- to three-year projects. To assure a strong industrial commitment and involvement, industry matches or in some cases exceeds DOE's funding for FPD technology. DOE had planned to increase FPD funding to $10 million for FY 1995, and to $20 million a year over the next several years to move technology out of the lab and into industry. For FY 1995, however, due to a tightening budget for its defense programs, DOE only expects to allocate $7.5 million for FPDs.

One program that crosses boundaries of Federal agencies is the National Center for Advanced Information Components Manufacturing (mentioned above), an ARPA funded, DOE managed, $60 million program at Sandia National Lab. This center develops manufacturing tools, materials, deposition methods, and reliability improvement techniques with applications to all FPD technologies.

The Department of Commerce is interested in FPDs not only as an important commercial product, but also as a contribution to the National

Information Infrastructure (NII). FPD technologies will be used to view data on computers and other electronic devices, an essential part of the NII. NIST's in-house efforts focus on developing tools to measure the characteristics and performance of FPDs, to facilitate testing and selection of competing products. These include measuring the optical performance, obtaining data on subjective viewing experiences using simulations, and developing testing and measurement standards. NIST's Advanced Technology Program (ATP) funded several projects in FPD technology in FY 1993. (See table 3.) For FY 1994, ATP funded four projects in FPD technology development totaling $13 million distributed over a two- to three-year period. Several projects involve FPD development or have basic applications to FPDs. All ATP projects will be cost-shared with industry partners, and some projects may have to compete for funding each year against projects in other technology areas.

TABLE 3.
ATP Projects in FPD Technology
(FY 1993) funds

Project Name	Duration (Years)	1st year request ($1000s)
Holographic Graded-Index Non-Lambertian Scattering Screens/Components with Light-Shaping	2	$430
Large Area Digital HDTV Field Emitter Display	3	$611
Advanced Cathode/Flat Florescent Light Sources	2.25	$243
Optically Controlled Alignment Materials for LCD	3	$651
Patterning Technology for Color FPDs	3	$3,000
Scalable High Density Electronics Based on Multifilm Modules	3	$979
Advanced Manufacturing Technology/Low-Cost FPD	5	$1,251

Source: U.S. Dept. of Commerce. National Institute of Standards and Technology, Advance Technology Program, Project Briefs.

The rescissions passed by the House of Representatives on February 22, 1995, may substantially alter these programs. A total of $107 million in rescissions for the ATP program was included for FY 1995. The Department of Commerce will have some discretion in deciding how the rescissions are applied; however, some, if not all, of the ATP money may already have been spent for FY 1995. For FY 1996, the Clinton Administration has requested a total of $491 million for the ATP. It is unlikely that the 104th Congress will approve this request. How much Congress will eventually appropriate for the ATP for FY 1996, and how that may affect current and proposed FPD projects, remains to be determined.

NASA's FPD technology work, conducted at Ames Research Center, Moffett Field, CA, is developing computer-aided design (CAD) tools for the engineering, performance optimization, and systems integration of FPDs. In this ARPA-funded project, NASA is working closely with DOD to study how FPD technologies would fit into some of the various defense systems being developed, as well as in NASA developed spacecraft. NASA scientists have developed models for how large, AMLCD FPD technologies would operate in aircraft cockpits. The models allow researchers to modify FPD technology parameters to optimize image properties and reduce manufacturing costs. NASA is also installing FPDs in the Space Shuttle, and some have counted the spending on this effort toward NASA's FPD program. This is, however, just an upgrade to an existing system, and does not contribute to the advancement of FPD technology since no new technology is being developed (and also is not included in table 1). To date, NASA has been modestly affected by the congressional rescissions for FY 1995. However, the agency may find its overall R&D budget affected for both FY 1995 and FY 1996 after congressional action is completed. This may also determine its future FPD program.

The National Science Foundation (NSF) funds basic university research in materials and processes that may be used in FPD development and manufacturing. However, several Members of the 104th Congress are seeking to redefine NSF's role in technology development and manufacturing. They may seek to redefine NSF's priorities, which may reduce the agency's involvement in FPD-related programs.

THE FPD INITIATIVE AND NATIONAL SECURITY ISSUES

The FPD Initiative: Building U.S. Capabilities in Flat Panel Displays

In September 1994, DOD released a report entitled *Industrial Capabilities for Defense,* part of a larger series of studies assessing DOD's technology and industrial needs and capabilities. Appendix B, "Building U.S. Capabilities in Flat Panel Displays, A Report of the Flat Panel Display Task Force," outlined the DOD strategy for maintaining FPD technology capability for military applications.[28]

This report (also known as the Flamm Report) recommends the Flat Panel Display (FPD) Initiative. The FPD Initiative is a proposed five-year, $587 million program focusing on existing Federal R&D efforts, primarily those being conducted by DOD's Advanced Research Projects Agency (ARPA). DOD's efforts would also be coordinated with the R&D being performed by other agencies, including the Departments of Energy and Commerce, the National Aeronautics

---

[28] U.S. Dept. of Defense. Industrial Capabilities for Defense. Washington: Office of the Assistant Secretary of Defense for Economic Security, Sept. 1994. Appendix B: chapters I - VIII. 102 p.

and Space Administration (NASA), and the National Science Foundation. Federal funding for the FPD Initiative would terminate at the end of FY 2000.

The keystone of the FPD Initiative is the emphasis on DOD's need for early, assured, and affordable access to FPD technology. Early access to the FPD design and manufacturing process allows defense suppliers to integrate FPD technology into the larger electronic systems of which they are a part. The armed forces must have early access to the most advanced systems to develop the strategy, tactics, and integration with command and control systems. Assured access to FPD technology is important because DOD has been unable to persuade foreign FPD manufacturers (who currently control 97 percent of the military FPD market) to build customized FPDs for military use. DOD has concluded that only by developing a substantial domestic FPD industry can the military meet its procurement needs during times of high demand.[29]

An important reason why DOD access to FPD is a problem is because Sharp Inc., the world's leading producer of LCD technology, reportedly has denied DOD this technology. According to the U.S. military, Sharp company officials have flatly stated in meetings with DOD officials that they will not sell to military users, because it is neither efficient nor profitable for their company[30] However, Sharp officials also have stated that their company already does sell to DOD, through subcontractors.[31] In this respect, Sharp officials are correct, although DOD officials also contend that receiving Sharp's technology from second and third sources may pose national security problems during wartime.[32]

DOD bases the need for the FPD Initiative on the following rationale. Although DOD will be only a small part of the market for FPDs, DOD's critical requirements for FPDs are sufficiently important to warrant the development of a domestic capability. DOD has determined that the traditional policy of supporting a defense-unique, captive industry, would be rendered ineffective since DOD funding could not possibly keep up with innovation in the international commercial industry. Instead, DOD has opted to invest in the creation of dual use capabilities, so that DOD's needs might be met by domestic commercial suppliers. An added advantage of fostering a domestic FPD infrastructure is that a lucrative commercial FPD industry might be generated that can create high paying jobs and enhance the American economy.

---

[29] Ibid., p. I-7.

[30] Interview with Kenneth Flamm, Special Assistant on Dual-Use Technology to the Secretary of Defense on July 6, 1994.

[31] Hess, Pamela. Sharp Will Open Domestic Production Facility for AMLCDs in October. Inside the Pentagon, v. 10, no. 38, Sept. 22, 1994. p. 4.

[32] Interview with Kenneth Flamm on July 6, 1994.

The Flamm Report contains five recommendations for U.S. Government action:

1. Continue to support FPD research and development (R&D), in developing product, process and manufacturing technology. These efforts would focus on developing lower cost displays, increasing display performance and functionality, and integrating additional capabilities into the display unit.

2. Build national expertise in a high volume FPD process technology. In addition to the AMLCD test bed established at OIS, a second display manufacturing test bed would be established (already underway with the AT&T/Xerox/Standish project).

3. Encourage U.S. industry investment in FPD production through focused R&D incentives. Since no domestic industry has made any significant investment in high volume FPD production facilities, the Government can stimulate such investment through competitive R&D awards on next generation products and processes to companies making a commitment to high volume manufacturing of current generation technology. It is contended that this would avoid the U.S. infringement of global trade rules that bar manufacturing subsidies. The total program would support up to four world class, high volume, production facilities. Competitions would be neutral with respect to technology, so that any FPD technology would be eligible for an award as long as the firm were to commit resources for manufacturing and match the Government support for R&D. According to DOD, the requirement that any Government funding be "technology neutral" assures that DOD attains the best capabilities available from a variety of technological approaches to meet a variety of needs.

4. Foster market development. To help establish a U.S. FPD industry, the Federal Government would consolidate its buying program for general purpose displays. This acquisition program would be managed by an executive agent, who would also organize an interagency working group to identify applications and products to improve agency performance and stimulate market demand for new products that the U.S. FPD industry could target. Possible products could include portable electronic blackboards, low energy computers, portable video conferences, high resolution imagery and data display, and other applications utilizing the National Information Infrastructure.

In addition, to stimulating external market development, the report calls for the Department of Commerce to design a program to promote the export of FPDs. The Department of Commerce would also study current tariff policies on products related to FPD technology to alter policies that impede domestic FPD manufacturing. The report recommends that the U.S. Trade Representative should make it a priority to gain access to foreign FPD markets. In addition, the report suggests that DOD should consider negotiating with foreign nations to exchange knowledge of certain U.S. technologies to gain access to knowledge on FPD technology in foreign countries. Finally, the FPD Initiative calls for an interagency team to assess competitive behavior in world FPD markets. The objective of this effort would be to identify trends, such as price discrimination, predatory pricing, denial of products, etc., that could affect the success of the U.S. FPD program.

5. Management and evaluation. The National Economic Council to the President would provide strategic oversight to ensure that the FPD Initiative is conducted coherently across Federal agencies, and would provide an annual review to evaluate progress toward stated goals. NEC review of the FPD Initiative may be problematic if its perspective is biased. Since the NEC and the Federal agencies implementing the FPD Initiative all report to the President, there may be a lack of objectivity needed to review the program.

The FPD Initiative would encourage the establishment of four high-volume FPD production facilities. DOD's rationale for supporting these sites is that, according to DOD's estimate, it would require at least four production-scale facilities to manufacture the multiple types of displays using several different technologies to meet DOD's needs.[33] These needs include displays for cockpit applications, helmet mounted displays, large-screen projection displays, and large workstation displays. DOD has determined that it would be impractical to expect a single supplier to develop all of the capabilities necessary to build each of these types of displays. ARPA has already identified three of the potential production facilities as proposals selected in the FY 1994 TRP competitions concluded in the fall 1994.

National Security Issues

There are three key issues supporting arguments for a larger and more vital domestic FPD industry. One is concern within DOD that the United States is **dependent** upon foreign sources for FPD technologies. A second is whether this dependency creates a **vulnerability** for U.S. national security. The third issue is whether **substitute** FPD technologies are available to supplant foreign sources. These three issues are important for defining the Federal interest and scope of support of FPD technologies.

Dependency, in a broad national context, means that a technology, component or product comes from foreign sources. The United States not only is dependent on foreign sources for many military technologies and commodities, but its defense industry also provides many military technologies to buyers in global markets. Clearly, in commercial, non-military markets, the flows of technologies and products across national borders is what spurs the growth of world trade. Some may argue that dependency on foreign sources for

---

[33] Personal conversation with D. Slobodin, Program Manager, Display Technology, ARPA, Sept. 8, 1994.

technologies is a critical part of the larger global market of technology flows and transfer.[34] Certainly, dependency on foreign sources for military technologies may highlight a nation's (and its industries') weaknesses in technology development and commercialization. But global access to technologies, products, services and skills is an asset within a context of world trade.[35] Multiple foreign sources can disperse the risk of relying on one source for funding R&D or for an uninterrupted supply of technology. They may enhance the pace of technology development as many competitors race to capture markets. Dependency in an open global economy may confer advantages of cost, price, variety and quality or technologies on the recipients--not all of the advantages may come from being the supplying nation.[36]

In this respect, dependency is <u>not</u> the same as vulnerability. Generally, nations tend to define vulnerability--or more properly, a vulnerable situation--when a loss or restriction of a foreign source would have a serious impact on national security.[37] A nation becomes vulnerable when its dependency upon a foreign source for technology also exposes it to potential disruptions and embargoes of technologies and commodities, threatening the very survival of the nation. In other words, a dependency can become a vulnerability, but not all dependent situations are vulnerable situations. DOD currently assesses the dependency on foreign sources for FPD technology--primarily Japan--as a vulnerability for U.S. national security interests.

Any number of issues may create a situation of vulnerability-- a political, economic, or military action may cause disruptions of critical military technologies from a source to a recipient. Natural disasters, trade wars, labor strikes and military conflicts can precipitate changes in nations' policies regarding technology dependency to that of perceived vulnerability. An example of a situation where the United States successfully addressed a situation of technology dependency which had the potential of national vulnerability was in semiconductor technology. The semiconductor manufacturing technology (SEMATECH) consortium, was started in late 1987 due to concern that

---

[34] Technology transfer has different meanings and applications, depending upon the situation or circumstances involved. It can mean the transfer of legal rights (e.g., patents), information, skills, or the actual technology itself. U.S. Library of Congress. Congressional Research Service. Technology Transfer: Use of Federally Funded Research and Development. CRS Issue Brief 85031, by Wendy H. Schacht. p. 1-2. [updated regularly]

[35] Alexander, Arthur J. Are U.S. Weapons Producers Dependent on Japanese and Other Foreign Suppliers? Japan Economic Institute [JEI]. Washington. Aug. 28, 1992. p. 3.

[36] Prior to World War II, the United States enjoyed the advantage of receiving important military technologies developed abroad, including radar and jet engines. JEI, p. 3.

[37] JEI, p. 3.

Japanese domination of the basic memory semiconductor chip markets--both in the United States and globally--was resulting in total foreign sources of semiconductor chips for U.S. weapons. According to its supporters, this posed a significant national security risk and would make a dependent situation vulnerable to disruptions of supplies of technology. An increase in the U.S. semiconductor industry's global market share since 1988 is often cited by SEMATECH supporters as proof that creation of this consortium was an appropriate way of addressing the potential vulnerability issue.

The third issue is substitution. If a nation or industry can substitute a different technology for a technology upon which it is either dependent, or even vulnerable, then dependence or vulnerability can be reduced or removed. But issues of dependency, vulnerability, and substitution are not just technology acquisition or access issues. They also are directly linked to manufacturing capability. A lack of a domestic manufacturing base or a substitute technology virtually implies dependency on a foreign source to produce a technology or product, and may also make a dependent situation a potential vulnerability. In other words, a nation's inability to manufacture an important technology, commodity or product means there are no choices--it must import the technology, if it is critical to that nation's security. DOD contends that currently there are no choices for high-volume, low cost FPD technology, save for what is available from abroad.

DOD is currently trying to address issues of technology dependence, vulnerability, substitution and the key issue of manufacturing capability through a variety of means. DOD manufacturing programs use Federal funding to help initiate and promote new manufacturing technologies and processes with U.S. industry. DOD's policy for supporting dual-use technologies is consistent with this policy of addressing domestic technology capability, not dependency or vulnerability. This policy stresses that, whenever possible, existing commercial technology should be purchased "off the shelf" by DOD, rather than buying or using technology which has military specifications ("mil-spec") and applications--which may be more expensive. The FPD Initiative would be part of the larger dual-use policy, and DOD contends that it addresses important issues of U.S. dependency, vulnerability, and substitution.

THE U.S. DISPLAY CONSORTIUM

In July 1993, 11 U.S. corporations formed an alliance to promote domestic development, an industry infrastructure and manufacturing of FPD technology.[38] This industry group is called the United States Display Consortium (USDC), and it receives part of its funding from ARPA's FPD program and the rest from industry membership fees. The USDC was created to provide a "common platform" so that FPD manufacturers and potential

---

[38] There are now 16 U.S. FPD manufacturers and developers as member of the USDC.

manufacturers, equipment and materials suppliers, and other interested parties can exchange information and improve communication.[39]

A key focus point of the USDC also will be support of U.S. suppliers of equipment and materials. In this respect, comparisons with the SEMATECH consortium are apparent: an industry-Government consortium, partially funded by the DOD, designed to improve U.S. manufacturing capability in an important sector of the electronics industry. Perhaps the most important comparison is that both consortia seek to enhance communication and technology transfer between suppliers and manufacturers. The USDC views the current (and growing) U.S. FPD industry as a single entity, not as disparate segments. This is important because the consortium is moving to link vertically the U.S. FPD industry through a "stream" of information and technology to form a standard body of knowledge.[40] This would start with suppliers of equipment and materials, and would continue through to manufacturers and producers of FPD technology, and ultimately to users of FPD technologies for commercial and military applications. Some of the broad objectives of the USDC include improving the quality of U.S. suppliers of materials, manufacturing equipment, and systems; enhancing U.S. FPD manufacturing processes; developing industry benchmarks and standards; and encouraging technology transfer and communication among industry members.[41]

However, unlike the SEMATECH consortium, there is no central USDC manufacturing facility or site. The USDC instead will establish "neutral forums" in which industry members identify issues critical to the industry.[42] Also, the size of the two respective industries mean that by definition each consortium serves different needs. While the U.S. semiconductor industry was at a distinct disadvantage from Japanese and other foreign semiconductor manufacturers, its total world market share never fell below 45 percent.[43] In contrast, the Japanese electronics firm Sharp alone holds just under 45 percent of the global market for LCD technology, and Japanese firms hold greater than 90 percent of that market in 1994.

---

[39] U.S. Display Consortium. USDC Fact Sheet and Membership List. San Jose, CA, 1994. p. 1. [unpublished]

[40] Interview with Peter Mills, Chief Executive Officer, USDC on Sept. 16, 1994.

[41] U.S. Display Consortium. Annual Report 1994. San Jose, CA. USDC, 1994. 16 p.

[42] These forums are neutral in that no specific technology or technologies are advocated or promoted by the USDC as an organization, but instead it is left to the individual members of the consortium to set technology priorities. U.S. Display Consortium. Fact Sheet and Membership List. p. 1.

[43] SEMATECH Accomplishments. Austin, TX, SEMATECH, Sept. 1993. p. 8.

In FY 1994 ARPA provided just under $15 million funding for FPD technology programs.[44] The contributing USDC industry members provided $11 million directly to FPD programs. In addition, the USDC members provided an additional $4 million to operate the consortium, staff project teams, help create the industry "roadmap" which will provide a strategy for the U.S. FPD industry, establish industry standards, and support additional activities and programs. Therefore of the nearly $30 million of Government and industry support already committed to the USDC, just over half is coming from the U.S. industry. This goal, intended to be reached in year five of the consortium, was achieved in the first year of the USDC.[45]

The USDC contends, that its objectives are consistent with the policy goals outlined in the Flamm Report.[46] Advocates of a strong U.S. public/private sector role in developing FPD technologies describe the efforts as complimentary. Support for the USDC is based on similar reasons for current public sector support of FPD: the impact which FPD technology has on the U.S. military and the potential growth of the commercial market. Because both sectors, in their own ways, are so vital to the United States, supporters contend that there is a strong need to create a broader and healthier domestic FPD industry. As a result, communication among U.S. FPD manufacturers, suppliers of capital equipment, and users of FPD technologies (a role of the USDC) becomes critical for developing a domestic FPD capability an objective of the FPD Initiative. Some in the U.S. electronics industry have called for additional Federal and industry support of the USDC.[47]

POLICY OPTIONS

The Federal role of supporting and funding FPD technologies, including (but not limited to) the FPD Initiative, may be revisited in the 104th Congress. Some policy options and considerations may include the following.

---

[44] The total funding commitment from ARPA is expected to be $20 million through FY 1995.

[45] Personal communication with Heidi M. Hoffman, Director of Government-Industry Relations, USDC on Oct. 25, 1994.

[46] U.S. Display Consortium, Annual Report 1994, p. 6-7.

[47] The American Electronics Association (AEA), in April 1994, released a task force report on the importance of the U.S. FPD industry. Among its recommendations is an industry-Government partnership in high-volume FPD manufacturing, creating an "infrastructure" for supporting all segments of the industry, and reducing barriers of entry for potential U.S. FPD manufacturers. The American Electronics Association. [A Critical Time for Industry] U.S. High-Volume Manufacturing of Flat Panel Displays. Washington: AEA Flat Panel Systems Task Force, Apr. 1994. 5 p.

Reexamine the DOD FPD Mission

Policymakers may wish to consider reexamining DOD's R&D role. Some argue that DOD should concentrate exclusively on its own technology needs for warfare. They argue that this is more consistent with DOD's, and particularly ARPA's, role rather than the establishment of an explicit dual-use technology policy and the proposed FPD Initiative. This may mean that DOD may have to selectively choose from a wide array of FPD technologies which specifically meets its needs, rather than creating a broader technology base which also will have commercial benefits. If DOD needs an FPD technology, it either supports a U.S. firm which can develop the FPD technology, or buys the best technology available in the global marketplace. In this scenario, the growth of the U.S. FPD industry would most likely be less relevant, since individual companies, laboratories or industry niches would provide DOD's technology needs. Also, there would be no Federal role to foster a commercial U.S. FPD industry.

Some contend that the benefit of this approach is that it would tailor DOD FPD R&D programs towards technology development most clearly defined by national security goals and objectives. The cost of this approach could likely be significantly lower than broader approaches which would attempt to increase the overall size of the U.S. FPD industry. The role of the Federal Government would be more clearly defined, and growth of a commercial U.S. industry would be less tied to DOD funding to keep the industry healthy. The argument is made that the commercial and defense industrial bases have different market demands, users' needs, and applications, and DOD should address its own technology needs in that context.

In February and March 1995, the House of Representatives and Senate of the 104th Congress considered FY 1995 budget rescissions for many Federal agencies, including DOD. On February 22, 1995. the House of Representatives approved significant rescissions in the DOD budget, including $502 million for FY 1994 and FY 1995 obligated funds for the TRP program. Since the TRP program supports a wide range of dual-use programs, including FPD technology, DOD's program may be significantly altered. To date, the Senate has considered, but not voted on these rescissions. Supporters of these rescissions contend that in an era of very tight budget constraints, support of a broad FPD technology Initiative is well beyond the Federal role.

But others have serious concern that a policy focussing more narrowly on DOD's support of technology may not be strategic, may not recognize the traditional role of DOD in supporting technologies which have a broad technology base, and may not recognize the importance of FPD technology both militarily or commercially. Many, if not most, observers of the old system of military specific, narrowly focused DOD technology funding and support contend that it was wasteful and inefficient. A more narrowly focused DOD FPD program may help provide the defense establishment with specific FPD technology requirements--but at what cost? While the overall funding levels for a DOD-only FPD program might be lower than the FPD Initiative, critics contend that it may result in other inefficiencies in procurement and acquisition.

A more narrowly focused DOD FPD program may not likely address concerns of the lack of U.S. FPD manufacturing capability, nor the continued reliance on Japanese firms for advanced FPD technology. Supporters of the FPD Initiative state that FPD technology is a vital part of weapons systems, and an important part of the military hardware used to support U.S. combat forces worldwide. DOD has emphasized that its entire dual-use technology strategy is tied to the agency's mission in a post-Cold War era, and argues that a narrowing of that mission might undermine the total military preparedness going into the next century.[48]

Support the FPD Initiative

Policymakers and industry leaders will be faced with considering support for the FPD Initiative. Again, this proposal would extend DOD funding for FPD technology through the end of the decade, would provide only "technology neutral" support for existing and prospective R&D, and perhaps most importantly, help develop the FPD manufacturing base in the United States, among other provisions. The FPD Initiative is also part of a larger DOD effort to support and develop a dual-use technology base for procurement and acquisition. The total cost of the FPD Initiative over the next five years would be $587 million, funded through ARPA.

The FPD Initiative could provide continuity and consistency of Federal support for the nascent U.S. industry. If the FPD Initiative is successful in nurturing a domestic FPD industry, it could very likely address DOD concerns of early, assured and affordable access to an important technology base. The "technology neutral" policy may mean that DOD truly will not pick "winners" and "losers" in developing domestic sources of FPD technology. "Exit ramps" may mean that DOD can move quickly from one technology to the next, as commercial developments warrant. The FPD Initiative could provide needed leverage to ensure that the "upstream" capital equipment suppliers, the manufacturers and producers of FPD components, and the final users of FPD products, will not be vulnerable to foreign sources. Kenneth Flamm and other supporters of the FPD Initiative contend that the United States faces fierce global competition and that DOD has *no choice* but to address national economic security problems where there are no competitive market alternatives.

The payoff from the FPD Initiative for the United States may be substantial. It could help create a U.S. FPD manufacturing industry which can address the potentially great commercial applications of FPD technology. It could help assure DOD of the technology it needs, when it needs it, to meet the security needs of the next century. It could be cost-effective in the long-term, since Federal funding would be leveraged with industry support. It could assist the entire U.S. electronics industry, from semiconductor to computer manufacturers, as new uses and applications of technology are developed.

---

[48] U.S. Dept. of Defense, Industrial Capabilities for Defense, p. 1-2.

The DOD dual-use strategy and the importance of FPD technology for military applications is at the heart of the FPD Initiative. DOD officials contend that DOD's support for fundamental technology research which eventually has both military and commercial applications has a long and productive history. During the 1940s and 1950s, DOD funded nearly all of the early R&D in the U.S. computer industry, and advances in aircraft design and jet propulsion had their genesis in DOD funding.[49] This support provided a significant technology base--the capacity to create, develop, improve, and manufacture--for both U.S. commercial and military markets.[50]

But many observers raise questions: is there a point at which military and commercial technology applications become so divergent that common manufacturing is no longer possible, and how can policymakers accurately assess that point of divergence? These are important questions, and relevant to the FPD technology issue. At some point in any R&D process, DOD's needs and the commercial markets' needs may be different, calling for differing approaches to technology development and commercialization. DOD officials contend that they understand when the technology base is the same as the commercial sector and when it is different. Supporters of the FPD Initiative argue that their policy approach reflects a core argument which is, and always has been, of the national security needs of DOD.

Congressional opposition to the FPD Initiative also could raise the following issues. DOD will have to address issues of whether the FPD Initiative would ultimately result in a Federal Government which has greatly expanded its scope and mandate. How much should DOD support commercial manufacturing, and how close is it to directly supporting commercial manufacturing? It also is unclear why 15 percent of the total FPD market has been set as the target for sustaining the U.S. FPD industry, or whether this market share could be sustained if DOD funding were to be withdrawn, since DOD purchases of FPD technology likely will never account for more than 5 percent of total market procurement. In addition, attempting to increase U.S. commercial FPD capabilities may alter or affect market developments. Critics contend that the FPD Initiative will broaden the Federal interest from national security to national economic security. They claim that this approach usually is inefficient and may actually delay commercial FPD developments. While the U.S. computer and semiconductor industries were nurtured by Government contracts in the late 1950s through early 1960s, there are significant differences between that era and today's: a lack of global competition with U.S. industry; U.S. firms' product life cycles are now measured in months, not years; and the cost of capital and manufacturing technology are less than 40 years ago.

---

[49] Deutch, John M. A Problem the Pentagon Will Solve. The Washington Post, June 21, 1994. p. A17.

[50] U.S. Library of Congress. Congressional Research Service. DOD's Dual-Use Strategy, by John D. Moteff. CRS Rept. 95-322. Washington, Mar. 2, 1995. 6 p.

There is also concern whether support of commercial FPD technologies may cross a policy line between R&D subsidy and production subsidy (particularly if test-bed facilities are supported by DOD). Under the new GATT agreement, the test-bed facilities are defined as pre-competitive technology, and therefore not restricted under GATT. Supporters of the FPD Initiative contend that there is little reason to assume that the proposed initiative will fall outside of the GATT agreement. But some would contend that the broader problem is that technology policy arguments--and they would include the FPD Initiative--attempt to redress trade and competitiveness issues, and in the end are a poor policy substitute.[51]

The FPD Initiative also may not address what is larger xthan a domestic technology situation. Is the current U.S. FPD situation a technology or trade issue--or both--and would the FPD Initiative be appropriate for either? Supporters at DOD would answer that the FPD Initiative is intended to redress an existing technology competitiveness and domestic manufacturing capability problem, and that, therefore it is good policy. But others would contend that, with the DOD emphasis on commercial, dual-use technology, trying to redress a foreign sourcing issue is really a trade problem. The Departments of State, Commerce, and the U.S. Trade Representative could negotiate with their counterparts in Japan to guarantee access, or the Japanese should expect similar retaliatory measures from the United States.

Any group which has its base of operations in the United States could be eligible to receive FPD Initiative funding. This would leave room for foreign firms which operate in the United States to be participants in FPD funding. Should (and can) commercial FPD technology developed with U.S. taxpayers' dollars be spent only on U.S. firms? DOD officials contend that allowing a broad range of potential participants would provide the necessary capital and expertise for developing domestic FPD technology. Advantages in FPD manufacturing knowledge may also be gained by U.S. firms working with foreign participants. But if a national economic security argument is made for Federal funding of FPD technology and creating a domestic industry, how is the U.S. national interest served by having foreign firms participate?

Finally, there are questions regarding evaluating and assessing the potential success of the FPD Initiative. Some FPD Initiative goals are clear and quantitative. Setting a 15 percent market share objective for the U.S. FPD industry by the end of the decade is a specific goal of the Initiative. Other provisions, such as assessing the technical merits of DOD FPD projects may ensure that there is a defined role for the Federal Government to enter and exit R&D programs. But policymakers may question the role of the National Economic Council in evaluating the FPD Initiative and its funding, and ask whether U.S. industry will be part of the evaluation process. Will the NEC evaluate the FPD Initiative's success solely on the size of U.S. market share--or will other criteria be needed? Can the NEC make objective policy decisions on

---

[51] Barfield, Claude. Flat Panel Displays: A Second Look. Issues in Science and Technology. Winter 1994-95. p. 21-25.

the FPD Initiative in conjunction with industry, or will political issues also be a part of the evaluation process? In the past, congressional policymakers have included investigatory and oversight provisions when constructing Government-industry technology programs (for example, the role of the General Accounting Office to examine the SEMATECH consortium), and may consider such an arrangement if the FPD Initiative is funded.

Shift Support to Civilian Agencies

Another approach would be to recognize a Federal interest in commercial FPD technology and give primary funding responsibility to a civilian agency or agencies. If the Federal interest is to ensure the development of a larger commercial domestic FPD industry, then funding programs may more properly reside in agencies with R&D missions more closely correlated with the needs of the commercial sector. Several civilian agencies do fund FPD programs, but DOD's FPD program, particularly the ARPA program, is much larger than all other Federal agencies' programs combined.

In previous Congresses, some asked whether the national interest in developing government-industry technology capability would be better served by having a single, civilian agency direct the Federal effort. They contended that the benefit of establishing a lead Federal civilian agency for important technologies like FPD technology is that it would more formally acknowledge a Federal interest in developing a healthy commercial FPD industry. Such a change could help clarify Federal R&D funding goals and would recognize per se the broader goals of supporting the U.S. FPD industry. For some, DOD's support of U.S. FPD technology as part of a larger national security argument may cloud the real reason why there is a Federal interest in supporting FPD technology and manufacturing.

A civilian agency like the Department of Commerce could take this role (its FPD efforts were described in the section on Federal FPD Programs). The Technology Administration within the Department of Commerce already provides broad policy outlines for encouraging government-industry cooperation in high technology development, as well as support for the NIST's ATP. In addition, NIST may have personnel familiar with commercial market and technology issues and could make policy and funding decisions based on that knowledge. The Clinton Administration has shown strong support for NIST and ATP. For FY 1996, the Administration has requested that the ATP receive $491 million, an increase of $60 million.

However, other policymakers are seeking to reduce, not enlarge, the role of civilian agencies in technology development. On February 22, 1995, the House of Representatives approved rescissions of $107 million reduction of the Advanced Technology Program (ATP) program for FY 1995, which the Senate is also considering. For FY 1996, funding for tne ATP, while currently under debate, may ultimately reflect congressional concern, rather than Administration support, for the program. In this environment, it may be unlikely that a larger funding role for a single civilian agency is established.

As the Federal agency providing the most support for FPD technology, ARPA has a long and successful history of supporting and nurturing technologies. It has demonstrated that its personnel is knowledgeable and sophisticated at selectively picking critical technologies which have an importance not only to the defense sector but to the commercial sector as well. ARPA has successfully supported technologies which had later commercial success and applications. There is no guarantee that a civilian agency could address national technology issues more competently than ARPA; nor is there any consensus that the Federal interest should be more broadly, and formally, expanded to include a national commercial technology policy.

Pursue a New Federal Role: Industrial Policy?

Another viewpoint is that U.S. policy needs to be extended beyond agency support of FPD technology. Should a broader policy FPD perspective include other issues, such as cost of capital, tax, investment, education and training as part of the policy debate? For many, this revives the decade-old debate on U.S. Industrial Policy and a large, expanded Federal Government making decisions for U.S. industry. Others contend that current U.S. technology policy is an amalgam of disconnected and uncoordinated policy decisions putting U.S. industry at a disadvantage in a competitive global economy.

It is not clear what would constitute a broader Federal role--and would be defined as Industrial Policy--for FPD technology. Some might argue that even current attempts to create a dual-use technology base, support critical U.S. technologies, and provide funding for the FPD Initiative, may constitute a form of Industrial Policy. Kenneth Flamm, Principal Deputy Assistant Secretary of Dual Use Policy and International Programs, responds that the FPD Initiative is not Industrial Policy, but instead Technology Policy, with the focus on technology and results, not on Federal programs and large-scale funding.

Over the last decade several U.S. industries (e.g., the semiconductor industry), have proposed to Federal policymakers several policy recommendations to assist firms competing in the global economy. Often, these recommendations are similar in broad intent, if not in the specific policy approaches. For example, U.S. industry groups have recommended instituting a permanent research and experimentation (or development) tax credit, accelerated depreciation of high technology equipment, stronger protection of U.S. intellectual property rights abroad, greater education and training efforts for future high technology workers, and a trade policy which is consistent with their view of U.S. domestic technology policy. To date, Federal policymakers generally have addressed these issues individually and separately. At issue is the question of whether these issues should be addressed more comprehensively and uniformly.

Any proposal for a form of Industrial Policy for FPD technology, however, would meet strong objections. Some contend that it could result in a greater concentration of decision making and authority in Federal agencies,with bureaucrats making decisions for the FPD industry. In the past, recommendations to create a Science and Technology Agency, to create a civilian ARPA or otherwise centralize U.S. S&T policy have met with resistance. At the heart of these proposals is the assumption that Federal policymakers have all the available information for creating the right policies. Critics believe that the Government cannot implement a better mix of technologies, management practices, work habits, culture and economics to ensure the growth of a healthy U.S. FPD industry than the market can.

Speaker of the House Gingrich has indicated that he supports applying information and telecommunication technologies in both public and private sector to provide information dissemination for the American people. In addition, it is likely that defense budget and policy debates in the 104th Congress will focus on issues of military readiness and DOD capability. The Federal role in developing FPD technologies is likely to be an important part of both policy debates. However, it is also likely that many Members of Congress will examine how changes in the marketplace--rather than Federal technology programs--can best provide commercial and military FPD technologies. The rescissions of the TRP, ATP and other Federal programs, offset by initiatives to enact tax, regulatory and other policy revisions, are one indication of this new direction. Many believe that a growing Federal role is unlikely, considering these changes.

Advocate a Free Market Approach

Many call for a free market approach for commercializing FPD technology. The key issues in the current FPD debate are access to the best new FPD technology and the manufacturing capability to make FPD products. Therefore, some make the argument that U.S. industry should be the primary driver of FPD technology. The Federal Government can neither substitute for an industry nor effectively create one in the absence of a strong technology and manufacturing base and initiative from the private sector. They contend that government funding of a commercial technology distorts both the consumer demand and the technology push for new products. According to one expert, government bureaucrats make bad business executives.[52]

Interestingly, some of the same policy recommendations brought forth by advocates of a broader Federal role--the Industrial Policy position--also are presented by those who favor the free market approach. For example, modification of the U.S. tax code is often supported by both positions. However, the supporters of a broader Federal role often call for a capital gains reduction or accelerated depreciation rates in addition to direct Federal support of industrial R&D, while many free market supporters contend that changes in tax policy should supplant current Federal technology programs.

Several case studies of how other nations support technology development and commercialization demonstrate how imperfect the role of governments can

---

[52] Beltz, Cynthia. Would You Invest in Bureaucracy Inc.? The New York Times, May 1, 1994. p. A12.

be in trying to direct, support or otherwise guide technology development. In Europe, both national and transnational (mainly European Union) programs to galvanize the semiconductor and high definition television industries, among others, have fallen far short of supporters' expectations.[53] In Japan, even the much-discussed "model" of technology development and policy--relying on a strong Japanese government presence to guide areas of technology development--is often misunderstood by foreign observers. Over 80 percent of all R&D undertaken in Japan is performed by Japanese industry. While the Japanese government does provide some coordination and sets over-arching national policy directives, it is Japanese industry which commercializes technology (like Sharp, Inc., in AMLCD technology). Even the Japanese have made significant mistakes when it comes to picking areas of technology "winners", as in HDTV. (See appendix A.) Similar experiences in government-industry cooperation, subsidy or support in other nations like Taiwan, Korea, Latin America and elsewhere indicate that there is no perfect model for the United States.

Instead, some observers contend that in a truly global economy, commercial technology flows across borders. The global free market is a fertile "hothouse" for high technology, with new products constantly introduced regardless of their source, meeting customers' demands and needs, and not the wants of political constituencies. Free marketeers would argue that, when left to their own devices, U.S. inventors, capitalists and entrepreneurs are inventive and innovative and can match any other nation when it comes to developing new technologies.

Several questions remain, however. Can a small U.S. FPD industry survive in a purely free market environment, and what is the cost to the United States if this industry cannot survive without some assistance? Some argue--including the supporters of the FPD Initiative--that a very small niche industry cannot meet the larger national economic and military needs of the United States. They contend that a situation of continued foreign dependency and potential vulnerability for imported FPD technology is untenable. While the perfect model for government-industry cooperation may not exist, some would argue that insisting on a pure free market approach (and leaving the U.S. FPD industry to go it alone in the global economy) is an inflexible approach to a serious problem for U.S. technology development. Some also are concerned that tax and regulatory reform may be advantageous for larger U.S. firms, which have the legal, financial and R&D resources to best take advantage of changes in these policies, while small and medium-sized firms, or new firms, may not be able to benefit as greatly.

Some point to the USDC as a good example of how the Federal Government--ARPA--can provide a relatively modest amount of Federal support to address a serious technology shortcoming in this country. They contend that,

---

[53] See Appendix A; and U.S. Library of Congress. Congressional Research Service. International Science and Technology: Issues for U.S. Policymakers, coordinated by Glenn J. McLoughlin. CRS Report 94-773. Washington, Sept. 16, 1994. 101 p.

usually, when U.S. firms together provide some of their own funding and initiative to support common industry goals, then matching Federal support is warranted. For them, the SEMATECH consortium is an example of successfully leveraged Federal support. To date, the corporate members of the USDC have demonstrated both the interest and wherewithal for taking the initiative to create a stronger U.S. FPD industry.

Still, some might contend that the very act of supporting near-commercial technology development and manufacturing may alter the market in such a way that cost, price, quality and innovation all are affected--and not always positively. Even the Federal Government's role in the success of SEMATECH and the revival of the U.S. semiconductor industry has not been definitively understood. The product innovation and manufacturing capability of U.S. semiconductor firms may have had as significant a role in this industry's revival over the last decade as the role of the consortium. Supporters would contend that these differences are a strong reason for continued, but limited, Federal support for USDC--the U.S. FPD industry is at an even greater disadvantage than a comparable industry in recent years, yet it is just as vital to the U.S. economy.

Policymakers may wish to consider at what point the role of government policy to support R&D ends and the benefit of the freedom of firms to innovate and compete begins. They may wish to consider when technology development and commercialization should be left to those willing to undertake the risk and reap the benefits of industrial R&D. Policymakers may consider, on a case-by-case basis, whether national or economic security concerns warrant a Federal role, and whether issues of dependency, vulnerability and technology substitution are at stake. Beyond these criteria, Congress may consider whether support for technology commercialization and industrial manufacturing is best left to the industrial, not Federal, sector.

CONCLUSION

Members of the 104th Congress will be asked to strike a balance between two national policy goals. The first goal is to address the concern that the United States be militarily prepared going into the 21st Century. Supporters of the FPD Initiative contend that FPD technology has, and will, play an important role in this readiness. During the Persian Gulf War, U.S. forces successfully deployed advanced weapons systems in jets, tanks, and ships that included FPD applications. For supporters, this demonstrated that the importance of FPD technology is not part of some abstract policy option but instead part of a real military strategy for using dual-use technology. Guaranteed and cost-effective access to FPD technology is a part of this DOD strategy through a domestic technology policy.

The second policy goal is to reduce the size and scope of the Federal Government, where appropriate. Many supporters of this perspective contend that the FPD Initiative has serious flaws. In their eyes, the Initiative calls for the Federal Government, and particularly DOD, to depart from its traditional role of supporting technology development only for military uses. Instead, DOD may undertake the more difficult task of creating--and perhaps managing--a private market serving not only military but commercial needs. In addition, critics contend the FPD Initiative comes at a time when congressional policymakers seek to rescind and otherwise reduce Federal spending in many areas. For them, the overriding imperative is that the Federal deficit must be reduced in those areas where the Government does not have a specific role to play for its citizens.

The solution may fall somewhere between these two perspectives. Concerns about military preparedness and the need to extend DOD support of FPD technology may be challenged by concerns to reduce the size of the Federal deficit by cutting those programs where either the domestic market or global trade can provide answers. Policymakers may seek to modify the FPD Initiative, or place it within the context of a reduced dual-use policy initiative. Some believe that failure to achieve such a balance may create policy missteps difficult to rectify in the future: the inability for the U.S. military to gain access to FPD technology when it is needed. Others believe that a broad FPD initiative may be counter to current efforts in Congress to limit the role and scope of the Federal Government.

APPENDIX A

The current public policy discussion about FPD technology should be taken within the context the many other initiatives considered by Federal policymakers. Several have redefined the role of Government and industry in technology development, while others either never went beyond the planning stage or, if implemented, did not seem to meet the goals and objectives of supporters. A partial list of these programs and initiatives follow.

VHSIC

In FY 1980, the Department of Defense ended an ambitious 10-year funding program for supporting high speed integrated circuits technology. This program--the Very High Speed Integrated Circuits, or VHSIC, program--was begun in response to Japanese market penetration in the global semiconductor market.[54] In particular, Japanese semiconductor manufacturers, with strong government support, had begun the Very Large-Scale Integrated Circuits (VLSIC) program to "leapfrog" Japanese semiconductor companies into the next generation of semiconductor devices. U.S. semiconductor and DOD officials were concerned that foreign firms would continue to gain market shares and drive U.S. firms out of the industry. If this occurred, the U.S. industrial base, particularly semiconductor manufacturers and suppliers of capital equipment, would erode and the DOD would be increasingly captive to foreign sources for its technology needs.[55]

The VHSIC program has had vocal supporters and critics. Supporters point to the long-term results which have come from DOD-sponsored R&D. While it is hard to point to any direct commercial technology benefits which have come from the VHSIC program, they contend that fundamental industrial R&D was maintained during a time in which many U.S. semiconductor firms could not (or would not) invest in R&D which did not have an immediate payoff. Supporters also contend that VHSIC R&D had a major effect of maintaining important defense-related semiconductor R&D, particularly in materials research and specific military applications. Critics just as strongly contend that the VHSIC program has been an inefficient and ineffective method to support the U.S. semiconductor industry. They contend that the focus of the program was too

---

[54] The VHSIC program was intended to last 10 years at a cost of about a $1 billion. Opposition to the VHSIC program, particularly during the congressional appropriations process, delayed the initial funding.

[55] U.S. Library of Congress. Congressional Research Service. The U.S. Semiconductor Industry and the Sematech Proposal, by Glenn J. McLoughlin and Nancy R. Miller. Washington, Apr. 23, 1987. p. 8. [archived]

much on the needs of the DOD and too little on industry strengths and capacity. They point out that if the VHSIC program had been as successful as its proponents had originally promised, there should not have been a need for the subsequent SEMATECH initiative (described below).

The issues raised during the VHSIC program are instructive as they relate to similar FPD technology issues. National security concerns, technology dependence and vulnerability, industry participation and global competition are apparent in both VHSIC and FPD R&D. However, the VHSIC program was not primarily intended for the broad commercial semiconductor industry. It was intended only for the technology needs for the Department of Defense, and should be evaluated on that criterion.

SEMATECH

In February 1987, the Defense Science Board (DSB) of the Department of Defense released a report, *Report of the Defense Science Board on Semiconductor Dependency*.[56] The authors of the report, representing U.S. industry, academia and Government, cited evidence that U.S. semiconductor manufacturers had lost significant market shares to foreign, particularly Japanese, semiconductor manufacturers. The report recommended a legislative initiative for creating a semiconductor manufacturing facility, jointly funded by the Federal Government and members of the U.S. semiconductor industry. This was called the Semiconductor Manufacturing Technology, or SEMATECH, initiative. The SEMATECH initiative was proposed in direct response to concerns that Japanese domination of the basic memory semiconductor chip markets--both in the United States and globally--might result in complete foreign sources of semiconductor chips for U.S. weapons, which would be a potential national security risk.[57]

By fall 1987, Congress provided $100 million in funding for SEMATECH through the Defense Advanced Research Projects Agency (DARPA) of the Department of Defense.[58] Congress has continued to provide DARPA with

---

[56] Office of the Under Secretary of Defense Acquisition. Defense Science Board. Report of the Defense Science Board on Semiconductor Dependency. Washington, Feb. 1987. 103 p.

[57] For a more comprehensive history of SEMATECH, see: Williams, Robert. SEMATECH Case Study N9-389-067. 32 p.; U.S. Library of Congress. Congressional Research Service. Semiconductor Manufacturing Technology Proposal: SEMATECH. CRS Issue Brief 87212, by Glenn J. McLoughlin. 12 p. [updated regularly]; and U.S. Congressional Budget Office. Using R&D Consortia for Commercial Innovation: SEMATECH, X-Ray Lithography, and High Resolutions Systems. 120 p.

[58] DARPA was renamed the Advanced Research Projects Agency (ARPA) in 1993. See: U.S. Library of Congress. Congressional Research Service. The

(continued...)

funding for the consortium. SEMATECH recently has announced that it will not take any Federal funding after FY 1996.[59] To date, it has been considered the most visible example of where Government and industry have been able to work together to address a national security and technology issue which has had both military and commercial applications. For some, SEMATECH has become the "model" for other types of Government-industry technology collaboration.

Generally, supporters of SEMATECH give the consortium high marks for its effectiveness and impact on the U.S. semiconductor industry. An important reason why Congress passed the SEMATECH initiative was that it agreed with industry supporters and outside observers that there is a national security issue at stake if the United States continues to rely heavily on imported semiconductor chips. SEMATECH appears to have addressed both the semiconductor concerns of DOD, as well as played a part in the revival of the U.S. semiconductor industry.

But SEMATECH also has been supported by a mature U.S. industry, which during its lowest point in the mid-1980s, never fell below 45 percent of the global semiconductor market. U.S. semiconductor firms, most notably led by the Intel Corporation, have continued to develop and commercialize advanced semiconductor devices (especially in the microprocessor and logic chip fields). Other factors also have contributed to the overall success of the U.S. semiconductor industry: a trade agreement with Japan which increased U.S. exports of semiconductor chips from approximately 7 percent to nearly 20 percent, a declining dollar in global markets, which has made U.S. exports cheaper and more competitive, a new resurgence in semiconductor technology innovation (coming directly from U.S. firms), and lower cost of capital.[60] In this respect, the U.S. semiconductor industry and participants in the SEMATECH consortium and the industry at large have enjoyed advantages which other nascent industries, especially the U.S. FPD industry, may not yet have.

High Definition Television (HDTV)

In 1988 through 1990, after the creation of SEMATECH and just before the current interest in FPD technology, another technology initiative which received

---

[58](...continued)
Defense Advanced Research Projects Agency: DARPA, by Michael E. Davey. CRS Rept. 93-27. Washington, Jan. 15, 1993. 15 p.

[59] Southerland, Daniel and Elizabeth Corcoran. Computer Chip Group to End Reliance on U.S. Funds. The Washington Post, July 18, 1994. p. A4.

[60] Stoner, Leigh. Electronics Industry Braces as U.S. Ponders Remedy in Japan Trade Dispute. New Technology Week, Feb. 22, 1994. p. 2; and U.S. Library of Congress. Congressional Research Service. U.S. Japanese Trade: Semiconductor Agreement by William H. Cooper. CRS Rept. 93-484. Washington, May 13, 1993. 6 p.

a great deal of attention was High Definition Television (HDTV). In many respects, FPD technology is a high definition technology; or, more accurately, a high resolution technology.[61] But support for HDTV emphasized the commercial applications of this technology in homes and businesses, particularly applying the new technology to television broadcast and reception. Experts also noted military applications, in targeting, missile guidance and command and control.[62]

For many, HDTV became, after SEMATECH, the largest and most important technology development issue linked to industrial and economic growth. Like SEMATECH and now FPD technology, industry and Government leaders viewed intense competition from Japan and other southeast Asian nations as a reason for increased Federal support. The Japanese had begun supporting HDTV technology development in the early 1970s, and by the mid-1980s had established a prototype of an analog satellite broadcast standard in Japan.[63]

In FY 1988 DARPA provided $30 million in "seed" money for HDTV research and Congress seriously considered several bills to provide additional Federal support.[64] In May 1989, the American Electronics Association (AEA) proposed increasing this support by recommending $1.35 billion to fund HDTV development in the United States, including $1.0 billion in Federal Government loan guarantees for manufacturing facilities, $300 million from 1989 through 1992 to expand DARPA's HDTV research, and $50 million to develop national broadcast standards.[65]

While there was discussion and interest by Congress on the merits of the AEA proposal,[66] it did not receive the same level of support as the SEMATECH

---

[61] U.S. Library of Congress. Congressional Research Service. High-Definition Television and Advanced Video Information Systems. CRS Issue Brief 89088, by David B. Hack. [archived]

[62] Ibid., p. 4.

[63] U.S. Library of Congress. Congressional Research Service. Japanese and Industrial Associations: Their Role in High-Technology Policymaking [A Japan Task Force Report], coordinated by Dick K. Nanto and Glenn J. McLoughlin. CRS Report 91-477. Washington, June 6, 1991. p. 26-48.

[64] Ibid., p. 71.

[65] American Electronics Association. Background Information: Development of a U.S.-based ATV Industry. Santa Clara, CA. AEA, May 9, 1989.

[66] U.S. Congress. House. Committee on Energy and Commerce. Subcommittee on Telecommunications and Finance. High Definition Television. Hearings, 100th Congress, 1st and 2d Sessions, Oct. 8, 1987, June 23, Sept. 7,

(continued...)

initiative. One issue was whether there was a basis for Federal support at the level proposed by the AEA. While the SEMATECH proposal included a substantial amount of Federal support--$500 million over five years--the AEA HDTV proposal struck many as beyond what the Federal Government should do to support a technology absent a commercial base. In this respect, the HDTV debate mirrored the current FPD technology debate--would there be a U.S. "industry" willing to fully commercialize and manufacture HDTV technology?

HDTV will likely remain of great interest to Federal policymakers in 1995. ARPA funding levels, while criticized by many as too low, did initiate several U.S. laboratory consortia, all of which were racing to develop a broadcasting standard to compete with the Japanese standard in the early 1990s. In mid-1994, the Grand Alliance consortium announced that it had successfully completed work on a digital HDTV standard--technically more advanced than the Japanese analog standard. By successfully creating a more technically advanced digital broadcasting standard, in effect the United States will force foreign HDTV researchers and manufacturers to comply with the U.S. standard. Ultimately, the more advanced U.S. HDTV broadcasting standard also will likely become the global standard as the technology becomes more commercialized.

For many, this clearly demonstrates that relatively smaller, focused and leveraged Federal funding is the most appropriate approach for technology development. In the HDTV example, this is particularly true where U.S. industry has a clear interest and can develop the technology with minimal Federal involvement. Still, some observers also caution not to read too much into the recent U.S. success in HDTV. While the broadcasting standard is critical, an HDTV manufacturing base is still lacking in the United States. As in the case of the U.S. FPD industry, the ability of a U.S. HDTV industry to compete globally will likely rest in an advanced manufacturing base and capability, not from setting broadcasting standards alone.

Multi-Chip Modules (MCMs)

Multi-chip module (MCM) technology comprises a compact package incorporating several high-density integrated circuits to form an electronic

---

[66](...continued)
1988. Washington, U.S. Govt. Print. Off., 1988. 425 p.; U.S. Congress. House. Committee on Science, Space and Technology. High Definition Television. Hearings, 101st Congress, 1st Session. Mar. 12, 1989. Washington, U.S. Govt. Print. Off., 1989. 458 p; U.S. Congress. House. Committee on Science, Space and Technology. Subcommittee on International Scientific Cooperation. The International HDTV Standard Setting Process and the Role of International Standards in U.S. Competitiveness. Hearings, 101st Congress, 1st Session. May 31, 1989. Washington, U.S. Govt. Print. Off., 1989. 157 p; and U.S. Congress. Senate. Committee on Government Affairs. Prospects for Development of a U.S. HDTV Industry. Hearing, 101st Congress, 1st Session. Aug. 1, 1989. Washington, U.S. Govt. Print. Off., 1989.

system or subsystem.[67] MCM technology is smaller in size, lighter in weight, faster in speed of operation, and has higher rates of reliability than conventional integrated circuit (IC) packaging used today. It also has the potential of lower cost than conventional IC packaging, if high-volume, high-quality production processes are developed for this technology.

MCM technology has been identified by both industry leaders and Federal policymakers as critical to the U.S. electronics industry. DOD has determined that there is a distinct strategic advantage to using MCM technology in advanced military equipment. ARPA has an MCM development program, which was started in FY 1988 at $2 million, and grew to approximately $80 million per year by FY 1993, spread across several projects. ARPA's FY 1995 funding for MCM development is approximately $75 million.

In addition to the military needs, many industry experts believe that MCMs will eventually be used for commercial electronics packaging. According to one estimate, the market for MCMs is expected to reach $1.8 billion by the year 2000, with MCMs expected to be used in 80 percent of all electronic systems by that time.[68] Therefore, the now nascent MCM industry stands to become a driver of the entire electronics industry. ARPA's MCM program is similar to its FPD technology program in that it is a dual-use technology with a critical defense application. DOD also is attempting to meet its military technology needs by fostering development of the commercial industry. The funding levels are also comparable. The only major difference is that there is no coordinated effort among Federal agencies for funding MCM development, as there is currently with FPD technology. In this respect, the ARPA MCM technology program has not had the visibility that the FPD program has received--nor the criticism. Its funding and support within ARPA have remained fairly consistent.

There has been some criticism of the Federal MCM technology program, just as there is with the FPD programs. Some contend that ARPA's strategy of funding the commercial technology development of MCM technology is ill-placed. They argue that the U.S. industry should be left completely alone to compete in the international market for new technologies. Others may agree that there is a proper role for the Government to play, but feel that perhaps DOD is not the right place to fund new technologies that will eventually have primarily commercial markets. Still others say that ARPA's MCM program is poorly conceived because the amount of Federal funding is not nearly enough to jump-start a whole new industry, These criticisms add to concerns whether a DOD dual-use strategy can effectively guarantee both U.S. military technology needs as well as an industrial base for commercial use.

---

[67] U.S. Library of Congress. Congressional Research Service. Is DOD the Place to Fund Dual-Use Technology? Coordinated by John D. Moteff. CRS Report 93-496. Washington, May 17, 1993. p. 89.

[68] Personal conversation with representative from the Electronic Industries Association.

MANUFACTURING, TECHNOLOGY, AND COMPETITIVENESS

Wendy H. Schacht

BACKGROUND AND ANALYSIS

Overview of Manufacturing

Introduction

A strong manufacturing sector is essential to the economic welfare of many communities and regions of the United States. In recent years, manufacturing has consistently generated just under one-fifth of the U.S. gross national product. According to information in the latest version of *Statistical Abstract of the United States 1997*, manufacturing companies employ almost 19 million people and add $1.7 trillion in value to the economy. Although the United States is now primarily a service economy, manufacturing remains a critical component. Manufactured products are used in the delivery of many, if not most, services. Maintenance of manufactured items constitutes a sizable share of the service economy.

Manufacturing is also an integral factor in achieving the mission responsibilities of the federal departments and agencies. For example, improved process technologies can conserve energy used in production, an objective of the Department of Energy. Changes in manufacturing can increase the quality of weapons and decrease their cost to the Department of Defense. Robotics are part of the National Aeronautics and Space Administration's space exploration capabilities.

In recognition of the value of manufacturing to the nation, the Congress has supported various initiatives to facilitate the development and application of new and improved process technologies, both in the government and in the industrial community. Those efforts include both direct research and development (R&D) funding and indirect measures to encourage increased private sector investment. This issue brief will highlight several federal activities in the manufacturing arena, thus providing a context within which to assist decisionmakers as they assess programmatic and budgetary choices.

Manufacturing and Competitive Success

In the face of international competition, increases in U.S. productivity must be continue if the nation is to maintain a high standard of living and generate economic growth while American firms engage in trade, retain market shares, and offer high quality goods and services. Industrial competitiveness is the ability under fair conditions to produce and deliver a given good at a lower price than other suppliers, or to deliver a better product at a given price. Firms can accomplish this in several ways, but advances in productivity are the only means that do not ultimately reduce living standards.

Technological advancement is an important means of increasing productivity. Technology allows for more efficient ways to produce existing products and for the development of new ones. Such advances can result in more effective use of resources and in quality improvements. And because technological progress is not necessarily determined by economic forces — it can also be influenced by advances in science, the organization and management of firms, government activity, or serendipity — it may affect productivity independently of economic factors such as the cost of capital or labor.

Trade in manufactured products is central to international competition, yet it is in manufacturing that the United States is experiencing a large trade imbalance. If U.S. industry responses to foreign competitors erode the standard of living, that can have a deleterious effect on the domestic economy. Thus, increasing productivity in manufacturing can be critical. In many areas, technological advancement has played an important role in allowing foreign firms to gain cost and quality advantages over American companies. While technological progress cannot be expected to solve U.S. trade problems in the short term, over the long run, enhanced technologies and manufacturing techniques can be very important for increasing productivity and economic growth. Domestic improvements in the design and manufacture of products can result in goods American consumers prefer to purchase over goods from abroad.

New technologies and their application can improve production capabilities in various ways. The factors that make a company competitive include quality and cost controls, rapid manufacture of goods, and the flexibility to meet changes in demand. Designing a new product for ease of manufacture — concurrent engineering — is often a particularly successful way to cut production expenses and make higher quality, more reliable goods. After an initial, often large, capital investment, automation also can reduce expenditures and speed production. The benefits of automation tend to be greatest in the reduction of indirect costs — those involving inventory needs, product flow, quality, facilities, and management — rather than direct labor costs.

Many markets are in a constant state of flux. Products quickly become obsolete, often necessitating numerous and prompt changes in the manufacturing process. New demands by customers require rapid alterations in production to reflect market needs. The flexibility that can be built into automated manufacturing permits changes to be made in a timely fashion. Some experts believe that with automation, economies of scale and standardized products are no longer the primary means to realize a profit. Experience thus far indicates that the *initial* unit cost for each type of product may not decrease, and may well increase with automation. However, the ability to produce numerous specialized products in smaller "batches" and at faster speeds can contribute to the increased profitability of manufacturing companies.

It appears that there are many cases where American companies have not remained competitive because they have not kept up with advances in manufacturing, as evidenced by such consumer goods as VCRs and televisions. While new markets are developed as a result of radical new technologies, market shares are maintained by constant refinements in products and improvements in the process of manufacturing them. Most innovations are actually enhancements of existing products and processes. As Don Kash of George Mason University notes, "70 percent of new ideas come out of manufacturing and marketing, not R&D." Studies by Edwin Mansfield (University of Pennsylvania) found that almost half of U.S. industrial R&D is devoted to the creation of *new* products and processes and only one-third of total spending is directed toward improved process technologies. That is in contrast to Japan, where the emphasis is on *refinements* in existing products and processes and on new manufacturing developments.

Constraints

Innovative firms often face contraints on activities related to the commercialization of new technologies and the use of new production techniques. Primary among these is the high cost of capital. That can discourage risk taking, R&D spending, purchase of new equipment, and long term business and investment perspectives. Lowering the cost of capital, however, does not necessarily ensure the increased use of improved manufacturing technology; the savings might be applied elsewhere (e.g., to augment salaries for management). The value of technology needs to be reflected in business decisions. The companies that are successful appear to operate in a similar manner; one that integrates research, development, engineering, design, manufacturing, marketing, and finance. However, American firms often see investments in technology as necessary, separate, and unavoidable costs, rather than as a principal means to achieve increased productivity and resource efficiency. Requests for assistance at centers set up to help small and mid-size companies tend to involve financial and managerial concerns rather than technical ones.

Additional constraints hamper the adoption of new automated manufacturing systems in American firms. In the United States, the relevant R&D typically is performed by companies that produce these systems, or at universities. Thus, the results tend to be generic in nature and often require extensive, and expensive, modification to address the particular needs of individual companies. That is in contrast to Japan, where work typically is performed by on-line engineers for specific use within the company. That allows the work to reflect the specialized design and production requirements of particular firms.

For automation to be effective, it is often necessary for a company to view its manufacturing process in a new way, considering greater integration of design, engineering, manufacturing, product, and marketing. The absence of such a perspective can lead to an inadequate amount of information on which to make informed decisions regarding the purchase and use of new techniques. A piecemeal approach to automation can sabotage expected and desired results. Problems are not generally failures of technology, but rather failures in implementation.

The expertise of the workforce is critical to the success of manufacturing firms. Increasingly, manufacturing requires the ability to operate computers and other sophisticated technological equipment. However, it appears that the educational and skill levels of available workers will be inadequate to meet the anticipated demands. Through the next decade, the total number of new workers is projected to decline and the pool of potential employees is expected to be less educated and less trained in the tasks necessary to meet job performance requirements.

Small Business

Problems associated with manufacturing are often exacerbated when small firms are involved. This is of concern because of the significant contribution of certain types of small firms to job creation, innovation, and manufacturing. Research by the National Science Foundation has shown that companies of fewer than 500 employees are about 2.5 times as innovative per employee as large firms. According to the Department of Commerce, the country's 381,000 small companies represent over 95% of the nation's total manufacturing firms. They also supply a significant portion of the parts used by large manufacturers.

Innovative small firms often find that they face additional difficulties because of limited resources. Fewer assets for collateral mean that banks are often unwilling to provide loans for new equipment. Automated systems available tend to meet the requirements of larger firms. In contrast to Japan, where suppliers have concentrated on the small business market, in the United States there are few vendors of equipment, information, and expertise specifically designed to the needs, organization, or operation of small manufacturing firms. Finally, in a situation where trained workers are scarce, employment competition may put small companies at another disadvantage, since they often pay lower wages and offer fewer benefits than large firms.

Congressional Initiatives

In recent years, Congress has generated legislative initiatives that directly or indirectly affect the manufacturing community. The government typically funds research and development to meet the mission requirements of the departments and agencies and to address national needs that otherwise would not be met without federal assistance. [For more information, see CRS Issue Brief 91132, *Industrial Competitiveness and Technological Advancement: Debate Over Government Policy*.] An overview of federal activities in manufacturing reflects this. The advanced manufacturing program at the National Institute of Standards and Technology, Department of Commerce (DOC), furthers the laboratory's mission to assist industry. The Mantech program of the Department of Defense (DOD) and various activities of the Defense Programs at the Department of Energy (DOE) aim to improve the manufacturing capabilities and efficiencies of the suppliers of weapons systems, as well as to clean up nuclear waste. However, as concerns were expressed over the continued competitive success of the U.S. manufacturing sector and the potential effects of any diminished capacity on the economy as well as on national security, different approaches to the federal role have been explored. Below are selected examples of the variety of governmental efforts in the manufacturing arena.

Technology Transfer

The major government-wide effort to move technologies, techniques, and expertise out of the federal R&D system is the Stevenson-Wydler Technology Innovation Act of 1980 (P.L. 96-480), as amended. The legislation requires that the transfer of technology from government laboratories be part of each federal agency's mission. That mandate is based on the assumption that there are technologies, techniques, and skills — including those used in manufacturing — developed in pursuit of each department's responsibilities that might have further commercial application. The legislation provides, among other things, the legal basis for collaborative research and development endeavors between government laboratories and private firms (through cooperative R&D agreements) and creates a system of rewards for promoting commercialization of the results of federally funded R&D. Small businesses are given special consideration in the creation of joint ventures under the law. [For additional information, see CRS Issue Brief 85031, *Technology Transfer: Use of Federally Funded Research and Development.*]

Defense Programs

The Department of Defense funds several manufacturing initiatives directed toward the development and use of new process technologies in defense related manufacturing. The service-wide manufacturing technology program, Mantech, which originated in the late 1950s, attempts to improve productivity, reduce acquisition and life cycle costs, and decrease the time lag between an invention and its application on the factory floor. According to the Department, the projects supported by Mantech, "... are beyond the normal risk or interest of industry to pursue in a timely manner." Companies receive seed funding to demonstrate the benefits of the new technologies in actual use and thus facilitate the widespread adoption of these techniques. Other collaborative efforts are carried out in conjunction with the National Aeronautics and Space Administration, the Department of Energy, the Department of Commerce, and the National Science Foundation.

Appropriations for the Mantech program in the three Services (Army, Navy, Air Force), as well as the Defense Logistics Agency totaled $228 million in FY1995, $185 million in FY1996, and $197 million in FY1997. The President's budget request for FY1998 is $68 million. The Defense Advanced Research Projects Agency (DARPA), which finances high-risk, leading-edge technology development, also participates in the Mantech initiative. Funding appropriated in FY1995 was $101 million, $103 million (estimated) in FY1996, and $64 million (estimated) in FY1997. The FY1998 budget request allowed $61 million for this activity. Authorizations contained in P.L. 105-85 include Mantech funding of $133.2 million. P.L. 105-66 appropriated $145.178 million for Mantech broken down as follows: $33.029 million for the Army, $55 million for the Navy, $48.429 for the Air Force, and $8.720 for the Defense Logistics Agency. For FY1999, P.L. 105-261 authorized spending at $172.6 million including $14.5 million for the Army, $59.0 million for the Navy, $50.9 million for the Air Force, and $26.2 million for the Defense Logistics Agency. Appropriations contained in P.L. 105-262 totaled $185.1 million for Mantech with $36.9 million for the Army, $69 million for the Navy, $53 million for the Air Force, and $26.2 million for the Defense Logistics Agency.

It should be noted that other activities related to manufacturing are also supported by DOD, including the defense-wide joint robotics program, which received $23 million in

FY1995, $22 million in FY1996, and $29 million in FY1997. P.L. 105-85 authorized $28.2 million for the joint robotics program in FY1998; P.L. 105-66 appropriated the same amount of funding. For FY1999, $17.3 million was authorized and $15.3 million was appropriated for this effort.

Department of Energy

The Department of Energy funds many programs that advance manufacturing R&D. In the civilian arena, the materials research program under Basic Energy Sciences involves work in process technologies. In Energy Conservation, the industry conservation program works with the private sector, developing new techniques to reduce the energy expended during manufacturing. However, the majority of manufacturing R&D conducted by the department is in the Defense Program area and is related to weapons and nuclear clean-up activities.

A significant portion of the manufacturing work supported by DOE in pursuit of the agency's missions in national security, science, energy resources, and environmental quality takes place within the department's laboratories. The 33 laboratories work together along with other federal agencies and the private sector to expand the application of new manufacturing techniques. [For more information on the DOE laboratories, see CRS Issue Brief 97012, *Restructuring DOE and Its Laboratories: Issues in the 105th Congress.*] For example, the National Machine Tool Partnership is a cooperative effort to augment the capabilities of the nation's machine tool industry. Participants include Sandia National Laboratories, Los Alamos National Laboratory, Lawrence Livermore National Laboratory, the Oak Ridge Y-12 Plant and the Kansas City Plant, as well as Department of Commerce's National Institute of Standards and Technology (NIST). Individual companies also take an active role in the various long-term projects and support those efforts with matched funding. An umbrella CRADA has been created between the laboratories and two manufacturing associations under which member firms can work with the relevant federal research facilities.

The Technologies Enabling Agile Manufacturing (TEAM) project is another cooperative effort created to "deploy integrated, validated, design-to-manufacture tools and to identify processes that streamline product development, reduce costs, enhance quality, and shorten cycle time." This activity involves Sandia, Oak Ridge, Lawrence Livermore, Los Alamos, and Kansas City, as well as NIST, the National Science Foundation, and the Defense Advanced Research Projects Agency. Over 20 individual companies are partners, as are academic institutions, industry consortia, and trade associations. Research and development are performed in five thrust areas that have been selected in conjunction with industry representatives. The areas are product design and enterprise concurrency, virtual manufacturing, manufacturing planning and control, intelligent closed-loop processing, and enterprise integration. In operation since 1994, this program is expected to terminate at the end of the current year as the participants believe the activity has achieved its goal of changing ideas about the way manufacturing is performed.

Also responding to industry requests for assistance, DOE created the American Textile Partnership (AMTEX), a cooperative industry/university/government effort to use R&D to "strengthen the competitiveness of the integrated U.S. textile industry — including the fiber, textile, apparel, and fabricated product sectors." Through cooperative research and development agreements (CRADAs), the energy laboratories are working to improve the industry's manufacturing capabilities, quality, and productivity. Over half the funding is provided by the private sector, with federal support remaining within the participating government laboratories. The eight main projects are computer-aided fabric evaluation, cotton biotechnology, demand activated manufacturing architecture, electronic embedded fingerprints, on-line process control, rapid cutting, sensors for agile manufacturing, and textile resource conservation.

There are many additional cooperative programs that involve the Department of Energy laboratories. Several laboratories have facilities with advanced equipment and experienced personnel available for use by the private sector. For example, the Manufacturing Technologies Center at Sandia is a designated technology deployment center and user facility open for joint projects with industry, universities, and other government agencies. Similarly, the Intelligent Systems and Robotics Center (ISRC) will work with companies through partnering and licensing of intellectual property. Within the Center, the Agile Manufacturing Prototyping System is a national testbed that allows academic and industry researchers to verify their ideas for flexible manufacturing improvements. The ISRC's Robotic Manufacturing Science and Engineering Laboratory is, according to Sandia, "the first centralized facility designed specifically to bring intelligent machine technologies and technologists together" for cooperative R&D.

Department of Commerce/National Institute of Standards and Technology

The National Institute of Standards and Technology, a laboratory of the Department of Commerce, is mandated to assist industry in the advancement of manufacturing (among other things). The Manufacturing Engineering Laboratory (MEL) is developing new techniques, including "intelligent machines; advanced sensors for real-time in-process measurements; software for precision control of machine tools; and information technology." MEL also operates the National Advanced Manufacturing Testbed, to undertake research into the technologies and standards necessary for distributed and virtual manufacturing. Funding for the Manufacturing Engineering program was $19 million in FY1996, FY1997, and FY1998. The Omnibus Consolidated Appropriations Act, P.L. 105-277, provided funding for FY1999 at the same figure.

Other programs at NIST that support manufacturing R&D include the advanced processing activities in the Materials Science and Engineering Laboratory. NIST also operates the Manufacturing Extension Partnership (MEP), a program of regional centers to help small and medium sized companies use the knowledge and technologies developed under the agency's auspices. Created by P.L. 100-418, the Omnibus Trade and Competitiveness Act of 1988, this activity now provides technical and managerial assistance to firms in all 50 states and Puerto Rico. Federal funding is matched by non-national sources including state and local governments and industry. Centers are selected in response to competitive solicitations, and awards are merit-based. The sponsor is expected to provide expertise reflecting the needs of the business community and the types of industries in that region. No direct financial support is available to companies. [For more information, see CRS Report 97-104, *Manufacturing Extension Partnership Program: An Overview*.] Funding for MEP in FY1996 was $80 million and $99.9 million in FY1997. The FY1998 budget request was for $123 million. No FY1998 authorizing legislation was enacted, but P.L. 105-119 appropriated $113.5 million for the program. This law also extended government funding for centers beyond the 6-year limit, at one-third of the facility's annual cost and for a one year period, if a positive evaluation is received. For FY1999, the Administration proposed funding of $106.8 million, a decrease of 6% over the previous year. P.L. 105-277 appropriated the $106.8 million. The decrease reflects declining federal obligations as the MEP centers mature. P.L. 105-309, the Technology Administration Act of 1998, amended the program's authorizing legislation and allowed federal funding past the original 6-year limit.

The Omnibus Trade and Competitiveness Act also established the Advanced Technology Program (ATP) within NIST. This activity provides seed funding, matched by private sector investments, to companies or consortia of universities, firms, and government laboratories for the purpose of accelerating the development of generic technologies that have broad applications across industries. Several of the focused program areas designated for long-range support (determined in conjunction with industry) involve manufacturing initiatives including technologies for the integration of manufacturing applications, manufacturing composite structures, materials processing for heavy manufacturing, motor vehicle manufacturing technology, advanced manufacturing control systems, photonics manufacturing, and microelectronics manufacturing infrastructure. Many of the awards made in the general competitions also relate to manufacturing and involve machine tools, ceramics, and materials processing. [For a discussion of this activity, see CRS Report 95-36, *The Advanced Technology Program.*] While targeted for elimination by the 104th Congress, the program had strong Administration support and was funded, although at reduced levels. Appropriations for ATP were $221 million in FY1996 and $225 million in FY1997, later reduced to $218 million by P.L. 105-18. The President requested funding for FY1998 at $276 million. As noted above, no FY1998 authorization legislation was enacted. However, P.L. 105-119 provided FY1998 appropriations of $192.5 million. The Administration's FY1999 budget proposal includes $259.9 million for ATP, an increase of 35% over FY1998. While not providing such a large increase, P.L. 105-277 appropriated $203.5 million, 6% over the previous year.

The National Institute of Standards and Technology administers the Malcolm Baldridge National Quality Award program. This effort recognizes American companies that practice effective quality control in the provision of goods and services and requires the dissemination of information on the successful strategies and programs that led to the award. Guidelines and other criteria have been established for use by industry, government, and organizations in evaluating their own quality improvement activities. Funding for this program was $3 million in FY1996 and in FY1997. The agency wanted to expand the awards to include health care and education. As such, the Administration requested $5 million in funding. The House authorized $4.1 million for FY1998 and $5.3 million for FY1999. However, P.L. 105-119 provided appropriations of $3 million for current activities but not for any new efforts. Again this year, the President's budget proposed an expansion of the program and requested $5.4 million to achieve this goal. P.L. 105-277 appropriated the $4.9 million to undertake the additional activities in health care and education.

Selected Cross-Agency Efforts

There are several major cross agency R&D initiatives designed to promote industrial competitiveness ana economic growth through improvements in manufacturing and materials. Among these is the Partnership for a New Generation Vehicle, a cooperative effort under the leadership of the Department of Commerce that involves seven federal agencies (including the Departments of Defense, Energy, and Transportation, the Environmental Protection

Agency, the National Aeronautics and Space Administration, and the National Science Foundation) and U.S. automobile manufacturers. One objective of this activity is to develop advanced industrial manufacturing methods that would reduce costs and improve quality while simultaneously meeting national goals of fuel economy and pollution control. [For more information, see CRS Report 96-191, *The Partnership for a New Generation of Vehicles*.]

The National Electronics Manufacturing Initiative (NEMI) brings together large electronic equipment manufacturers, key suppliers, and government agencies to undertake strategic planning and partnering with the goals of making the most effective use of federal investments in electronics R&D. No specific programs or projects are expected, but several technology roadmaps have been developed and collaborative opportunities identified. The Manufacturing Infrastructure Initiative has as its objective the support of research and development in areas of interest to the U.S. manufacturing sector. Through government-industry partnerships, work at various federal departments and agencies is focusing on projects in advanced manufacturing systems, engineering tools for design and manufacture, advanced manufacturing processes and equipment, manufacturing education and training, and manufacturing technology deployment.

The Departments of Energy, Defense, and Commerce and the National Science Foundation supported the Next Generation Manufacturing project, an industry-led program designed to "develop a broadly accepted model of future manufacturing enterprises [and] recommend actions that manufacturers, working individually and in partnership with government, industry, and the academic community, can use to attain world-class status." In the spring of 1997, a report was issued detailing the results of a study involving the various interested parties. Efforts are now focused on implementing the recommendations made using the auspices of the Agility Forum, the Leaders for Manufacturing Program at MIT, and the Technologies Enabling Manufacturing (TEAM) program. In response, the National Institute of Standards and Technology, the Defense Advanced Research Projects Agency, the Department of Energy, and the National Science Foundation have begun a one-year project to develop "integrated manufacturing technology roadmaps" for use in directing further research investments.

Policy Perspectives

The Clinton Administration has made technology development a cornerstone of its economic growth policies. As such, the President has supported efforts to increase the generation of new manufacturing technologies and their adoption in industry. To that end, the Administration has promoted programs to increase the transfer of technology from the federal laboratories to the private sector; encouraged the formation of R&D partnerships between and among universities, industry, and the government; expanded the reach of manufacturing extension (the Manufacturing Extension Partnership); and provided direct funding for private sector efforts to undertake high-risk, broad-based manufacturing research and development (the Advanced Technology Program). [For more information, see CRS Issue Brief 91132, *Industrial Competitiveness and Technological Advancement: Debate Over Government Policy*.]

On a national level, the development of any government effort to facilitate technological progress in all areas, including manufacturing, has been difficult because of the absence of a consensus on the need for an articulated policy. Technology development and commercialization traditionally have been considered private sector functions in the United States. A variety of relevant programs and policies are in place, but the approach generally has been ad hoc. Much of what has been accomplished focuses on research or on agency mission requirements, which typically do not reflect the needs of the commercial marketplace.

The congressional support that has evolved for technological advancement tends to reflect a balance between legislatively mandated incentives for private sector investment and direct government financing of R&D. However, the 104th Congress called into question many of the established programs, particularly those that involved federal funding for technology development in industry. Unsuccessful attempts were made to terminate the Advanced Technology Program at the National Institute of Standards and Technology. The Technology Reinvestment Project in DOD — an activity that included manufacturing education programs and other efforts to develop dual-use (military/civilian) technologies — was eliminated. Supporters of indirect measures argue that the market is superior to government in deciding which technologies are worthy of investment. They prefer mechanisms that enhance market opportunities and the ability of the private sector to make choices. Those include tax credits, changes to the antitrust laws as they pertain to cooperative R&D, and liability or regulatory reform, among others. Proponents of direct government assistance maintain, conversely, that indirect methods can be wasteful and ineffective and may compromise other goals of public policy in the hope of stimulating innovative performance. Advocates of a direct approach argue that it is important to apply the nation's scarce resources in technological areas that have the greatest promise as determined by industry and supported by the business community's willingness to match federal funds.

The 105th Congress continued to explore manufacturing research and technology development within the context of budget decisions. There appears to be agreement with the basic premise that the government does have a role in fostering technological advancement in the private sector; the debate is over how that is best achieved within existing and future budget constraints. Research and development funding constitutes the largest single portion of total federal discretionary spending and is therefore subject to cuts as legislators move toward a balanced budget. As a result, it is unrealistic to expect significant increases in government funding for such work. Thus, priorities concerning the allocation of financing may need to be established.

The legislative initiatives expected to be examined and debated during the 106th Congress include on-going efforts to generate new advanced manufacturing technologies and to augment their adoption in the private sector. However, while the federal government can support programs to encourage improvements in the manufacturing sector, the final responsibility for action remains in the business community. Some would suggest that a significant segment of industry does not assign a sufficiently high value to technological progress in manufacturing, planning, and operations. While federal efforts can prove beneficial, the decision to commit greater resources to technological means of improving competitive success still resides largely with industry.

LEGISLATION

P.L. 105-66, H.R. 2266

A bill making FY1998 appropriations for the Department of Defense. Introduced July 25, 1997; referred to the House Committee on Appropriations. Reported out as an original measure on July 25. Passed House, amended, on July 29, 1997. Received in Senate the same day. Senate substituted language of S. 1005, amended, and passed bill July 29, 1997. Conference held September 16. House and Senate agreed to conference report September 25, 1997. Signed into law October 8, 1997.

P.L. 105-85, H.R. 1119

A bill to authorize FY1998 and FY1999 appropriations for the Department of Defense and the defense activities of the Department of Energy. Introduced March 19, 1997; referred to Committee on National Security. Reported to House, amended, June 16. Passed House, amended, June 25, 1997. Received in Senate July 7; Senate inserted text of S. 936, as amended, and passed the bill July 11, 1997. Conference report filed October 23, 1997. House agreed to conference report October 28 and the Senate agreed November 6, 1997. Signed into law November 18, 1997.

P.L. 105-119, H.R. 2267

A bill to provide FY1998 appropriations for the Departments of Commerce, Justice, State, and the Judiciary, and related agencies. Funds the Advanced Technology Program at $185 million and the Manufacturing Extension Program at $113.6 million. Introduced July 25, 1997; referred to Committee on Appropriations. Reported July 25, 1997. Passed House, amended, September 30, 1997. Received in Senate October 1. Senate substituted language of S. 1022, and passed the bill. Conference held November 7, 1997. House and Senate agreed to conference report November 13, 1997. Signed into law November 26, 1997.

P.L. 105-261, H.R. 3616

Department of Defense FY1999 Authorizations. Authorizes Mantech funding of $172.6 million, among other things. Introduced April 1, 1998; referred to the House Committee on National Security. Reported to House, amended, May 12. Passed House, amended, May 21, 1998 and received in Senate the following day. Language of S. 2057 substituted; passed Senate, amended, June 25, 1998. Conference report filed September 22. House agreed to conference report September 24, 1998. Senate agreed October 1, 1998. Signed into law October 17, 1998.

P.L. 105-262, H.R. 4103

Department of Defense FY1999 Appropriations. Appropriates $185.1 million for the Mantech program of the Department of Defense. Reported as an original measure from the House Appropriations Committee on June 22, 1998. Passed House, amended, June 24. Language of S. 2132 substituted and passed Senate in lieu of S. 2132 July 30, 1998. Conference report filed September 25, 1998. House agreed to conference report September 28, 1998; Senate agreed the following day. Signed into law October 17, 1998.

P.L. 105-277, H.R. 4328

Omnibus Consolidated Appropriations Act. Provides FY1999 appropriations of $203.5 million for the Advanced Technology Program and $106.8 million for the Manufacturing Extension Partnership at the National Institute of Standards and Technology. Extends the research and experimentation tax credit through June 30, 1999. Reported as an original measure from the House Committee on Appropriations July 24, 1998. Passed House, amended, on July 30, 1998; passed Senate, amended, in lieu of S. 2307 the same day. Conference held October 2, 1998. House agreed to conference report October 20; Senate agreed to conference report October 21, 1998. Signed into law October 21, 1998.

P.L. 105-309, H.R. 1274

The Technology Administration Act of 1998. Allows the federal government to provide financing for the Manufacturing Extension Partnership centers past the current six-year limit if a positive, independent evaluation is received at least every two years. Federal funding is limited to one-third the centers costs. Introduced April 10, 1997; referred to Committee on Science. Committee markup held April 16. Passed Committee, amended, April 21, 1997. Passed House April 24, 1997. Received in Senate April 24, 1997; referred to Committee on Commerce, Science, and Transportation. Passed Senate, amended, October 10, 1998. House agreed to Senate amendment October 13, 1998. Signed into law October 30, 1998.

Excerpted from Congressional Research Service Report Number IB 97041

COOPERATIVE R&D: FEDERAL EFFORTS TO PROMOTE INDUSTRIAL COMPETITIVENESS

Wendy H. Schacht

MOST RECENT DEVELOPMENTS

The promotion of cooperative research and development among industry, universities, and the federal R&D establishment has played an important role in President Clinton's articulated technology policy. Various congressional initiatives have supplemented this approach, augmenting support for various collaborative efforts including technology transfer from federal laboratories as well as the Advanced Technology Program (ATP) and the Manufacturing Extension Partnership (MEP) at the National Institute of Standards and Technology (NIST). However, many of these activities were revisited in the 104th Congress given the Republican majority's statements in favor of indirect measures such as tax policies, intellectual property rights, and antitrust laws to promote technology development; increased government support for basic research; and decreased direct federal funding for private sector technology initiatives. Although the cooperative programs were not terminated during the 104th Congress, several were funded at reduced levels. The 105th Congress continued support for most on-going federal programs that facilitate joint R&D and increased funding for ATP. The Taxpayer Relief Act of 1997, P.L. 105-34, extended the research and experimentation tax credit for 13 months from June 1, 1997 through June 30, 1998. Allowed to expire again, P.L. 105-277, the Omnibus Consolidated Appropriations Act of 1998, reinstated the tax credit through June 30, 1999. The pilot Small Business Technology Transfer (STTR) program was reauthorized through FY2001 by P.L. 105-135. P.L. 105-309, the Technology Administration Act of 1998, extends federal funding for the technology transfer centers of the Manufacturing Extension Partnership at one-third of costs after the initial 6 years. In the current Congress, H.R. 1744, as introduced May 10, 1999, would authorize $190.7 million in FY2000 funding for the Advanced Technology Program and $106.8 million for the Manufacturing Extension Partnership. The Senate authorization bill, S. 1407, introduced July 21, 1999, provides $209.6 million for ATP and $106.8 million for MEP. On October 20, 1999, the House and Senate agreed to the conference on H.R. 2670, which provides FY2000 appropriations of $142.6 million for ATP and $104.8 million for MEP (although the original House version of the bill included no funding for ATP). However, the President vetoed this legislation on October 25, 1999. H.R. 209, the Technology Transfer Commercialization Act, passed by the House on May 11, 1999, would make changes in the technology transfer legislation to improve the ability of the government to license federally owned inventions. S. 804 and S. 999 are similar. H.R. 2488, the Financial Freedom Act of 1999, as passed by the Congress but vetoed by the President, would have extended the research and experimentation tax credit through June 30, 2004. However, H.R. 2923, as reported to the House on September 28, 1999, was introduced to achieve the same objectives.

BACKGROUND AND ANALYSIS

Rationale

In response to concerns over competition from foreign firms, the U.S. Congress has increasingly looked for ways the federal government can stimulate technological innovation in the private sector. This technological advancement is critical in that it contributes to economic growth and long term increases in our standard of living. New technologies can create new industries and new jobs; expand the types and geographic distribution of services; and reduce production costs by making more efficient use of resources. The development and application of technology also plays a major role in determining patterns of international trade by affecting the comparative advantages of industrial sectors. Since technological progress is not necessarily determined by economic conditions, it can have effects on trade independent of shifts in macroeconomic factors that may affect the marketplace.

Joint ventures are an attempt to facilitate technological advancement within the industrial community. Academia, industry, and government can play complementary roles in technology development. While opponents argue that cooperative ventures stifle competition, proponents assert that they are designed to accommodate the strengths and responsibilities of these sectors. Collaborative projects attempt to utilize and integrate what the participants do best and to direct these efforts toward the goal of generating new goods, processes, and services for the marketplace. They allow for shared costs, shared risks, shared facilities, and shared expertise.

The lexicon of current cooperative activity covers various different institutional and legal arrangements. These ventures might include industry-industry joint projects involving the creation of a new entity to undertake research, the reassignment of researchers to a new effort, and/or hiring new personnel. Collaborative industry-university efforts may revolve around activities in which industry supports centers (sometimes cross-disciplinary) for research at universities, funds individual research projects, and/or exchanges personnel. Cooperative activities with the federal government might include projects that use federal facilities and researchers, federal funding for industry-industry or industry-university efforts, or financial support for centers of excellence at universities to which the private sector has access.

There are many different types of cooperative arrangements. The flexibility associated with this concept can allow for the development of institutional and organizational plans tailored to the specific needs of the particular project. Issues of patent ownership, disclosure of information, licensing, and antitrust are to be resolved on an individual basis within the general guidelines established by law governing joint ventures.

Collaborative ventures can be structured either "horizontally" or "vertically." The former involves efforts in which companies work together to perform research and then use the results of this research within their individual organizations. The latter involves activities where researchers, producers, and users work together. Both approaches are seen as ways to address some of the perceived obstacles to the competitiveness of American firms in the marketplace.

Joint Industrial Research

Traditionally, the federal government has funded research and development to meet mission requirements; in areas where the government is the primary user of the results; and/or where there is an identified need for R&D not being performed in the private sector. Most government support is for basic research which is often long-term and highly risky for individual companies; yet research can be the foundation for breakthrough achievements which can revolutionize the marketplace. Studies have shown that inventions based on R&D

are the more important ones. However, the societal benefits of research tend to be greater than those that can be captured by the firm performing the work. Thus the rationale for federal funding of research in industry.

The major emphasis of legislative activity has been on augmenting research in the industrial community. This focus is reflected in efforts to encourage companies to undertake cooperative research arrangements and expand the opportunities available for increases in research activities. Collaboration permits work to be done which is too expensive for one company to fund and also allows for R&D that crosses traditional boundaries of expertise and experience. A joint venture makes use of existing, and supports development of new resources, facilities, knowledge, and skills.

The concentration on increased research as a prelude to increased technological advancement was based upon the "pipeline model" of innovation. This process was understood to be a series of distinct steps from an idea through product development, engineering, testing, and commercialization to a marketable product, process, or service. Thus increases at the beginning of the pipeline — in research — were expected to result in analogous increases in innovation at the end. However, this model is no longer considered valid. Innovation is rarely a linear process and new technologies and techniques often occur that do not require basic or applied research or development. Most innovations are actually incremental improvements to existing products and processes. In some areas, particularly biotechnology, research is closer to a commercial product than this conception would indicate. In others, the differentiation between basic and applied research is artificial. The critical factor is the commercialization of the technology. Economic benefits accrue only when a technology or technique is brought to the marketplace where it can be sold to generate income and/or applied to increase productivity.

In the recent past, it was increasingly common to find that foreign companies were commercializing the results of U.S. funded research at a faster pace than American firms. In the rapidly changing technological environment, the speed at which a product, process, or service is brought to the marketplace is often a crucial factor in its competitiveness. The recognition that more than research needs to be done has lead to other approaches at cooperative efforts aimed at expediting the commercialization of the results of the American R&D endeavor. These include industry-university joint activities, use of the federal laboratory system by industry, and industry-industry development efforts where manufacturers, suppliers, and users work together.

Industry-University Cooperative Efforts

Industry-university cooperation in R&D is one important mechanism intended to facilitate technological innovation. Traditionally, universities perform much of the basic research integral to certain technological advancements. They are generally able to undertake fundamental research because it is part of the educational process and because they do not have to produce for the marketplace. The risks attached to work in this setting are fewer than those in industry where companies must earn profits. Universities also educate and train the scientists, engineers, and managers employed by companies.

Academic institutions do not have the commercialization capacity available in industry and necessary to translate the results of research into products and processes that can be sold in the marketplace. Thus, if the work performed in the academic environment is to be integrated into goods and services, a mechanism to link the two sectors must be available. Prior to World War II, industry was the primary source of funding for basic research in universities. This financial support helped shape priorities and build relationships. However, after the war the federal government supplanted industry as the major financial contributor and became the principal determinant of the type and direction of the research performed in academic institutions. This situation resulted in a disconnection between the university and industrial communities. Because industry and not the government is responsible for commercialization, the difficulties in moving an idea from the research stage to a marketable product or process appear to have been compounded.

Efforts to encourage increased collaboration between the academic and industrial sectors might be expected to augment the contribution of both parties to technological advancement. Company support for research within the university provides additional funds and information on the concerns and direction of industry. For many companies, access to expertise and facilities outside of the firm expands or complements available internal resources. Yet, such cooperation should not necessarily be seen as a panacea. Oftentimes, collaborative ventures fail because of various factors including conflicting goals, differing research cultures, and financial disagreements.

Federal Laboratory-Industry Interaction

The federal government can share its extensive facilities, expertise, knowledge, and new technologies with partners in a cooperative venture. In certain cases, the government laboratories have scientists and engineers with experience and skills, as well as equipment, not available elsewhere. The government also has a vested interest in technology development. It does not have the mandate or resources to manufacture goods but has a stake in the availability of products and processes to meet mission requirements. In addition, technological advancement contributes to the economic growth vital to the health and security of the nation.

Collaboration between government laboratories and industry is not, however, just a one way street. In several technological areas, particularly electronics and computer software, the private sector is more advanced in technologies important to the national defense and welfare of this country. Interaction with industry offers federal scientists and engineers valuable information to be used within the government R&D enterprise.

Federal Initiatives in Cooperative R&D

The cooperative venture concept is not new. In the early 1970s, the National Science Foundation established its Industry-University Cooperative Research Centers program. The Electric Power Research Institute, a research organization supported by the electric power utilities, has been in operation since 1973. In the private sector, the Microelectronics and Computer Technology Corporation (MCC), which performs research for its member firms, and the Semiconductor Research Corporation (SRC), which funds research in universities, were created in the early 1980s. The difference today is the number of projects and the scope of legislative activity designed to promote cooperative ventures.

Faced with pressures from foreign competition, the government's interest appears to be expanding beyond that of funding R&D, to meeting other critical national needs including the economic growth that flows from new commercialization in the private sector. While acknowledging that the commercialization of technology is the responsibility of the business community, in the past several years the government has attempted to stimulate innovation and technological advancement in industry. These activities often involve the removal of barriers to technology development in the private sector, thereby permitting market forces to operate and the provision of incentives to encourage increased innovation related efforts in industry. Cooperative R&D efforts are a part of both these trends.

The National Cooperative Research Act (P.L. 98-462) is designed to encourage companies to undertake joint research which is typically long-term, risky, and often too expensive for one company to finance. This legislation clarifies the antitrust laws and requires that the "rule of reason" standard be applied in determinations of violations of these laws; that cooperative research ventures are not to be judged illegal "per se". It also eliminates treble damage awards for those research ventures found in violation of the antitrust laws if prior disclosure (as defined in the law) has been made. In addition, P.L. 98-462 makes some changes in the way attorney fee awards are made to discourage frivolous litigation against joint research ventures without simultaneously discouraging suits of plaintiffs with valid claims. Over 700 joint ventures have filed with the Justice Department since this law was enacted.

P.L. 103-42, the National Cooperative Production Amendments Act of 1993, amends the National Cooperative Research Act by, among other things, extending the original law's provisions to joint manufacturing ventures. These provisions are only applicable, however, to cooperative production when (1) the principal manufacturing facilities are "located in the United States or its territories, and (2) each person who controls any party to such venture...is a United States person, or a foreign person from a country whose law accords antitrust treatment no less favorable to United States persons than to such country's domestic persons with respect to participation in joint ventures for production."

The Omnibus Trade and Competitiveness Act of 1988 (P.L. 100-418) created the Advanced Technology Program (ATP) at the Department of Commerce's National Institute of Standards and Technology. This program provides seed funding, matched by private sector investment, to companies or consortia comprised of universities, companies, and/or government laboratories for the development of generic technologies that have broad application across industrial sectors. To date, 431 projects have been funded representing approximately $1,384 million in federal financing matched by over $1,419 million in private sector support. Eleven initial R&D programs were selected for funding, almost half of which involved consortia. Twenty-seven awards were made to programs in the second year; approximately one-third were *consortia*. In December 1992, 21 new ATP awards were made, including three joint ventures. Thirty additional projects were funded in 1993, and, in October 1994, 41 awards were made in four key technology areas: information infrastructure for healthcare; tools for DNA diagnostics; component-based software; and computer-integrated manufacturing for electronics. Fourteen are cooperative efforts. In November 1994, 47 additional awards were made in the general competition and in the area of manufacturing composite structures. Twenty-four involve collaborative R&D. Of the 24 awards announced on July 13, 1995, 35% of the projects in the general competition were joint ventures and 29% in the focused competition. The following month 21 additional awards were made of which 9 were cooperative efforts. In early September, another 44 grants were awarded including 19 joint ventures. Later in that month, 10 more awards were made of which three were to cooperative efforts. On January 25, 1996, an additional four projects received awards; three involved multiple firms. In March 1997, NIST announced that it would fund 8 new proposals from the FY1996 general competition of which 2 were collaborative projects. Sixty-four awards were made in October 1997; 15 involving multiple companies. In October 1998, NIST awarded funding for 79 new projects involving more than 150 companies, 11 universities, and several federal laboratories. This reflects changes in the ATP selection criteria designed to encourage large companies to participate in joint ventures with small firms and academic institutions. (For more information, see CRS Report 95-36 SPR, *The Advanced Technology Program*.)

Appropriations for the Advanced Technology Program were $35.9 million in FY1991, $47.9 million in FY1992, and $67.9 million in FY1993. FY1994 appropriations expanded significantly to $199.5 million and even further to $431 million in FY1995. However, P.L. 104-6, the DOD Emergency Supplemental Appropriations and Rescissions Act, rescinded $90 million of this amount. The President's FY1996 budget request for ATP was $490.9 million. There were no authorizations. The original appropriations bill, H.R. 2076, which passed the Congress but was vetoed by the President, provided no financing for ATP. The final appropriations legislation, P.L. 104-134, funded the Advanced Technology Program at $221 million for FY1996. The following year, FY1997, the Omnibus Consolidated Appropriations Act (P.L. 104- 208) provided support levels of $225 million, but $7 million was rescinded by P.L. 105-18. While no authorization legislation was enacted for FY1998, P.L. 105-119 funded ATP at $192.5 million. The President's FY1999 budget request included $259.9 million for this program, an increase of 35%. However, P.L. 105-277, the Omnibus Consolidated Appropriations Act, funded ATP at $197.5 million, 3% above the previous year. This figure reflected a $6 million recission to account for "deobligated" funds resulting from prior projects that had been terminated early. In the FY2000 budget, the Administration has requested $238.7 million for ATP, an increase of 21% over the current year. H.R. 1744, introduced May 10, 1999, would authorize funding of $190.7 million for the program, 3% below FY1999. The Senate authorization bill, S. 1407, provides $209.6 million, a 6% increase. S. 1217, passed by the Senate on July 22, 1999, would appropriate $226.5 million, 15% more than the previous year. H.R. 2670, as passed by the House on August 5, 1999, contained no appropriated funding for ATP. The report accompanying the House bill states that ". . .the program has not produced a body of evidence to overcome those fundamental questions about whether the program should exist in the first place." The final version of H.R. 2670, as passed by the House and Senate on October 20, 1999, appropriates $142.6 million for the program, a decrease of 28% from the previous year. However, the President vetoed this legislation on October 25, 1999.

Several laws have attempted to facilitate industry-university cooperation. Title II of the Economic Recovery Tax Act of 1981 (P.L. 97-34) provided, in part, a temporary 25% tax credit for 65% of all company payments to universities for the performance of basic research. Firms were also permitted a larger tax deduction for charitable contributions of equipment used in scientific research at academic institutions. The Tax Reform Act of 1986 (P.L. 99-514) kept this latter provision, but reduced the credit for university basic research to 20% of all corporate expenditures for this work over the sum of a fixed research floor plus any decrease in non-research giving.

The 1981 Act also provided an increased charitable deduction for donations of new equipment by a manufacturer to an institution of higher education. This equipment must be used for research or training for physical or biological sciences within the United States. The tax deduction was equal to the manufacturer's cost plus one-half the difference between the manufacturer's cost and the market value, as long as it does not exceed twice the cost basis.

This research and experimentation tax credit expired in June 1992 when an extension contained in H.R. 11, the Enterprise Zone Tax Act, was vetoed by former President Bush. The Omnibus Budget Reconciliation Act, P.L. 103-66, reinstated the credit through July 1995 and made it retroactive to the date of its previous expiration. The credit again expired. However, P.L. 104-188, the Small Business Job Protection Act, reinstated the tax credit for application between July 1, 1996 and May 31, 1997. The Taxpayer Relief Act of 1997, P.L. 105-34, extended the credit for 13 months from June 1, 1997 through June 30, 1998. The tax credit expired once again but was reinstated through June 30, 1999, by P.L. 105-277. Several bills also were introduced that would have permitted the research tax credit to be applied to support for certain collaborative research consortia. In the current Congress, H.R. 2488, the Financial Freedom Act of 1999, as passed by both the House and Senate but vetoed by the President on September 23, 1999, would have extended the research and experimentation tax credit through June 30, 2004 and increased the credit rate applicable under the alternative incremental research credit by one percentage point per step. In response, H.R. 2923 has been introduced to accomplish the same objectives.

Amendments to the patent and trademark laws contained in P.L. 96-517 also were designed to foster interaction between academia and the business community. This law provides, in part, for title to inventions made by contractors receiving federal R&D funds to be vested in the contractor if it is a university, not-for-profit institution, or a small business. Certain rights to the patent are reserved for the government and these organizations are required to commercialize within a predetermined and agreed upon time frame. Providing universities with patent title is expected to encourage licensing to industry where the technology can be manufactured or utilized, thereby creating a financial return to the academic institution. University patent applications and licensing have increased since this law was enacted. (For more discussion on this topic see CRS Report 94-501, *The Bayh-Dole Act: Patent Policy and the Commercialization of Technology* and CRS Report 98-862, *R&D Partnerships and Intellectual Property: Implications for U.S. Policy*.)

Many cooperative industry-industry or industry-university programs are supported and/or organized by the federal departments and agencies. These include, but are not limited to, the National Science Foundation's Engineering Research Centers, the approximately 40 Industry-University Cooperative Research Programs, and the more recent Science and Technology Centers. A program to match small businesses interested in joint manufacturing technology efforts has been created in the Department of Commerce.

While most legislative activities are intended to facilitate technological advance across industries, there have been several recent efforts to provide direct assistance for cooperative ventures in a particular industry. These initiatives are based, in part, on national defense and economic security concerns over specific technologies that are, or are perceived as, potentially critical to a wide range of businesses. Among the joint ventures, funded primarily by the Department of Defense, have been SEMATECH (a joint private sector semiconductor manufacturing research effort which is now privately financed), the National Center for

Manufacturing Sciences, and the steel initiative. In addition, DOD supports the Software Engineering Institute and the Department of Energy assists in the Partnership for a New Generation Vehicle initiative that, among other things, encourages joint R&D between federal laboratories and private firms leading to commercialization.

Cooperation between industry and the federal R&D enterprise is another facet of the effort to increase industrial competitiveness through joint ventures. The federal government will spend an estimated $79 billion for research and development in FY1999 to meet the mission requirements of the federal departments and agencies. This has led to many technologies and techniques, as well as to the generation of knowledge and skills, which may have applications beyond their original intent. To foster their development and commercialization in the industrial community, various laws have established institutions and mechanisms to facilitate the movement of ideas and technologies between the public and private sectors.

The Stevenson-Wydler Technology Innovation Act (P.L. 96-480), as amended by the Federal Technology Transfer Act (P.L. 99-502) and the Department of Defense FY1990 Authorizations (P.L. 101-189), provides, in part, a legislative mandate for technology transfer from the federal government to the private sector, establishes a series of offices in the agencies and/or laboratories to administer transfer efforts, provides incentives for federal laboratory personnel to actively engage in technology transfer, and creates new contractual means for industry to work with the laboratories including cooperative research and development agreements. P.L. 104-113, the National Technology Transfer and Advancement Act, attempts to clarify existing policy with respect to the dispensation of intellectual property under a CRADA by amending the Stevenson-Wydler Act. (For additional information see CRS Issue Brief 85031, *Technology Transfer: Use of Federally Funded Research and Development*.) H.R. 209, the Technology Transfer Commercializaiton Act, passed by the House on May 11, 1999, makes changes in current practices concerning patents held by the government to make it easier for federal agencies to license such inventions to the private sector for commercialization. S. 804, introduced April 14, 1999, and S. 999, introduced May 11, 1999, are similar.

The Omnibus Trade and Competitiveness Act (P.L. 100-418) established a program of regional Centers for the Transfer of Manufacturing Technology (now part of the Manufacturing Extension Partnership effort) to facilitate the movement to the private sector of knowledge and technologies developed under the aegis of the National Institute of Standards and Technology. (For more discussion, see CRS Report 97-104, *Manufacturing Extension Partnership Program*.) In addition, the law required that NIST provide technical assistance to state technology extension programs in an effort to improve private sector access to federal technology. (For additional Information, see CRS Issue Brief 91132, *Industrial Competitiveness and Technological Advancement: Debate over Government Policy*.) Government-industry collaboration is further facilitated by a provision of the FY1991 National Defense Authorization Act (P.L. 101- 510) that amends Stevenson-Wydler to allow government agencies and laboratories to develop partnership intermediary programs to augment the transfer of laboratory technology to the small business community. A pilot activity under the Small Business Development Act of 1992, the Small Business Technology Transfer program, facilitates cooperative work between small companies and federal labs leading to the commercialization of new technology. Scheduled to sunset in FY1996, the program was extended for one year until P.L. 105-135 reauthorized it through FY2001 (see CRS Report 96-402, *Small Business Innovation Research Program*).

Issues

It is not yet known whether federal support of cooperative ventures signals a long-term commitment to the development of technology. However, statements issued by President Clinton and Vice President Gore during their first term in office (see *A Vision of Change for America and Technology for America's Economic Growth, A New Direction to Build Economic Strength*) set out a policy to actively promote joint development activities as well as joint research. The proposed initiatives discussed above are part of the Administration's strategy to provide both direct and indirect federal support for expanded cooperative R&D leading to commercialization. Given concerns over the federal budget, it is unlikely that large sums of government money will be forthcoming for such efforts in the future. However, other actions may reflect federal interest in the process of technological advancement. The use of the extensive government R&D system, with its expensive state-of-the-art facilities, can provide both academia and industry with resources that may be beyond their financial ability. And despite the often short-term focus of budget decisions, federal funds and non-monetary contributions to cooperative ventures may be leveraged by contributions from state and local agencies and the private sector.

If the proliferation of programs is any indication, state and local jurisdictions have been in the forefront of cooperative endeavors. Many state and local economic development activities focus on increasing innovation and the use of technology in the private sector. Instead of competing for companies to relocate, many of these jurisdictions now see additional benefits accruing from the creation of new firms and the modernization of existing ones through the application of new technology. Various states and localities are attempting to foster an entrepreneurial climate by undertaking the development and support of a variety of programs to assist existing high technology businesses, to promote the establishment of new companies, and to facilitate the use of new technologies and processes in traditional industries. While these efforts vary by state and locality, many of them include industry-university-government cooperation. Several of the President's proposals for increasing cooperative ventures build upon existing state and local activities in these areas. (For additional discussion, see CRS Report 96-958 SPR, *Technology Development: Federal-State Issues* and CRS Report 98-859, *State Technology Development Strategies: The Role of High Tech Clusters.*)

Proponents of cooperative work argue that certain benefits are associated with joint ventures. The increased popularity of this concept, and expanding federal support for this approach, however, might suggest some questions be raised to assess whether cooperative ventures are meeting expectations. Are there drawbacks to this effort in general and in specific instances? Are cooperative projects addressing the problems associated with the competitiveness of U.S. industry? Are they moving technology development in the right direction?

It is perhaps too soon to determine whether the cooperative R&D venture is an effective, or the most effective, mechanism to increase technological advancement. The time between research and the availability of a marketable product or process is generally thought to be from 10-15 years; less in those areas such as biotechnology where research and commercial commodity tend to be more closely linked. Many of the cooperative activities fostered by the federal government are of recent origin and therefore have not had sufficient time to generate measurable results. However, raising certain issues and/or questions might serve to develop a framework within which future, near term decisions concerning technology development and cooperative R&D might be addressed.

It might be expected that an increasing number of industries and/or companies will come to the federal government for assistance in supporting cooperative R&D activities. Despite opposition by some to what has been described as "picking winners or losers," various sectors of the government have chosen to provide funding for cooperative ventures in specific industries while requiring that the private sector generate matching funds. At the same time, there are programs and policies that attempt to facilitate cooperative efforts across industry in general. Decisions might need to be made whether one approach is better than the other, or if both should continue.

If part of government policy is to respond to individual industry requests for assistance, Congress may wish to consider developing procedures to select between industries and/or companies competing for limited federal funds. Can, and should, federal guidelines be established? In addition, is it possible to determine at this time what type of cooperative ventures are the most effective and efficient? Is there, in fact, one best model or should each venture be tailored to the specific situation? And finally, what are the implications of these decisions for policymaking in Congress?

Development

As noted above, innovation is a dynamic process that can involve idea origination, research, development, commercialization, and diffusion throughout the economy. However, it is not a linear process and an innovation may occur without developing through these steps. In fact, most innovations are actually incremental changes in existing goods and services in response to unmet market needs. The most crucial factor is the availability or use of the technology or technique in the marketplace.

In the recent past, the commercialization and diffusion of products and processes often stood out as significant problems in terms of the ability of U.S. industries to compete. Firms in several other countries, particularly Japan and the East Asian newly industrializing countries, have been successful in commercializing the results of R&D. In various instances, this was research initially performed in the United States, as evidenced by the VCR and semiconductor chips. Basic research and the pursuit of science are done successfully in the United States as indicated, in part, by the number of Nobel prizes awarded to Americans. However, excellence in science does not necessarily assure leadership in world markets. It has been noted that the United States was the world's premiere economic power in the 1920s when this nation was far from being in the forefront of science. Instead, market leadership is significantly affected by the development and application of technology to make the goods and services the consumers want to purchase.

Thus, questions may be raised as to whether programs and policies encouraging increased cooperative research, without concomitant efforts to facilitate the development and commercialization of technologies and techniques, can be effective mechanisms to increase the competitiveness of American industry. Do we need to know more about how to encourage the application of the research resulting from joint ventures in the manufacture of products and processes and in the delivery of services? Do these cooperative activities include mechanisms to facilitate the effective and timely transfer of the results back to the companies where they can be developed into goods for the marketplace? Since the major portion of the costs associated with bringing out a new product occur at the development and marketing stages, not in the research phase, should there be additional government incentives to encourage companies to spend funds for commercialization in addition to research?

Manufacturing

It is in the manufacturing arena where American companies appear to be the most vulnerable to foreign competition. Process technologies (those used in manufacturing) can significantly lower the costs of production and increase the quality of goods and services. In *Global Competition*, the President's Commission on Industrial Competitiveness (under former President Reagan) concluded that ". . . competitive success in many industries today is as much a matter of mastering the most advanced manufacturing processes as it is in pioneering new products."

The costs associated with the development and purchase of new manufacturing equipment are high. This is particularly true for the 381,000 small companies which make up a major segment of the manufacturing community. Several of the cooperative efforts supported by the federal government address these manufacturing concerns. The Manufacturing Technology program of the Department of Defense, the Advanced Manufacturing Technical Initiative of the Department of Energy, and the Manufacturing Technology Centers operated by the National Institute of Standards and Technology, although all different, are examples of government activities devoted to facilitating the development of new manufacturing techniques and their use in industry.

Considering the importance of manufacturing, the existing cooperative programs may not be sufficient to increase the competitiveness of American industry. Are there more effective types of joint ventures? Cooperative efforts, where resources could be pooled and the equipment shared, may be one way to improve the manufacturing capability of U.S. firms, large or small. Will joint manufacturing prove to be a viable option? Should existing cooperative manufacturing programs in certain agencies be expanded or should new efforts in other departments be developed? Should one government agency have the lead in policy determinations; if so, which federal department?

Defense vs. Civilian Support

Many of the industries interested in cooperative ventures with federal financial support have approached the Department of Defense and, to a lesser extent, the Department of Energy's Defense Programs because these agencies have the greatest amount of available resources and/or funding. They also tend to have the expertise to operate large-scale programs and maintain close ties with certain industrial sectors which could be encouraged to increase cooperation. In addition, both DOD and DOE have a vested interest in the availability of certain technologies which could be provided by a healthy domestic commercial market. However, questions remain whether sponsorship of certain cooperative ventures by

DOD and the Department of Energy's defense-related programs will lead to increased commercialization in the civilian marketplace.

Critics argue that defense spending is not an effective mechanism to increase industry's ability to innovate and develop new technologies. Much of the research and development in the defense arena may be too specialized, overdesigned, and/or too costly to have value for commercial markets. The R&D also tends to concentrate on weapon systems and other defense hardware rather than on process technologies that are often necessary to improve manufacturing productivity. One reason cited for the competitive problems of the machine tool industry was its focus on defense needs rather than on the commercial market which is larger in the aggregate.

On the other hand, the U.S. commitment to military R&D has contributed to a favorable balance of trade in the defense and aerospace industries. In the SEMATECH effort, the purpose of DOD support was to facilitate the *commercial* development of technologies with critical defense applications. The companies involved in SEMATECH are experienced semiconductor manufacturers and are knowledgeable about the markets' needs and operations. Thus, although the initial work performed by this semiconductor consortium may have been partially funded by the Defense Advanced Research Project Agency, it was designed to result in new products and processes in the civilian marketplace where both defense and commercial demand can be met. SEMATECH now operates without direct federal financing.

The issue of cooperative work between the Defense Department and the private sector leading to commercial technologies was addressed in the former Technology Reinvestment Project and is part of the more recent Dual-Use Partnership Project. The Department of Energy has been expanding cooperative R&D activities in Defense Program laboratories in conjunction with an increase in all DOE collaborative efforts with industry. Recent significant decreases in the technology transfer budgets may impeded this effort, but several DOE defense laboratories are actively pursuing joint ventures with industry. (See CRS Report 98-81, *Cooperative Research and Development Agreements and Semiconductor Technology: Issues Involving the "DOE-Intel CRADA".*)

Access by Foreign Firms

With worldwide communications systems, it is virtually impossible to prevent the flow of scientific and technical information. What is critical to competitiveness is the speed at which this knowledge is used to make products, processes, and services for the marketplace. However, it appears that many foreign firms are willing and able to take the results of research performed both in the United States and their own countries and rapidly make high quality commercial goods. Many of these companies are purchasing American businesses or establishing U.S. subsidiaries to access American expertise. With the increased activity in research consortia, particularly those with federal support, questions might be asked as to whether or not foreign companies should or could be barred from access to the results. A larger issue is how to define an "American company." Is it determined by majority ownership, manufacturing, location, value added to the U.S. economy, or by some other definition? In addition, since technology is most effectively transferred by person-to-person interaction, would cooperative activities between American industry and foreign firms produce an outflow of information which could be used to increase competitive pressures?

Direct vs. Indirect Support

Government efforts to facilitate cooperative ventures have included both indirect supports and direct federal funding. Indirect measures include such things as tax policies, intellectual property rights, and antitrust laws that create incentives for the private sector. Other initiatives include government financing (on a cost shared basis) of joint efforts such as the Advanced Technology Program and Manufacturing Extension Partnerships. In the past, participants in the legislative process generally did not make definite (or exclusionary) choices between these two approaches. However, these activities were revisited in the 104th Congress given apparent Republican preferences for the funding of basic research and not technology development. For example, efforts to eliminate the Advanced Technology Program, funding for flat panel displays, and agricultural extension reflected concern over the role of government in developing commercial technologies and generally resulted in reductions of direct federal financing for such public-private partnerships. Issues were again raised in the 105th Congress although no relevant, on-going program was terminated. As the 106th Congress begins its budget deliberations, the future of cooperative R&D may be expected to be explored further. (For more information, see CRS Report 95-50, *The Federal Role in Technology Development*.)

LEGISLATION

H.R. 209 (Morella)
The Technology Transfer Commercialization Act of 1999. Amends the Stevenson-Wydler Technology Innovation Act to improve the ability of government agencies to license federally-owned inventions. Introduced January 6, 1999; referred to Committees on Science and on the Judiciary. Reported from Committee on Science, with amendments, May 6, 1999. Discharged from Committee on the Judiciary the same day. Passed House May 11, 1999. Referred to Senate Committee on Commerce, Science, and Transportation September 29, 1999.

H.R. 1744 (Morella)
A bill to authorize appropriations for the National Institute of Standards and Technology for FY2000, among other things. Authorizes $190.7 million in funding for the Advanced Technology Program and $106.8 million for the Manufacturing Extension Partnership. Introduced May 10, 1999; referred to Committee on Science. Committee mark-up held May 26, 1999.

H.R. 2488 (Archer)
Financial Freedom Act of 1999. Amends the Internal Revenue Code of 1986 to extend the research and experimentation tax credit through June 30, 2004 and increases the credit rate applicable under the alternative incremental research credit by one percentage point per step, among other things. Introduced July 13, 1999; referred to the House Committee on Ways and Means. Reported from Committee, amended, July 16, 1999. Passed House on July 22, 1999. Senate substituted language of S. 1429, amended, and passed H.R. 2488 in lieu. Conference report agreed to by House and Senate August 5, 1999. Vetoed by President September 23, 1999.

H.R. 2670 (Rogers)

Departments of Commerce, Justice, and State, the Judiciary, and Related Agencies Appropriations Bill, FY2000. Makes FY2000 appropriations for the National Institute of Standards and Technology. Provides $99.8 million for the Manufacturing Extension Partnership and no funding for the Advanced Technology Program. Reported as an original measure from the House Committee on Appropriations August 2, 1999. Passed House, amended, August 5, 1999. On September 8, 1999, Senate substituted language of S. 1217 and passed H.R. 2670 in lieu with an amendment. Conference report filed October 19, 1999. In its final version, the bill appropriates $142.6 million for ATP and $104.8 million for MEP. House and Senate agreed to conference report October 20, 1999. Vetoed by President October 25, 1999

H.R. 2923 (Archer)

Amends the Internal Revenue Code of 1986 to extend the research and experimentation tax credit through June 30, 2004, among other things. Introduced September 23, 1999; referred to Committee on Ways and Means. Reported, amended, to House on September 28, 1999.

S. 804 (Rockefeller)

Technology Transfer Commercialization Act of 1999. A bill to improve the ability of federal agencies to license federally owned inventions. Introduced April 14, 1999; referred to Committee on Commerce, Science and Transportation.

S. 999 (Hatch)

Technology Transfer Act of 1999. Amends chapter 18 of title 35 U.S.C. to improve the ability of federal agencies to patent and license federally owned inventions. Introduced May 11, 1999; referred to Committee on the Judiciary.

S. 1217 (Gregg)

Departments of Commerce, Justice, and State, the Judiciary, and Related Agencies Appropriation Bill, 2000. Makes FY2000 appropriations for the National Institute of Standards and Technology. Provides $226.5 million for the Advanced Technology Program and $109.8 million for the Manufacturing Extension Partnership. Introduced June 14, 1999; reported as an original measure by Committee on Appropriations. Passed Senate, amended, July 22, 1999. On September 8, 1999, Senate vitiated previous passage, incorporated S. 1217 into H.R. 2670 as an amendment, and passed H.R. 2670 in lieu.

S. 1407 (Frist)

A bill to authorize appropriations for the National Institute of Technology for FY2000, among other things. Provides $209.6 million for the Advanced Technology Program and $106.8 million for the Manufacturing Extension Partnership. Introduced July 21, 1999; referred to Committee on Commerce, Science, and Transportation.

Excerpted from Congressional Research Service Report Number IB 89056

INDUSTRIAL COMPETITIVENESS AND TECHNOLOGICAL ADVANCEMENT: DEBATE OVER GOVERNMENT POLICY

Wendy H. Schacht

MOST RECENT DEVELOPMENTS

President Clinton has adopted a strategy for technological advancement as part of a defined national economic policy. This approach was initially supported by various congressional initiatives that supplemented funding for various technology development activities including the Advanced Technology Program (ATP) and the Manufacturing Extension Partnership (MEP) at the National Institute of Standards and Technology. However, many of these efforts were revisited in the 104th Congress given the Republican majority's statements in favor of (1) indirect measures such as tax policies, intellectual property rights, and antitrust laws to promote technological advancement; (2) increased government support for basic research; and (3) decreased direct federal funding for private sector technology initiatives. While no program was eliminated, several were financed at reduced levels. The 105th Congress continued to explore the proper role of the federal government in promoting private sector innovation. The Taxpayer Relief Act of 1997, P.L. 105-34, extended the research and experimentation tax credit for 13 months from June 1, 1997 through June 30, 1998. Allowed to expire again, P.L. 105-277 reinstated the credit through June 30, 1999. The pilot Small Business Technology Transfer (STTR) program was reauthorized through FY2001 by P.L. 105-135. For FY1999, P.L. 105-277, the Omnibus Consolidated Appropriations Act, funded ATP at $197.5 million and MEP at $106.8 million. The Technology Administration Act of 1998, P.L. 105-309, extends federal financing for the centers of the Manufacturing Extension Partnership past the original 6 years under certain conditions. In the current Congress, H.R. 1744, as introduced May 10, 1999, would authorize FY2000 funding for ATP at $190.7 million and $106.8 million for MEP. S. 1407, introduced July 21, 1999, authorizes $209.6 million for ATP and $106.8 million for MEP. On October 20, 1999, both the House and Senate agreed to the conference report on H.R. 2670, which provides FY2000 appropriations of $142.6 million for ATP and $104.8 million for MEP (although the original House version included no support for the Advanced Technology Program). However, on October 25, 1999, the President vetoed the legislation. H.R. 209, passed by the House on May 11, 1999, would make changes to existing laws to improve the ability of the government to license federally owned inventions. Similar legislative initiatives include S. 804 and S. 999. The Financial Freedom Act, H.R. 2488, passed by Congress, but vetoed by the President on September 23, 1999, would have extended the research and experimentation tax credit through June 30, 2004. However, H.R. 2923, reported to the House on September 28, 1999, has been introduced to achieve the same objectives. S. 296, a bill that provides for doubling the federal civilian R&D over the next 11 years, was passed by the Senate on July 26, 1999.

BACKGROUND AND ANALYSIS

Technology and Competitiveness

Interest in technology development and industrial innovation increased as concern mounted over the economic strength of the nation and over competition from abroad. For the United States to be competitive in the world economy, U.S. companies must be able to engage in trade, retain market shares, and offer high quality products, processes, and services while the nation maintains economic growth and a high standard of living. Technological advancement is important because the commercialization of inventions provides economic benefits from the sale of new products or services; from new ways to provide a service; or from new processes that increase productivity and efficiency. It is widely accepted that technological progress is responsible for up to one-half the growth of the U.S. economy, and is one principal driving force in long-term growth and increases in living standards.

Technological advances can further economic growth because they contribute to the creation of new goods, new services, new jobs, and new capital. The application of technology can improve productivity and the quality of products. It can expand the range of services that can be offered as well as extend the geographic distribution of these services. The development and use of technology also plays a major role in determining patterns of international trade by affecting the comparative advantages of industrial sectors. Since technological progress is not necessarily determined by economic conditions — it also can be influenced by advances in science, the organization and management of firms, government activity, or serendipity — it can have effects on trade independent of shifts in macroeconomic factors. New technologies also can help compensate for possible disadvantages in the cost of capital and labor faced by firms.

Federal Role

In the recent past, American companies faced increased competitive pressures in the international marketplace from firms based in countries where governments actively promote commercial technological development and application. In the United States, the generation of technology for the commercial marketplace is primarily a private sector activity. The federal government traditionally becomes involved only for certain limited purposes. Typically these are activities which have been determined to be necessary for the "national good" but which cannot, or will not, be supported by industry.

To date, the U.S. government has funded research and development (R&D) to meet the mission requirements of the federal departments and agencies. It also finances efforts in areas where there is an identified need for research, primarily basic research, not being performed in the private sector. Federal support reflects a consensus that basic research is critical because it is the foundation for many new innovations. However, any returns created by this activity are generally long term, sometimes not marketable, and not always evident. Yet the rate of return to society as a whole generated by investments in research is significantly larger than the benefits that can be captured by the firm doing the work.

Many past government activities to increase basic research were based on a "linear" model of innovation. This theory viewed technological advancement as a series of sequential steps starting with idea origination and moving through basic research, applied research, development, commercialization, and diffusion into the economy. Increases in federal funds in the basic research stage were expected to result in concomitant increases in new products and processes. However, this linear concept is no longer considered valid. Innovations often occur that do not require basic or applied research or development; in fact most innovations are incremental improvements to existing products or processes. In certain areas, such as biotechnology, the distinctions between basic research and commercialization are small and shrinking. In others, the differentiation between basic and applied research is artificial. The critical factor is the *commercialization* of the technology. Economic benefits accrue only when a technology or technique is brought to the marketplace where it can be sold to generate income or applied to increase productivity. Yet, while the United States has a strong basic research enterprise, foreign firms appear more adept at taking the results of these scientific efforts and making commercially viable products. Often U.S. companies are competing in the global marketplace against goods and services developed by foreign industries from research performed in the United States. Thus, there has been increased congressional interest in mechanisms to accelerate the development and commercialization processes in the private sector.

The development of a governmental effort to facilitate technological advance has been particularly difficult because of the absence of a consensus on the need for an articulated policy. Technology demonstration and commercialization have traditionally been considered private sector functions in the United States. While over the years there have been various programs and policies (such as tax credits, technology transfer to industry, and patents), the approach had been ad hoc and uncoordinated. Much of the program development was based upon what individual committees judged appropriate for the agencies over which they have jurisdiction. Despite the importance of technology to the economy, technology-related considerations often have not been integrated into economic decisions.

There have been attempts to provide a central focus for governmental activity in technology matters. P.L. 100-519 created within the Department of Commerce a Technology Administration headed by a new Under Secretary for Technology. In November 1993, the President established a National Science and Technology Council to coordinate decisionmaking in science and technology and to insure their integration at all policy levels (see more below). However, technological issues and responsibilities remain shared among many departments and agencies. This diffused focus has sometimes resulted in actions which, if not at cross purposes, may not have accounted for the impact of policies or practices in one area on other parts of the process. Technology issues involve components which operate both separately and in concert. While a diffused approach can offer varied responses to varied issues, the importance of interrelationships may be underestimated and their usefulness may suffer.

Several times, Congress has examined the idea of an industrial policy to develop a coordinated approach on issues of economic growth and industrial competitiveness. Technological advance is both one aspect of this and an altogether separate consideration. In looking at the development of an identified policy for industrial competitiveness, advocates argue that such an effort could ameliorate much of the uncertainty with which the private sector perceives future government actions. It has been argued that consideration and delineation of national objectives could encourage industry to engage in more long-term planning with regard to R&D and to make decisions as to the best allocation of resources. Such a technology policy could generate greater consistency in government activities. Because technological development involves numerous risks, efforts to minimize uncertainty regarding federal programs and policies may help alleviate some of the disincentives perceived by industry.

The development of a technology policy, however, would require a new orientation by both the public and private sectors. There is widespread resistance to what could be and has been called national planning, due variously to doubts as to its efficacy, to fear of adverse effects on our market system, to political beliefs about government intervention in our economic system, and to the current emphasis on short- term returns in both the political and economic arenas. Yet proponents note that planning can be advisory or indicative rather than mandatory. The focus provided by a technology policy could arguably provide a more receptive or helpful governmental environment within which business can make better decisions. Advocates assert that it could also reassure industry of government's ongoing commitment to stimulating R&D and innovation in the private sector.

Consideration of what constitutes government policy (both in terms of the industrial policy and technology policy) covers a broad range of ideas from laissez-faire to special government incentives to target specific high-technology, high-growth industries. Suggestions have been made for the creation of federal mechanisms to identify and support strategic industries and technologies. Various federal agencies and private sector groups have developed critical technology lists. However, others maintain that such targeting is an unwanted, and unwarranted, interference in the private sector which will cause unnecessary dislocations in the marketplace or a misallocation of resources. The government does not have the knowledge or expertise to make business-related decisions. Instead, they argue, the appropriate role for government is to encourage innovative activities in all industries and to keep market related decisionmaking within the business community that has ultimate responsibility for commercialization and where such decisions have traditionally been made.

The relationship between government and industry is a major factor affecting innovation and the environment within which technological development takes place. This relationship often has been adversarial, with the government acting to regulate or restrain the business community, rather than to facilitate its positive contributions to the nation. However, the situation is changing; it has become increasingly apparent that lack of cooperation can be detrimental to the nation as it faces competition from companies in countries where close government-industry collaboration is the norm. There are an increasing number of areas where the traditional distinctions between public and private sector functions and responsibilities are becoming blurred. Many assumptions have been questioned, particularly in light of the increased internationalization of the U.S. economy. The business sector is no longer be viewed in an exclusively domestic context; the economy of the United States is often tied to the economies of other nations. The technological superiority long held by the United States in many areas is being challenged by other industrialized countries in which economic, social, and political policies and practices foster government-industry cooperation in technological development.

The Clinton-Gore Approach

A major divergence from the past was evident in the approach suggested by President Clinton and Vice President Gore in two reports issued in February 1993; *A Vision of Change for America and Technology for America's Economic Growth, A New Direction to Build Economic Strength*. Their proposals called for a national commitment to, and a strategy for, technological advancement as part of a defined national economic policy. This detailed strategy offers a policy agenda for economic growth in the United States, of which technological development and industrial competitiveness are critical components.

In articulating a national technology policy, the approach initially recommended and subsequently followed by the Administration is multifaceted and provides a wide range of options while for the most part reflecting current trends in congressional efforts to facilitate industrial advancement. This policy increases federal coordination and augments direct government spending for technological development. While many past activities focused primarily on research, the new initiatives shift the emphasis toward *development* of new products, processes, and services by the private sector for the commercial marketplace. In addition, a significant number of the proposals aim to increase both government and private sector support for R&D leading to the commercialization of technology.

To facilitate technological advance, the Clinton-Gore approach focuses on increasing investment; *investment* in research, primarily civilian research, to meet the Nation's needs in energy, environmental quality, and health; investment in the development and commercialization of new products, processes, and services for the marketplace; investment in improved manufacturing to make American goods less expensive and of better quality; investment in small, high technology businesses in light of their role in innovation and job creation; and investment in the country's infrastructure to support all these efforts. To make the most productive use of this increased investment, the Administration has supported increased *cooperation* between all levels of government, industry, and academia to share risk, to share funding, and to utilize the strengths of each sector in reaching common goals of economic growth, productivity improvement, and maintenance of a high living standard.

On November 23, 1993, President Clinton issued Executive Order 12881 establishing a National Science and Technology Council (NSTC), a cabinet-level body to "...coordinate science, space, and technology policies throughout the federal government." According to the Fact sheet issued by the White House, the NSTC is to help establish clear national goals for federal R&D, assist in the implementation of the President's science and technology agenda, and insure that R&D considerations be integrated into decisionmaking at all policy levels. Membership includes the President, as chair; the Vice President; the Assistant to the President for Science and Technology; the Secretaries or Directors of the major R&D agencies; the Director of the Office of Management and Budget; the National Security Advisor; the Assistant to the President for Economic Policy; and the Assistant to the President for Domestic Policy. In addition, the National Economic Council offers expert advice and helps formulate policy proposals on issues related to economic growth, international competitiveness, and technological advancement. Thus, it appears that the Administration is attempting to view the activities of the federal R&D agencies and the executive branch advisory bodies comprehensively and not independently by mission.

Despite the continuing debate on what is the appropriate role of government and what constitutes a desirable government technology development policy, it remains an undisputed fact that what the government does or does not do affects the private sector and the marketplace. The various rules, regulations, and other activities of the government have become de facto policy as they relate to, and affect, innovation and technological advancement. It has been argued that these actions are not sufficiently understood or analyzed with respect to the larger context within which economic growth occurs. According to critics, these actions also are not coordinated in any meaningful way so that they promote an identifiable goal, whether that goal is as general as the "national welfare" or as specific as the growth of a particular industry.

Legislative Initiatives and Current Programs

Over the past several years, legislative initiatives have reflected a trend toward expanding the government's role beyond traditional funding of mission-oriented R&D and basic research toward the facilitation of technological advancement to meet other critical national needs, including the economic growth that flows from new commercialization and use of technologies and techniques in the private sector. An overview of recent legislation shows federal efforts aimed at (1) encouraging industry to spend more on R&D; (2) assisting small high-technology businesses; (3) promoting joint research activities between companies; (4) fostering cooperative work between industry and universities; (5) facilitating the transfer of technology from the federal laboratories to the private sector; and (6) providing incentives for quality improvements. These efforts tend toward removing barriers to technology development in the private sector (thereby permitting market forces to operate) and providing incentives to encourage increased private sector R&D activities. While most focus primarily on research, some also involve policies and programs associated with technology development and commercialization.

Increased R&D Spending

To foster increased company spending on research, the 1981 Economic Recovery Tax Act (P.L. 97-34) mandated a temporary incremental tax credit for qualified research expenditures. The law provided a 25% tax credit for the increase in a firm's qualified research costs above the average expenditures for the previous three tax years. Qualified costs included in-house expenditures such as wages for researchers, material costs, and payments for use of equipment; 65% of corporate grants towards basic research at universities and other relevant institutions; and 65% of payments for contract research. The credit applied to research expenditures through 1985.

The Tax Reform Act of 1986 (P.L. 99-514) extended the research and experimentation (R&E) tax credit for another 3 years. However, the credit was lowered to 20% and is applicable to only 75% of a company's liability. The 1988 Tax Corrections Act (P.L. 100-647) approved a 1-year extension of the research tax credit. The Omnibus Budget Reconciliation Act (P.L. 101-239) extended the credit through September 30, 1990 and made small start-up firms eligible for the credit. The FY1991 Budget Act (P.L. 101-508) again continued the tax credit provisions through 1992. The law expired in June 1992 when former President Bush vetoed H.R. 11 that year. However, P.L. 103-66, the Omnibus Budget Reconciliation Act of 1993, reinstated the credit through July 1995 and made it retroactive to the former expiration date. The tax credit again was allowed to expire until P.L. 104-188, the Small Business Job Protection Act, restored it from July 1, 1996 through May 31, 1997. P.L. 105-34, the Taxpayer Relief Act of 1997, extended the credit for 13 months from June 1, 1997 through June 30, 1998. Although it expired once again at the end of June, the Omnibus Consolidated Appropriations Act, P.L. 105-277, reinstated the tax credit through June 30, 1999. During the 105th Congress, various bills were introduced to make the tax credit permanent; other bills would have allowed the credit to be applied to certain collaborative research consortia. On August 5, 1999, both the House and Senate agreed to the conference report for H.R. 2488, the Financial Freedom Act, which would extend the credit for 5 years through June 30, 2004. This bill would also increase the credit rate applicable under the alternative incremental research credit by one percentage point per step. While President Clinton vetoed this bill on September 23, 1999, H.R. 2923, as reported September 28, 1999 from the Committee on Ways and Means, would accomplish the same objectives.

The Small Business Development Act (P.L. 97-219), as extended (P.L. 99-443), established a program to facilitate increased R&D within the small-business, high- technology community. Each federal agency with a research budget was required to set aside 1.25% of its R&D funding for grants to small firms for research in areas of interest to that agency. P.L. 102-564, which reauthorized the SBIR program, increased the set-aside to 2.5%, phased in over a 5-year period. Funding is, in part, dependent on companies obtaining private sector support for the commercialization of the resulting products or processes. The authorization for the program is set to terminate October 1, 2000. H.R. 2392, as passed by the House on September 27, 1999, would extend the SBIR effort through September 30, 2007. A pilot activity, the Small Business Technology Transfer (STTR) program, also was created to encourage firms to work with universities or federal laboratories to commercialize the results of research. This activity is funded by a 0.15% (phased in) set-aside. Set to expire in FY1997, the STTR originally was extended for one year until P.L. 105-135 reauthorized this activity through FY2001. (See CRS Report 96-402, *Small Business Innovation Research Program*.)

The Omnibus Trade and Competitiveness Act of 1988 (P.L. 100-418) created the Advanced Technology Program (ATP) at the Department of Commerce's National Institute of Standards and Technology. This program provides seed funding, matched by private sector investment, to companies or consortia of universities, industries, and government laboratories to accelerate the development of generic technologies that have broad application across industries. The first awards were made in 1991. To date, 431 projects have been funded representing approximately $1,384 million in federal dollars matched by more than $1,419 million in financing from the private sector. Over one-half of the awardees (including 68% in the latest competition) are small businesses or cooperative efforts led by such firms.

The first four competitions (through August 1994) were all general in nature. However, in response to large increases in federal funding, NIST, in conjunction with industry, identified various key areas for long-range support including: information infrastructure for healthcare; tools for DNA diagnostics; component-based software; manufacturing composite structures; computer-integrated manufacturing for electronics; digital data storage; advanced vapor-compression refrigeration systems; motor vehicle manufacturing technology; materials processing for heavy manufacturing; catalysis and biocatalysis technologies; advanced manufacturing control systems; digital video in information networks; tissue engineering; photonics manufacturing; premium power; microelectronics manufacturing infrastructure;

selective-membrane platforms; and adaptive learning systems. A general competition also was held each year. For FY1999, NIST has decided to curtail the focused programs and hold one competition open to all areas of technology. (For more information, see CRS Report 95-36, *The Advanced Technology Program*.)

Appropriations for the ATP include $35.9 million in FY1991, $47.9 million in FY1992, and $67.9 million in FY1993. FY1994 appropriations increased significantly to $199.5 million and even further in FY1995 to $431 million. However, P.L. 104-6, rescinded $90 million from this amount. There was no FY1996 authorization. The original FY1996 appropriations bill, H.R. 2076, which passed the Congress was vetoed by the President, in part, because it provided no support for ATP. The appropriations legislation finally enacted, P.L. 104-134, did fund the Advanced Technology Program at $221 million. For FY1997, the President's budget request was $345 million. Again, there was no authorizing legislation. However, P.L. 104-208, the Omnibus Consolidated Appropriations Act, provided $225 million for ATP, later reduced by $7 million to $218 million by P.L. 105-18, the FY1997 Emergency Supplemental Appropriations and Rescission Act. For FY1998, the Administration requested $276 million in funding. P.L. 105-119, appropriated FY1998 financing of ATP at $192.5 million, again at a level less than the previous year. The Administration's FY1999 budget proposal included $259.9 million for this program, a 35% increase. While not providing such a large increase, P.L. 105-277 does fund ATP for FY1999 at $197.5 million, 3% above the previous year. This figure reflects a $6 million recission contained in the same law that accounts for "deobligated" funds resulting from early termination of certain projects. In the FY2000 budget, the President has requested $238.7 million for ATP, an increase of 21% over the current year. H.R. 1744, introduced May 10, 1999 and considered by the House Committee on Science, would authorize funding of $190.7 million for the program in FY2000, 3% below the previous year. The Senate authorization legislation, S. 1407, introduced July 21, 1999, provides $209.6 million, a 6% increase. S. 1217, as passed by the Senate on July 22, 1999, would appropriate $226.5 million for ATP, 15% over FY1999. H.R. 2670, as passed by the House on August 5, 1999, provided no funding for the activity. The report to accompany the bill stated that there was insufficient evidence to ". . .overcome those fundamental questions about whether the program should exist in the first place." The final version of the bill, passed by both the House and Senate on October 20, 1999, funds ATP at $142.6 million for FY2000, a 28% decrease from the previous year. However, on October 25, 1999 the President vetoed the legislation.

Industry-University Cooperative Efforts

The promotion of cooperative efforts among academia and industry is aimed at increasing the potential for the commercialization of technology. (For more information, see CRS Issue Brief 89056, *Cooperative R&D: Federal Efforts to Promote Industrial Competitiveness*.) Traditionally, basic research has been performed in universities or in the federal laboratory system while the business community focuses on the manufacture or provision of products, processes, or services. Universities are especially suited to undertake basic research. Their mission is to educate and basic research is an integral part of the educational process. Universities generally are able to undertake these activities because they do not have to produce goods for the marketplace and therefore can do research not necessarily tied to the development of a commercial product or process.

Subsequent to World War II, the federal government supplanted industry as the primary source of funding for basic research in universities. It also became the principal determinant of the type and direction of the research performed in academia. This resulted in a disconnect between the university and industrial communities. The separation and isolation of the parties involved in the innovation process is thought to be a barrier to technological progress. The difficulties in moving an idea from the concept stage to a commercial product or process are compounded when several entities are involved. Legislation to stimulate cooperative efforts among those involved in technology development is viewed as one way to promote innovation and facilitate the international competitiveness of U.S. industry.

Several laws have attempted to encourage industry-university cooperation. Title II of the Economic Recovery Tax Act of 1981 (P.L. 97-34) provided, in part, a 25% tax credit for 65% of all company payments to universities for the performance of basic research. Firms were also permitted a larger tax deduction for charitable contributions of equipment used in scientific research at academic institutions. The Tax Reform Act of 1986 (P.L. 99-514) kept this latter provision, but reduced the credit for university basic research to 20% of all corporate expenditures for this over the sum of a fixed research floor plus any decrease in non-research giving.

The 1981 Act also provided an increased charitable deduction for donations of new equipment by a manufacturer to an institution of higher education. This equipment must be used for research or research training for physical or biological sciences within the United States. The tax deduction is equal to the manufacturer's cost plus one-half the difference between the manufacturer's cost and the market value, as long as it does not exceed twice the cost basis. These provisions were extended through July 1995 by the Omnibus Budget Reconciliation Act of 1993, but then expired until restored by the passage of P.L. 104-188 and P.L. 105-277 as noted above.

Amendments to the patent and trademark laws contained in P.L. 96-517 (commonly called the "Bayh-Dole Act") also were designed to foster interaction between academia and the business community. This law provides, in part, for title to inventions made by contractors receiving federal R&D funds to be vested in the contractor if they are small businesses, universities, or not-for-profit institutions. Certain rights to the patent are reserved for the government and these organizations are required to commercialize within a predetermined and agreed upon time frame. Providing universities with patent title is expected to encourage licensing to industry where the technology can be manufactured or used thereby creating a financial return to the academic institution. University patent applications and licensing have increased siginificantly since this law was enacted. (See CRS Report 94-501, *The Bayh-Dole Act* and CRS Report 98-862, *R&D Partnerships and Intellectual Property: Implications for U.S. Policy.*)

Joint Industrial Research

Private sector investments in basic research are often costly, long term, and risky. Although not all advances in technology are the result of research, it is often the foundation of important new innovations. To encourage increased industrial involvement in research, legislation was enacted to allow for joint ventures in this arena. It is argued that cooperative research reduces risks and costs and allows for work to be performed that crosses traditional boundaries or expertise and experience. Such collaborative efforts make use of existing and support the development of new resources, facilities, knowledge, and skills.

The National Cooperative Research Act (P.L. 98-462) encourages companies to undertake joint research. The legislation clarifies the antitrust laws and requires that the "rule of reason" standard be applied in determinations of violations of these laws; cooperative research ventures are not to be judged illegal "per se." It also eliminates treble damage awards for those research ventures found in violation of the antitrust laws if prior disclosure (as defined in the law) has been made. In addition, P.L. 98-462 makes changes in the way attorney fee awards are made. Defendants can collect attorney fees in specified circumstances, including when the claim is judged frivolous, unreasonable, without foundation, or made in bad faith. However, the attorney fee award to the prevailing party may be offset if the court decides that the prevailing party conducted a portion of the litigation in a manner which was frivolous, unreasonable, without foundation, or in bad faith. These provisions were included to discourage frivolous litigation against joint research ventures without simultaneously discouraging suits of plaintiffs with valid claims. Over 700 joint research ventures have filed with the Department of Justice since passage of this legislation.

P.L. 103-42, the National Cooperative Production Amendments Act of 1993, amends the National Cooperative Research Act by, among other things, extending the original law's provisions to joint manufacturing ventures. These provisions are only applicable, however, to cooperative production when (1) the principal manufacturing facilities are "...located in the United States or its territories, and (2) each person who controls any party to such venture...is a United States person, or a foreign person from a country whose law accords antitrust treatment no less favorable to United States persons than to such country's domestic persons with respect to participation in joint ventures for production."

Commercialization of the Results of Federally Funded R&D

Another approach to encouraging the commercialization of technology involves the transfer of technology from federal laboratories and contractors to the private sector where commercialization can proceed. Because the federal laboratory system has extensive science and technology resources and expertise developed in pursuit of mission responsibilities, it is a potential source of new ideas and knowledge which may be used in the business community. (See CRS Issue Brief 85031, *Technology Transfer: Utilization of Federally Funded Research and Development*, for more details.)

Despite the potential offered by the resources of the federal laboratory system, however, the commercialization level of the results of federally funded R&D remained low. Studies indicated that only approximately 10% of federally owned patents were ever utilized. There are many reasons for this low level of usage, one of which is the fact that some technologies and/or patents have no market application. However, industry unfamiliarity with these technologies, the "not-invented-here" syndrome, and perhaps more significantly, the ambiguities associated with obtaining title to or exclusive license to federally owned patents also contribute to the low level of commercialization.

Over the years, several governmental efforts have been undertaken to augment industry's awareness of federal R&D resources. The Federal Laboratory Consortium for Technology Transfer was created in 1972 (from a Department of Defense program) to assist in transferring technology from the federal government to state and local governments and the private sector. To expand on the work of the Federal Laboratory Consortium, and to provide added emphasis on the commercialization of government technology, Congress passed P.L.

96-480, the Stevenson-Wydler Technology Innovation Act of 1980. Prior to this law, technology transfer was not an explicit mandate of the federal departments and agencies with the exception of the National Aeronautics and Space Administration. To provide "legitimacy" to the numerous technology activities of the government, Congress, with strong bipartisan support, enacted P.L. 96-480 which explicitly states that the federal government has the responsibility, "...to ensure the full use of the results of the nation's federal investment in research and development." Section 11 of the law created a system within the federal government to identify and disseminate information and expertise on what technologies or techniques are available for transfer. Offices of Research and Technology Applications were established in each federal laboratory to distinguish technologies and ideas with potential applications in other settings.

Several amendments to the Stevenson-Wydler Technology Innovation Act have been enacted to provide additional incentives for the commercialization of technology. P.L. 99-502, the Federal Technology Transfer Act, authorizes activities designed to encourage industry, universities, and federal laboratories to work cooperatively. It also establishes incentives for federal laboratory employees to promote the commercialization of the results of federally funded research and development. The law amends P.L. 96-480 to allow government-owned, government-operated laboratories to enter into cooperative R&D agreements (CRADAs) with universities and the private sector. This authority is extended to government-owned, contractor-operated laboratories by the Department of Defense FY1990 Authorization Act, P.L. 101-189. (See CRS Report 95-150, *Cooperative Research and Development Agreements (CRADAs)*.) Companies, regardless of size, are allowed to retain title to inventions resulting from research performed under cooperative agreements. The federal government retains a royalty-free license to use these patents. The Technology Transfer Improvements and Advancement Act (P.L. 104- 113), clarifies the dispensation of intellectual property rights under CRADAs to facilitate the implementation of these cooperative efforts. The Federal Laboratory Consortium is given a legislative mandate to assist in the coordination of technology transfer. To further promote the use of the results of federal R&D, certain agencies are mandated to create a cash awards program and a royalty sharing activity for federal scientists, engineers, and technicians in recognition of efforts toward commercialization of this federally developed technology. These efforts are facilitated by a provision of the National Defense Authorization Act for FY1991 (P.L. 101-510), which amends the Stevenson-Wydler Technology Innovation Act to allow government agencies and laboratories to develop partnership intermediary programs to augment the transfer of laboratory technology to the small business sector.

Amendments to the Patent and Trademark law contained in Title V of P.L. 98- 620 make changes which are designed to improve the transfer of technology from the federal laboratories — especially those operated by contractors — to the private sector and increase the chances of successful commercialization of these technologies. This law permits the contractor at government-owned, contractor-operated laboratories (GOCOs) to make decisions at the laboratory level as to the granting of licenses for subject inventions. This has the potential of effecting greater interaction between laboratories and industry in the transfer of technology. Royalties on these inventions are also permitted to go back to the contractor to be used for additional R&D, awards to individual inventors, or education. While there is a cap on the amount of the royalty returning directly to the lab in order not to disrupt the agency's mission requirements and congressionally mandated R&D agenda, the establishment of discretionary funds gives contractor-operated laboratories added incentive to encourage technology transfer.

Under P.L. 98-620, private companies, regardless of size, are allowed to obtain exclusive licenses for the life of the patent. Prior restrictions allowed large firms use of exclusive license for only 5 of the 17 years (now 20 years) of the life of the patent. This should encourage improved technology transfer from the federal laboratories or the universities (in the case of university operated GOCOs) to large corporations which often have the resources necessary for development and commercialization activities. In addition, the law permits GOCOs (those operated by universities or nonprofit institutions) to retain title to inventions made in the laboratory within certain defined limitations. Those laboratories operated by large companies are not included in this provision.

In the 105th Congress, the House passed H.R. 2544, the Technology Transfer Commercialization Act, a bill that would have made alterations in current practices concerning patents held by the government to make it easier for federal agencies to license such inventions. On May 6, 1999, similar legislation again passed the House. This bill, H.R. 209, amends the Stevenson-Wydler Technology Innovation Act and the Bayh-Dole Act to decrease the time delays associated with obtaining an exclusive or partially exclusive license. Agencies now must publicize the availability of technologies for 3 months using the *Federal Register* and then provide an additional 60 day notice of intent to license by an interested company. H.R. 209 would shorten the period to 15 days in recognition of the ability of the internet to offer widespread notification and the necessity of time constraints faced by industry in commercialization activities. Certain rights would be retained by the government. The bill also would allow licenses for existing government-owned inventions to be included in CRADAs. In the Senate, S. 804, introduced on April 14, 1999, and S. 999, introduced on May 11, 1999, are intended to accomplish similar objectives.

The Omnibus Trade and Competitiveness Act (P.L. 100-418) mandated the creation of a program of regional centers to assist small manufacturing companies to use knowledge and technology developed under the auspices of the National Institute of Standards and Technology and other federal agencies. Federal funding for the centers is matched by non-federal sources including state and local governments and industry. Originally, seven Regional Centers for the Transfer of Manufacturing Technology were selected and are operational: the Great Lakes Manufacturing Technology Center at the Cleveland Advanced Manufacturing Program in Ohio; the Northeast Manufacturing Technology Center at Rensselaer Polytechnic Institute in Troy, New York (now called the New York Manufacturing Extension Partnership); the South Carolina Technology Transfer Cooperative based at the University of South Carolina in Columbia; the Midwest Manufacturing Technology Center at the Industrial Technology Institute in Ann Arbor, Michigan; the Mid-American Manufacturing Technology Center at the Kansas Technology Enterprise Corporation of Topeka; the California Manufacturing Technology Center at El Camino College in Torrance; and the Upper Midwest Manufacturing Technology Center in Minneapolis.

The initial program was expanded in 1994 to create the Manufacturing Extension Partnership (MEP) to meet new and growing needs of the community. In a more varied approach, the Partnership involves both large centers and smaller, more dispersed organizations sometimes affiliated with larger centers as well as the NIST State Technology Extension Program which provides states with grants to develop the infrastructure necessary to transfer technology from the federal government to the private sector (an effort which was also mandated by P.L. 100-418) and a program which electronically ties the disparate parties together along with other federal, state, local, and academic technology transfer organizations. There are now centers in all 50 states and Puerto Rico. Since the manufacturing extension activity was created in 1989, awards made by NIST have resulted in the creation of approximately 400 regional offices. [It should be noted that the Department of Defense also funded 36 centers through its Technology Reinvestment Project (TRP) in FY1994 and FY1995. When the TRP was terminated, NIST took over support for 20 of these programs in FY1996 and funded the remaining efforts during FY1997.]

Funding for this program was $11.9 million in FY1991, $15.1 million in FY1992, and $16.9 million in FY1993. In FY1994 support for the expanded Manufacturing Technology Partnerships was $30.3 million. The following fiscal year, P.L. 103-317 appropriated $90.6 million for this effort, although P.L. 104-19 rescinded $16.3 million from this amount. While the original FY1996 appropriations bill, H.R. 2076, was vetoed by the President, the $80 million funding for MEP was retained in the final legislation, P.L. 104-134. The President's FY1997 budget request was $105 million. No FY1997 authorization legislation was enacted, but P.L. 104-208 appropriated $95 million for Manufacturing Extension while temporarily lifting the six-year limit on federal support for individual centers. The Administration requested FY1998 funding of $123 million. Again no authorizations were passed. However, the FY1998 appropriations bill, P.L. 105-119, financed the MEP program at $113.5 million. This law also permitted government funding, at one-third the centers total annual cost, to continue for additional periods of one year over the original six-year limit, if a positive evaluation is received. The President's FY1999 budget included $106.8 million for the MEP, a 6% decrease from current funding. The Omnibus Consolidated Appropriates Act, P.L. 105-277, appropriated the $106.8 million. The decrease in funding reflects a reduced federal financial commitment as the centers mature, not a decrease in program support. In addition, the Technology Administration Act of 1998, P.L. 105-309, permits the federal government to fund centers at one-third the cost after the 6 years if a positive, independent evaluation is made every two years. For FY2000, the Administration has requested $99.8 million in support for the MEP. Again, the lower federal share indicates a smaller statutory portion required of the government. H.R. 1744 authorizes spending of $106.8 million for the program, the same amount as in the Senate authorization bill, S. 1407. The original appropriations bill that passed the Senate, S. 1217, would have provided $109.8 million for the Manufacturing Extension Partnership, an increase of 3% over FY1999. H.R. 2670, as passed initially by the House, would appropriate $99.8 million for this activity. The final version of the bill passed by both House and Senate provides FY2000 appropriations of $104.8 million. However, the President has vetoed the legislation. (For additional information see: CRS Report 97-104, *Manufacturing Extension Partnership Program: An Overview*.)

A New Approach?

As indicated above, the laws affecting the R&D environment have included both direct and indirect measures to facilitate technological innovation. In general, direct measures are those which involve budget outlays and the provision of services by government agencies. Indirect measures include financial incentives and legal changes (e.g., liability or regulatory reform; new antitrust arrangements). Supporters of indirect approaches argue that the market is superior to government in deciding which technologies are worthy of investment. Mechanisms that enhance the market's opportunities and abilities to make such choices are preferred. Advocates further state that dependency on agency discretion to assist one technology in preference to another will inevitably be subjected to political pressures from entrenched interests. Proponents of direct government assistance maintain, conversely, that indirect methods can be wasteful and ineffective and that they can compromise other goals of public policy in the hope of stimulating innovative performance. Advocates of direct approaches argue that it is important to put the country's scarce resources to work on those technologies that have the greatest promise as determined by industry and supported by its willingness to match federal funding.

In the past, while Republicans tended to prefer reliance on free market investment, competition, and indirect support by government, participants in the debates generally did not make definite (or exclusionary) choices between the two approaches, nor consistently favor one over the other. For example, some proponents of a stronger direct role for the government in innovation are also supporters of enhanced tax preferences for R&D spending, an indirect mechanism. Opponents of direct federal support for specific projects (e.g., SEMATECH, flat panel displays) may nevertheless back similar activities focused on more general areas such as manufacturing or information technology. However, the 104th Congress directed their efforts at eliminating or curtailing many of the efforts which previously had enjoyed bipartisan support. Initiatives to terminate the Advanced Technology Program, funding for flat panel displays, and agricultural extension reflected concern about the role of government in developing commercial technologies. The Republican leadership stated that the government should directly support basic science while leaving technology development to the private sector. Instead of federal funding, changes to the tax laws, proponents argue, will provide the capital resources and incentives necessary for industry to further invest in R&D. During the 105th Congress many of the same issues were considered. While funding for several programs decreased, particularly in FY1998, support for most on-going activities continues. How the debate over federal funding evolves in the 106th Congress may serve to redefine thinking about the government's efforts in promoting technological advancement in the private sector.

LEGISLATION

H.R. 209 (Morella)

The Technology Transfer Commercialization Act of 1999. Amends the Stevenson-Wydler Technology Innovation Act to improve the ability of government agencies to license federally owned inventions. Introduced January 6, 1999; referred to Committees on Science and on the Judiciary. Reported from Committee on Science, with amendments, May 6, 1999. Discharged from Committee on the Judiciary the same day. Passed House May 11, 1999. Referred to Senate Committee on Commerce, Science, and Transportation September 29, 1999.

H.R. 1744 (Morella)

A bill to authorize appropriations for the National Institute of Standards and Technology for FY2000, among other things. Authorizes $190.7 million in funding for the Advanced Technology Program and $106.8 million for the Manufacturing Extension Partnership. Introduced May 10, 1999; referred to House Committee on Science. Committee mark-up held May 26, 1999.

H.R. 2392 (Talent)
Amends the Small Business act to extend the authorization for the Small Business Innovation Research Program through September 30, 2007. Introduced June 30, 1999; referred to Committees on Small Business and Science. Reported from Committee on Small Business September 23, 1999; discharged from Committee on Science the same day. Passed House, amended, September 27, 1999.

H.R. 2488 (Archer)
Financial Freedom Act of 1999. Amends the Internal Revenue Code of 1986 to extend the research and experimentation tax credit through June 30, 2004 and increases the credit rate applicable under the alternative incremental research credit by one percentage point per step, among other things. Introduced July 13, 1999; referred to the House Committee on Ways and Means. Reported from Committee, amended, July 16, 1999. Passed House July 22, 1999. Senate substituted language of S. 1429, amended, and passed H.R. 2488 in lieu. House and Senate agreed to conference report August 5, 1999. Vetoed by President September 23, 1999.

H.R. 2670 (Rogers)
Departments of Commerce, Justice, and State, the Judiciary, and Related Agencies Appropriations Bill, FY2000. Makes FY2000 appropriations for the National Institute of Standards and Technology. Provides $99.8 million for the Manufacturing Extension Partnership and no funding for the Advanced Technology Program. Reported as an origianl measure from the House Committee on Appropriations August 2, 1999. Passed House, amended, August 5, 1999. On September 8, 1999, Senate substituted the language of S. 1217 and passed H.R. 2670 in lieu with an amendment. Conference report filed October 19 and House and Senate agreed to report October 20, 1999. The final version of the bill provides $142.6 million for ATP and $104.8 million for MEP. Vetoed by President October 25, 1999.

H.R. 2923 (Archer)
Amends the Internal Revenue Code of 1986 to extend the research and experimentation tax credit through June 30, 2004, among other things. Introduced September 23, 1999; referred to the House Committee on Ways and Means. Reported, amended, to House September 28, 1999.

S. 296 (Frist)
Federal Research Investment Act. Provides for doubling of the federal civilian research budget over the next 11 years. Introduced January 22, 1999; referred to Committee on Commerce, Science, and Transportation. Ordered reported by Committee on Commerce, Science, and Transportation, with amendment, May 5, 1999. Passed Senate, amended, July 26, 1999. Referred to House Committee on Science July 27, 1999.

S. 804 (Rockefeller)
Technology Transfer Commercialization Act of 1999. A bill to improve the ability of federal agencies to license federally owned inventions. Introduced April 14, 1999; referred to Committee on Commerce, Science and Transportation.

S. 999 (Hatch)
Technology Transfer Act of 1999. Amends chapter 18 of title 35 U.S.C. to improve the ability of federal agencies to patent and license federally owned inventions. Introduced May 11, 1999; referred to Committee on the Judiciary.

S. 1217 (Gregg)

Departments of Commerce, Justice, and State, the Judiciary, and Related Agencies Appropriation Bill, 2000. Makes FY2000 appropriations for the National Institute of Standards and Technology. Provides $226.5 million for the Advanced Technology Program and $109.8 million for the Manufacturing Extension Partnership. Introduced June 14, 1999; reported as an original measure by Committee on Appropriations. Passed Senate, amended, July 22, 1999. On September 8, 1999, Senate vitiated previous passage, incorporated S. 1217 into H.R. 2670 as an amendment, and passed H.R. 2670 in lieu.

S. 1407 (Frist)

A bill to authorize appropriations for the National Institute of Technology for FY2000, among other things. Provides $209.6 million for the Advanced Technology Program and $106.8 million for the Manufacturing Extension Partnership. Introduced July 21, 1999; referred to Committee on Commerce, Science, and Transportation.

Excerpted from Congressional Research Service Report Number IB 91132

ENTREPRENEURIAL HIGH-GROWTH COMPANIES: A FRAMEWORK FOR THINKING ABOUT THE FEDERAL POLICY ROLE

Alan S. Gregerman

Entrepreneurship and its Key Characteristics

In recent years "entrepreneurs" have emerged as the popular heroes of the U.S. economy. The media sing their praises, literally hundreds of books tell their stories, business schools rush to create courses and programs designed to teach "their" skills, big companies try to act more like them, and individuals from all walks of life set out in record numbers to start and build their own new companies. At a time when the U.S. economy is very strong, there is widespread belief that entrepreneurs and their new high-growth companies are a very important part of the equation.

Defining Entrepreneurship

When we think of "entrepreneurs" we often think of people like Fred Smith, the founder of Federal Express, who created an entirely new industry by identifying the need to send information and products "absolutely, positively overnight"...or John Johnson, of Johnson Publishing Company, who created a media empire by responding to the aspirations of middle-class black America...or Bill Gates, of Microsoft, who transformed his vision of a personal computer in every home and workplace into the world's dominant information technology company...or Lilian Vernon, of the Lilian Vernon Company who turned a small home-based handbag company into a model for selling products by mail order...or Pleasant Rowland, the creator of the Pleasant Company, who has built a dynamic educational business by celebrating the experience of growing up as an American girl...or Ray Kroc, the now legendary founder of MacDonald's, who reinvented the concept of restaurants and food service efficiency. These people and others like them have changed the business landscape and in the process changed the way that we live, work and buy things.

This report defines "entrepreneurs" as those people who see new opportunities in new or existing markets and build, or at least try to build, growing companies to tap them. Entrepreneurs add value to the marketplace by offering new or improved products and services, or by creating new ways of doing business that enhance the choices that customers have. In this era of individual initiative there is a tendency to call everyone who starts or runs a business an entrepreneur. But that tends to miss the real point. Entrepreneurs are innovators who build ventures that have the potential to grow in size and value. Peter Drucker, the renowned business thinker, defines entrepreneurial firms as enterprises that have "special characteristics over and above being new and small. Indeed, entrepreneurs are a minority among new businesses. They create something new, something different; they change or transmute values."[1]

The Importance of Entrepreneurs to the Economy. The start-up and growth of truly "entrepreneurial" firms is critical to the overall vitality of the U.S. economy. While their precise impact is a subject of considerable debate,[2] it is clear that these enterprises are an important and dynamic source of economic growth, the nonstop innovation of new products, services and ways of doing business, new jobs and training opportunities for people at all levels of the workforce.

The focus of this report is not intended to downplay the role of small businesses in our economy. After all, the vast majority of U.S. firms are small—having less than 20 employees.[3] These enterprises as a group form a critical part of the backbone of the national economy, providing a wide range of goods and services to customers and a reliable source of jobs and income for their owners and employees. They also receive a great deal of attention from public policies and programs and have a major federal advocate in the Small Business Administration. But most have limited potential for growth.[4] Hicks contends that newness—defined as the age of a venture—is a critical and overlooked determinant of the potential for firms to grow and innovate. Rarely do firms that have remained small for an extended period shift into a high-growth mode.[5]

Can Entrepreneurship Be Taught? There is considerable debate in the literature about whether the founders and builders of high-growth companies are born or made. In other words, can individuals be taught entrepreneurial skills that will enable them to see real business opportunities and build successful ventures? Some analysts suggest that there is a clear entrepreneurial "profile" that can be used to describe successful entrepreneurs.[6] This profile includes a set of personal traits such as persistence, flexibility, an independent nature, boredom with routine tasks, an often aggressive style, difficulty working for other people and a willingness to take moderate risks. It can also include characteristics that reflect a person's childhood development, school performance, family history and birth order. There are even tests

---

[1] Drucker, Peter F., *Innovation and Entrepreneurship: Practice and Principles.* New York: Harper & Row Publishers, 1985, p. 22.

[2] See the discussion below.

[3] Office of Advocacy, U.S. Small Business Administration, *United States: Employer Firms and Employment, 1988-1995.* In fact, most are very small. It is estimated that of the 25 million business establishments in America roughly 75 percent have only one full-time employee.

[4] See Birch, David L. *Who's Creating Jobs?,* Cambridge, MA: Cognetics, Inc., 1995. Birch's analysis suggests that only about three percent of firms have high-growth potential.

[5] Hicks, Donald A., *A Nanoeconomic Perspective on the Growth and Development of the Texas Manufacturing Base, 1970-1991,* Office of the Comptroller, State of Texas, 1993.

[6] See Harrell, Wilson, *For Entrepreneurs Only,* Franklin Lakes, New Jersey: Career Press, 1995.

that would-be entrepreneurs can take to see how they measure up against a set of these "key" characteristics.[7]

Other experts believe that entrepreneurship is a systematic process that can be taught and learned.[8] As evidence they describe the step-by-step process by which many very successful companies have been built, widespread interest in entrepreneurial education programs at colleges and universities and changes in the workplace that are enticing a much broader group of people into entrepreneurship.[9] This group includes more women—who are the fastest growing segment of the entrepreneurial economy—and now account for more than half of the new businesses being started in the U.S.[10] It also includes growing numbers of minorities,[11] new immigrants,[12] young people—many of whom are on the cutting edge of the hottest technologies,[13] and corporate executives seeking new challenges.[14] McClelland placed great emphasis on "achievement motivation" as the key ingredient in entrepreneurial success and attempted to demonstrate that this behavior that could be learned and developed.[15]

Clearly, the answer is not cut and dry. There is something unique about many of the people who end up starting and building high-growth companies.[16] But there

---

[7] One widely used test is "The Entrepreneurial Quiz" developed by Joseph Mancuso at the Center for Entrepreneurial Management in New York. This test is discussed in detail in the *Checklist for Starting Your Own Business,* New York: Center for Entrepreneurial Management, 1996.

[8] See Drucker, Peter F., *Innovation and Entrepreneurship: Practice and Principles,* and Vesper, Karl H., *New Venture Mechanics,* Englewood Cliffs, NJ: Prentice Hall, 1993. There are also literally hundreds of books that are designed to teach people how to build a new and thriving business from the ground up.

[9] There are over 1,100 schools of higher education in the U.S. that currently offer an entrepreneurship curriculum, and several hundred not-for-profit and for-profit organizations teaching skills to people of all ages.

[10] See Henry, Shannon, "The New Girl's Network: Welcome to the Club," *TECH Capital,* Fall 1997, pp. 38-48. Also Brown, Carolyn M., "They've Got the Power," *Black Enterprise,* August 1996, pp. 63-88.

[11] See Woodard, Michael D., *Black Entrepreneurs in America: Stories of Struggle and Success,* New Brunswick, NJ: Rutgers University Press, 1997.

[12] Engardio, Pete and Peter Burrows. "Where Immigrants Find a Melting Pot of Gold," *Business Week,* August 25, 1997, p. 123.

[13] See Schlender, Brent, et al. "Cool Companies," *Fortune,* July 7, 1997, pp. 84-110.

[14] See Useem, Jerry, "The New Entrepreneurial Elite," *Inc.,* December 1997, pp. 50-57.

[15] McClelland, David, *The Achieving Society,* New York: Irvington Publishers, 1976.

[16] For a detailed discussion see White, Sammis B. and Paul D. Reynolds, "What Can the Public Sector Do To Increase New Business Starts?," *Frontiers of Entrepreneurship Research* (1994 edition), Center for Entrepreneurial Studies, Babson College, and Reynolds, Paul D. and Sammis B. White, "Government Programs and High Growth New Firms," *Frontiers of Entrepreneurship Research* (1996 edition), Center for Entrepreneurial Studies, (continued...)

are also a lot of successful entrepreneurs who did not begin their careers with visions of entrepreneurship. Whether they were "latent" entrepreneurs is unclear. And while there is very little analysis that links participation in entrepreneurial training programs to the start-up of fast growing companies, the fact that many successful entrepreneurs came out of other high-growth businesses speaks to the importance of "on-the-job" education. So the truth is probably somewhere in between.

Can Entrepreneurs Be Grown Locally? It is worth noting that some places have a greater concentration of entrepreneurs and high-growth companies than other places. The leading example of this is the Silicon Valley in northern California which is widely viewed as the most entrepreneurial local economy in the world. The history of its development is a virtual family-tree of business innovation beginning in 1939 in a garage in Palo Alto where William Hewlett and David Packard built an oscillator for Walt Disney, and continuing with the invention of the integrated circuit in 1959 at Fairchild Semiconductor and the microprocessor in 1972 at Intel.

Most experts contend that entrepreneurial "cultures" evolve over time, spurred initially by the presence of a limited number of innovation-intensive firms and often supported by the people and research of local universities and research centers. Over time, individuals leave these firms or institutions to launch new businesses capitalizing on local networks of talent, ideas, financing and other helpful resources.[17] It is encouraging to note that these cultures are emerging, in varying degrees, in a wide range of communities across the U.S. These include major metropolitan areas like Boston, Washington, D.C., Seattle, Austin and Las Vegas, and mid-size metropolitan areas such as Salt Lake City, Lincoln, Nebraska, Ann Arbor, Michigan, and Boise, Idaho, where key industries are spurring large numbers of entrepreneurial ventures.[18]

Keys to Building a Successful and Growing Company

It is not an easy task to start and build a high growth company. Even when an entrepreneur has a great idea that meets an established or emerging market need, business success still relies on a combination of skillful execution, the right timing, imagination, passion and a measure of luck. A good way to understand how the process unfolds is to look at the stages that a typical entrepreneur goes through in turning his or her idea into a growing concern.[19] But it is important to emphasize that

---

[16](...continued)
Babson College, pp. 621-632.

[17] See Rebello, Kathy (editor), "Silicon Valley: How It Really Works," *Business Week*, Special Issue, August 18-25, 1997.

[18] See "Almanac: Where the Start-Ups Are," *Inc.*, May 19, 1998, pp. 131-136, and Hyatt, Joshua, "The Most Entrepreneurial City in America: Vegas," *Inc.*, May 19, 1998, pp. 66-81.

[19] See Vesper, Karl H., *New Venture Mechanics*, Englewood Cliffs, NJ: Prentice Hall, 1993, for an in-depth guide to the stages that entrepreneurs go through in conceiving their ideas and turning them into succes_ful business ventures. Other excellent sources include Timmons, Jeffry A., *New Venture Creation: Entrepreneurship for the 21st Century* (fourth (continued...)

the process is dynamic, with many mid-course changes and "corrections" along the way, and each individual brings his or her own unique variation to starting and building a company.

Concept Development. In the initial stage, the entrepreneur (or entrepreneurial team) identifies a sound market opportunity and the nature of a specific business that will address the opportunity.

Business Planning. In this stage the entrepreneur (or entrepreneurial team) analyzes the market opportunity in detail and develops a plan for building the business. A complete plan will typically address the following set of issues:

- Market and competition;
- Products and/or services to be offered;
- Strategy for winning a large and sustainable share of the market;
- Selling plan and sales activities;
- Business structure and management;
- Operational plan for producing the company's products/services—this includes staffing and the use of technology;
- Financial plan for funding the start-up and growth of the business;
- Assessment of the key risks involved.

In addition to testing the feasibility of the business idea, the business plan also serves as the principal tool in attracting any outside financing needed to build the business. It should be pointed out that many successful businesses have been started without a formal business plan and relied, at least initially, on simply the intuition and resourcefulness of the entrepreneur.

Product or Service Development. In this stage the entrepreneur (or entrepreneurial team) designs and tests their products and/or services, including appropriate research and development activities and the building of prototypes.

Start-Up/Launch. In this stage the entrepreneurial team is assembled (if it does not exist already), financing is organized, appropriate business relationships are formed, initial sales are made, business operations—i.e., the way that the company either produces its products or delivers its services—are designed, tested and refined, and the administrative side of the business is set up.

Growth and Expansion. In this stage, or set of stages, the business grows into a more established and growing concern that must balance its entrepreneurial spirit with the appropriate degree of structure to assure quality, customer satisfaction, employee satisfaction and profitability/an appropriate return on investment.

---

[19](...continued)
edition), Burr Ridge, IL: Irwin, 1994, and Sahlman, William A. and Howard H. Stevenson, *The Entrepreneurial Venture*, Boston: Harvard Business School, Practice of Management Series, 1992.

Each stage of business development requires a special set of skills, know-how, commitment, resources and relationships. And each stage comes with its own share of challenges, opportunities and risks. Much of the literature and discussion on entrepreneurship emphasizes the "risk-taking" nature of entrepreneurs and their ventures. After all, a large percentage of new businesses either fail or cease operations within the first three years after start-up.[20] In reality, successful entrepreneurs and their backers see themselves taking a moderate degree of risk in putting their reputations and resources on the line to advance their ideas. They try to place highly-informed bets based on a combination of educated intuition (i.e., industry knowledge) careful analysis (i.e., homework and business planning) and manageable risks. Having said this, it is important to note that many entrepreneurs failed in business at least once before building a successful company. The act of failing is obviously a powerful and necessary learning experience for some people in understanding what it takes to grow a business.

Most venture capitalists (i.e., people who make a living investing in companies with high-growth potential) would argue that there are only two key ingredients to starting and building a successful company: a very strong business idea and the management talent to implement the idea effectively in order to maximize the business opportunity.[21] And, if forced to choose between the two, they would place greater emphasis on the strength of the management team. In their minds, skilled people can probably make more of an imperfect idea than less skilled people can of a great one. In fact, most of the literature on business success points directly to a wide range of management mistakes and/or miscalculations as the principal reasons for failure.[22]

One entrepreneurial consulting firm with extensive experience across a wide range of industries has identified ten keys to building a successful growing company.[23] These are:

- A sound idea that meets a large enough market need;
- A solid game plan for succeeding in the market that includes a clear and compelling "value proposition" to customers;
- A strong management team with the skills to grow the business and to surround itself with the right expertise and partners;
- A true sense of passion and enthusiasm for the business and the value that it provides;

---

[20] For a detailed discussion of business failures see Fullenbaum, Richard F. and Mariana A. McNeill, *The Function of Failure,* U.S. Small Business Administration, 1994.

[21] For a leading venture capitalist's assessment see Malone, Michael S., "John Doerr's Startup Manual," *Fast Company,* 1996, pp. 82-85. See also Silver, A. David, *Up Front Financing: The Entrepreneur's Guide,* New York: John Wiley & Sons, 1988. His discussion of "demonstrable economic justification" provides a good framework for thinking about the elements of a strong business idea and management's role in executing it.

[22] Gerber, Michael E., *The E-Myth Revisited: Why Most Small Businesses Don't Work and What to Do About It,* New York: Harper Business, 1995.

[23] VENTURE WORKS Inc., *Building a Successful New Venture: A Guidebook for Entrepreneurs,* 1996.

- The ability to sell the company's products/services with skill;
- The ability to produce the company's products/services with skill;
- A unending commitment to quality and customer satisfaction;
- The ability to raise sufficient capital during each stage of growth and manage operations profitably on an ongoing basis;
- The ability to innovate consistently;
- The ability to deal with adversity.

These ten "keys" underscore the critical role of management in turning a strong idea into a growing and profitable enterprise. They also provide a framework for thinking about what entrepreneurs and their firms need to survive and prosper.

Improving Entrepreneurial Success.. Given these keys to starting and building a high-growth company, it is clear that in order to succeed entrepreneurs need access to four critical things. These are:

- Knowledge
- Markets
- Talent
- Capital

Access to Knowledge. As noted above, entrepreneurs and their teams must have a sound business idea, access to in-depth knowledge of the marketplace and the right "expertise" to build their business.[24] This includes the appropriate technical and industry knowledge plus the strategic and practical know-how to organize, manage and build the enterprise. It also means the ability to stay as smart as possible as the business grows and the industry and world around it keep changing.

Access to Markets. Entrepreneurs and their teams must also have the ability to gain access to the market and to influence the buying choices of decision-makers. This includes participating in the right business networks and having the right contacts and relationships to leverage new business opportunities. It also means forming the right alliances with partners, suppliers and, at times, competitors. And as more entrepreneurial firms compete globally, it means having access foreign markets.

Access to Talent. Entrepreneurs and their teams must be able to find, train, motivate and keep the best people possible. This becomes an even bigger challenge as the pace of technological change accelerates and its influence expands in all areas of business operations. It also becomes a bigger challenge as companies must compete in the market for technically-trained workers.

Access to Capital. Entrepreneurs and their teams must be able to pull together sufficient capital at the right times to launch their business and fund it through each stage of development and growth. Gaining access to capital is probably the hottest topic in entrepreneurial circles. Interestingly, most entrepreneurs start their businesses

---

[24] Bhide, Amar, "The Questions Every Entrepreneur Must Answer," *Harvard Business Review*, November-December 1996, pp. 120-130.

using personal resources and funds from family, friends and "informal" investors.[25] The skill of piecing together financing from different sources and getting by on a limited amount of up-front money (i.e., "bootstrapping") is a key characteristic of many very successful entrepreneurs.[26] A relatively small percentage of firms ever receive formal venture capital.[27]

Later sections of this report examine whether the private market satisfies these needs and what issues might be appropriate for the public policy consideration.

The Economics of Entrepreneurship

The entrepreneur has wandered in and out of economic theory during the past 200 years receiving relatively little attention from many mainstream economists. More recently, however, a new generation of thinkers is redefining our understanding of the dynamic and entrepreneurial nature of the U.S. economy.

The Entrepreneur's Place in Economic Theory

Early Thinking About Entrepreneurs. Richard Cantillon, an Irish economist, is credited with introducing the term "entrepreneur" into economic theory in the mid-1700's.[28] But it was the French economist Jean-Baptiste Say who gave the term greater prominence in the early 1800's.[29] He argued that entrepreneurs played a critical role in producing useful value by "shifting economic resources out of an area of lower and into an area of higher productivity and greater yield." John Stuart Mill popularized the term in England in the mid-1800's using it widely in his thinking and writing.[30] But by the turn of the century the concept of the entrepreneur had literally disappeared from the economics literature. This was due in large part to the development of neoclassical theory, with its emphasis on perfect information and perfect competition.

Alfred Marshall's neoclassical theory of economics, which was developed further by many other leading economists, provided a basis for general equilibrium theory and

---

[25] Mamis, Robert A., "Great Companies Started With $10,000 or Less," *Inc.*, August 1997, pp. 40-51. This article debunks the notion that it takes a lot of money to start and build a high-growth company...providing case examples and insight on businesses like Nike, Land's End and The Limited that were started on a shoestring.

[26] Bhide, Amar, "Bootstrap Finance: The Art of Start-Ups," *Harvard Business Review*, November-December 1992, pp. 109-117.

[27] Bylinsky, Gene, "Who Will Feed the Startups?," *Fortune*, June 26, 1995, pp. 98-106.

[28] Cantillon, Richard, *Essai sur la nature du commerce en general*, ed. H. Higgs, London: Macmillan, 1931 (1755).

[29] Say, Jean-Baptiste, *A Treatise on Political Economy*, New York: Augustus M. Kelley, 1964 (1803).

[30] Mill, John Stuart, *Principles of Political Economy*, London: John W. Parker, 1848.

a relatively static view of an economy dominated by large firms.[31] This theory regarded entrepreneurs merely as fixed factors of production thus negating any role that they might play as innovators. In a world of perfect information and perfect markets there was no added value provided by individual initiative, or more effective entrepreneurial strategy and execution. The entrepreneur did appear in the work of John Maynard Keynes, who acknowledged explicitly the role of entrepreneurs and their decision-making in influencing the level of production and employment.[32] In particular, he focused on thinking about the effect of changes in money-wages on entrepreneurial decisions about pricing and capital investment.

Schumpeter and "Creative Destruction". In the 1930's and 1940's Joseph Schumpeter provided much of the foundation for current thinking about the role of entrepreneurs in the economy. His theory of "creative destruction" emphasized the importance of entrepreneurs as innovators who allocate resources in order to exploit new inventions and "carry out new combinations."[33] Schumpeter thought that innovation was the critical driver of capitalism and that entrepreneurs were the people responsible for "the doing of new things or the doing of things that are already being done in a new way."[34] In the process their efforts acted to create entirely new industries and "destroy" existing ones. (He also expressed the belief that over time large firms would tend to dominate the economy.)

But Schumpeter was really alone for many years. It was not until the late 1960's that mainstream economists started to focus on the important role of entrepreneurs in the economy. Baumol, in particular, suggested the need for a more flexible (and less precise) model of the economy that factored in the important role of entrepreneurial initiative and decision-making.[35] Leibenstein acknowledged that two critical assumptions underlying the theory of the production function were to blame. First, "that the complete set of inputs are specified and known to all actual or potential firms" in an industry and, second, "that there is a fixed relation between inputs and outputs."[36]

One might have assumed that microeconomics would pay more attention to entrepreneurs. Until recently, however, microeconomic theory has largely viewed

---

[31] Marshall, Alfred, *Principles of Economics*, London: Macmillan, 1961 (1890).

[32] Keynes, John Maynard, *The General Theory of Employment, Interest and Money*, New York: Harcourt, Brace & World, 1935. In particular see Chapter 19 on "Changes in Money-Wages," pp. 257-279.

[33] Schumpeter, Joseph A., *The Theory of Economic Development*, Trans. by R. Opie, Cambridge, MA: Harvard University Press, 1934. See also *Capitalism, Socialism and Democracy*, New York: Harper Colophon, 1975 (1942), especially Chapter VII: "The Process of Creative Destruction."

[34] Schumpeter, Joseph A., "The Creative Response in Economic History," *Journal of Economic History*, November 1947, p. 151.

[35] Baumol, William J., "Entrepreneurship in Economic Theory," *The American Economic Review*, Volume 58, Issue 2, Papers and Proceedings, May 1968, pp. 64-71.

[36] Leibenstein, Harvey, "Entrepreneurship and Development," *The American Economic Review*, Volume 58, Issue 2, Papers and Proceedings, May 1968, p. 73.

business managers as a group of equally capable people who make "rational" choices in maximizing the profits of their firms. The fact that they could take distinct and creative actions beyond making simple tradeoffs does not fit neatly into the equation. But it is precisely the ability to innovate, and rethink the possible combinations of people and technology (in particular) that truly sets apart the most successful firms in their marketplaces.

In their study of leading companies, Treacy and Wiersema identify three distinct "value disciplines" that are perfected by market leaders.[37] These are operational excellence, product leadership and customer intimacy. Operational excellence means providing the right mix of quality, price and service. Product leadership means providing the best or "state-of-the-art" offerings. And customer intimacy means building close relationships with customers in order to anticipate their needs and provide the best possible solutions. Within each of these three categories are a wide range of options for entrepreneurial initiative aimed at creating a competitive advantage in the marketplace. Entrepreneurs, acting at a point in time in response to a dynamic market, are continually looking for the right (and unique) combinations of resources and actions.

New Growth Theory, Dynamic Capitalism. Today researchers are focusing more attention on entrepreneurs and their role in economic growth. Among current thinkers, Kirchhoff's theory of "Dynamic Capitalism" and Romer's "New Growth Theory" are receiving the most attention. These theories run in a different direction than more traditional economic thinking about relatively static economies in equilibrium dominated by large and highly efficient firms. Kirchhoff is a leading proponent of the view that entrepreneurship is the driving force of capitalism. His theory of "dynamic capitalism" argues that the "dynamics of firm formation, growth, decline, and termination" create a cycle of renewal and growth in the economy.[38]

Romer contends that the world economy today is not constrained by a scarcity of resources and limits to growth. Instead he argues that there is "unbounded opportunity, where new ideas beget new products, new markets, and new possibilities to create wealth."[39] In his formulation, human capital and knowledge have become the most influential ingredients in economic growth. This is demonstrated by the emergence of whole new industries based on information technology and the selling of information, including the rapid commercialization of the Internet. It remains to be seen if these industries, while they clearly have great growth potential, will break the mold of economic theory. There are already real limits in the availability of human capital that are beginning to constrain growth.

---

[37] Treacy, Michael and Fred Wiersema, *The Discipline of Market Leaders*, Reading, MA: Addison-Wesley, 1995.

[38] Kirchhoff, Bruce A., *Entrepreneurship and Dynamic Capitalism: The Economics of Business Formation and Growth,* Westport, Connecticut: Praeger, 1994.

[39] See Romer, Paul M., "Beyond Classical and Keynesian Macroeconomic Policy," *Policy Options,* July-August 1994. Also, "Economic Growth," *The Fortune Encyclopedia of Economics,* David R. Henderson (editor), New York: Warner Books, 1992.

Many economists and others have built on Schumpeter's thinking. Hicks, for example, in his in-depth analysis of the Texas economy, describes a "churning" process in local/entrepreneurial economies that is characterized by a high number of business births and deaths.[40] He argues that this churning is vital to economic growth because it produces, in healthy economies, a larger number of young and rapidly-growing firms. These economies in essence "regenerate" themselves. Others like Useem and Richman have used this analysis to suggest that the dynamic process of new business formation is the critical driving force in our growing economy.[41]

The Role of Entrepreneurial Firms in the Economy

In recent years economists and others have focused more attention on the role of small and entrepreneurial firms as contributors to economic growth, innovation and jobs. A catalyst for this discussion (and subsequent controversy) was research done by David Birch of MIT and funded initially by the U.S. Department of Commerce.

With respect to innovation, traditional economic thinking has emphasized the role of large firms given their market power. Arrow argues that entrepreneurs do not play as significant a role in innovation and commercialization of new ideas as one might believe from the literature.[42] He focuses on "giant" firms, though he acknowledges that they often develop the ideas generated by small companies. But there is a growing body of research that shows that smaller firms also have an important role to play. The leading proponents here include Acs and Audretsch[43] and Scherer[44] who argue that small firms have "an innovation-related advantage" over large firms that enables them to be more effective leveraging their internal R&D and university-based support. The core of this advantage includes a relative lack of bureaucracy that often constrains the development of new ideas in larger and more established companies, the central place that innovation holds in the strategies of small and growing firms, a greater level of comfort with risk-taking, an ability to move more quickly in pursuing opportunities, a willingness to develop new products and processes that may have too small a payoff to entice larger firms, and a stronger spirit of internal collaboration.

---

[40] Hicks, Donald A., *A Nanoeconomic Perspective on the Growth and Development of the Texas Manufacturing Base, 1970-1991,* Office of the Comptroller, State of Texas, 1993.

[41] "The State of Small Business," *Inc.* Special Issue, May 1997. See especially "Opening Argument: Churn, Baby, Churn," by Jerry Useem, pp. 25-32, and "Creators of the New Economy," by Tom Richman, pp. 44-48.

[42] Arrow, Kenneth J., "Innovation in Large and Small Firms: An Empirical Analysis," in *Entrepreneurship,* Joshua Ronen (ed.), Lexington, MA: Lexington Books, 1983, pp. 15-28.

[43] Acs, Zoltan J. and David B. Audretsch, "Innovation and Firm Size: The New Learning," in M. Dodgson and R. Rothwell (eds.), *International Journal of Technology Management,* Special Publication on Small Firms and Innovation, 1993, pp. 23-35.

[44] Scherer, F.M., "Changing Perspectives on the Firm Size Problem," in Zoltan Acs and David Audretsch (eds.), *Innovation and Technological Change: An International Comparison,* Ann Arbor, MI: University of Michigan Press, 1991, pp. 24-38.

It is important to point out that many of today's major technology firms started as new entrepreneurial firms not that long ago. And a growing number of established companies are opting to acquire entrepreneurial firms and their technologies rather that trying to "innovate" internally.[45]

Nelson focuses on technical advance as the key to economic growth, suggesting that a diverse group of firms—of different sizes and stages of development—is the prime mover in growing the economy.[46] In fact, large firms have a distinct advantage when innovation requires the ability to invest a very high level of financial, technical and human resources.

With respect to economic growth and job creation, Birch caused considerable debate in 1979 by arguing that small firms—defined as businesses with less than 100 employees—created most of the net new jobs in the U.S. economy.[47] His findings and methodology have been questioned by other researchers such as Armington and Odle[48] and Brown, Hamilton and Medoff,[49] and supported by others such as the Small Business Administration[50] and a more in-depth analysis by Kirchhoff and Greene.[51]

Birch's most recent research suggests that only about 3% of small firms, which he has labeled "gazelles," actually do very much growing at all.[52] This assessment is reinforced by Inc. magazine's annual review of the fastest growing private companies in the U.S.[53]

---

[45] See Henkoff, Ronald, "Growing Your Company: Five Ways To Do It Right," Fortune, November 25, 1996, especially pp. 84-88. A recent and much publicized example of the trend to acquire entrepreneurial firms is Lucent's purchase of Yurie Systems to gain access to its innovative technology. This is even more noteworthy given Lucent's history of innovation.

[46] Nelson, Richard R., *The Sources of Economic Growth,* Cambridge, Massachusetts: Harvard University Press, 1996.

[47] Birch, David L., *The Job Generation Process,* Report prepared for the Economic Development Administration, U.S. Department of Commerce, Washington, D.C., 1979. See also *Job Creation in America,* New York: The Free Press, 1987.

[48] Armington, Catherine, and Marjorie Odle, "Small Business, How Many Jobs?" *Brookings Review,* 20:14-17, 1982.

[49] Brown, Charles, James Hamilton and James Medoff, *Employers Large and Small,* Cambridge, MA: Harvard University Press, 1990.

[50] U.S. Small Business Administration, *Handbook of Small Business Data,* Washington, D.C.: U.S. Government Printing Office, 1988.

[51] Kirchhoff, Bruce A. and Patricia G. Greene, "Response to Renewed Attacks on the Small Business Job Creation Hypothesis," *Frontiers of Entrepreneurship Research* (1994 edition), Center for Entrepreneurial Studies, Babson College.

[52] Birch, David L., *Who's Creating Jobs?*, Cambridge, MA: Cognetics Inc., 1997.

[53] Various authors, "The Inc. 500: America's Fastest Growing Private Companies," *Inc.*, 1990-1997. This publication is a widely read and cited standard on the state of entrepreneurship in America.

Summary

While entrepreneurs have not held a central role in mainstream economic theory, there is growing interest in improving our understanding of their contributions to economic growth. In particular, we know that successful entrepreneurs and their companies are dynamic innovators with great skill in envisioning new combinations of resources and creating new products, services, new ways of doing business and entirely new industries. They improve the choices that customers have by shifting the production function through the use of knowledge, technology, the power of new ideas and the design of new business practices.

As economic thinking evolves there is probably sound analysis that could be done to integrate entrepreneurial vision and initiative into existing theories. This means being able to model the unique decision-making processes of entrepreneurs in a more dynamic way as they identify, screen and exploit business opportunities. Microeconomic theory, however, is basically static and looks at the marketplace at a particular point in time, rather than looking dynamically at how entrepreneurs innovate, create new products and services, and adapt new technologies. It is this special ability to look at the world, information and markets in a different way that makes entrepreneurs critical drivers in business building and economic growth.

Are There Gaps Affecting Entrepreneurship in the Private Economy?

In the general case, economic theory suggests that the private sector economy allocates resources efficiently producing the right level and mix of goods and services to meet demand. In this framework the role of the public sector is simply to facilitate the process. There are cases, however, in which the private economy does not work efficiently and public policy may have an important role to play in addressing the short-coming.[54] One of the purposes of this report is to look at whether or not that is the case with respect to entrepreneurship and high-growth firms and, if so, to identify those areas warranting consideration of public policies to address the inefficiencies.

The literature typically views shortcomings in the marketplace as "gaps" such as the "credit gap" that is widely discussed as a barrier to business formation and growth. In thinking about the appropriate federal policy role vis-a-vis high-growth companies, there are three important questions to ask. First, are there gaps in the private sector economy that adversely affect all entrepreneurs and growing firms? Second, are there gaps in the private sector economy that adversely affect specific groups of entrepreneurs and growing firms? Third, if gaps do exist are there appropriate federal policies that could successfully address them?

These are tough questions to answer given the relative lack of available data. One could make the case, especially in today's strong economy, that the private

---

[54] One example here is the case of monopolies.

market—combined with existing public policies—does an effective job of nurturing entrepreneurs and their high-growth companies. After all, the U.S. is viewed widely as a center of entrepreneurial initiative, and our economy continues to grow fueled to a significant extent by the start-up and growth of new firms. And, people from all walks of life are entering the ranks of entrepreneurs in record numbers. There are, however, data that suggest that the ranks of high-growth companies still include a relatively small number of women and minorities—in particular African-Americans and Hispanics. One could also make the case that even more people would start and build growing companies if the "environment" were more supportive. Add to this the fact that our current economic fortunes might not continue unabated, and we are left with a less than definitive answer.

Gaps Affecting All Entrepreneurial Firms

As discussed above, entrepreneurs need four things in order to start and build successful companies. These are access to knowledge, access to markets, access to talent and access to capital. In an important sense we can view entrepreneurship as a test to pull these four ingredients together to start and build a growing company. To pass the test entrepreneurs must combine skill, ingenuity, persistence, good timing and a measure of luck. And they must go up against larger and more established firms that often have greater access to these resources. But it is for the most part, at least at the outset, a good system for screening people and their ideas.

Access to Knowledge. Access to knowledge is the real foundation for entrepreneurial initiative. An entrepreneur sees what appears to be a compelling market opportunity and sets out to determine the best way to seize it. This means understanding the marketplace and the target customers in as much detail as possible. It also means looking at the key trends and forces that are influencing the market and the likely competition for the idea. This information provides the basis for developing a sound game plan and effectively designing products and services that really meet the market need. But that is just the start of building a business, because entrepreneurs and their teams must also have the knowledge and talents to win customers, the ability to develop, produce and deliver their products and services skillfully—including conducting any necessary R&D, and knowledge to manage operations profitably on an ongoing basis. And they must continue to stay as smart as possible, in line with the pace of change in their industry and the overall economy, in order to guide the growth of their companies.

In today's economy information about starting and building a business is more widely available than ever before. Growing interest in entrepreneurship has spawned a range of business resources, organizations, national and local networks, publications and educational offerings that are intended to support would-be entrepreneurs. The rapidly expanding use of the Internet is bringing a world of valuable information right to everyone's desktop and helping to level the knowledge playing field. There are, of course, differences in the ability of people to tap information in pursuing a business opportunity. Keeley and Knapp contend that differences in business performance are tied to a set of factors that include subtle personal qualities such as the ability to learn and adapt knowledge.[55] But there is plenty of information out there to access, for many types of businesses, if someone has the ability and motivation to tap it. Reynolds and White contend that there is no clear "formula" for determining whether or not a new firm will grow rapidly. Rather they suggest that it is a function of some key characteristics of the people involved including their strength of motivation, perseverance, creativity, energy, luck and timing.[56] One could argue that both creativity and timing are related directly to skillfully developing and utilizing knowledge.

Having said this it is important to point out that many entrepreneurs learn about the market and basic business skills by working for someone else. As an employee they gain specific industry and technical expertise along with general knowledge about how to run a business. When they are fortunate enough to work for a strong company and a talented mentor they have a clear advantage in being prepared for entrepreneurial opportunity. Even when they haven't worked directly for a successful firm or entrepreneur, the founders of many young firms have found mentors to coach them. But is access to the right mentors and the right entrepreneurial "proving grounds" uneven? While there are no definitive data available, it is reasonable to assume that some potential and emerging entrepreneurs have less access to these resources by virtue of their background and geography. It is widely acknowledged that some communities have strong cultures that facilitate the mentoring and networking process.[57]

Colleges and universities are becoming important participants in entrepreneur development by offering curriculum, supporting mentoring and networking, helping to facilitate technology transfer and enhancing access to basic market information. Some are even trying to link emerging entrepreneurs with local investors and sources of capital. Programs like the Center for Entrepreneurial Studies at Babson College, the Dingman Center for Entrepreneurship at the University of Maryland and the IC2 Institute at the University of Texas are strong examples of the evolving role.

There is even growing interest in providing access to entrepreneurial education for children at an early age as a way to seed the next generation of entrepreneurs. This effort is being sponsored by a wide range of not-for-profit organizations.

---

[55] Keeley, Robert H. and Robert W. Knapp, "Founding Conditions and Business Performance: High Performers vs. Small vs. Venture Capital-Backed Start-Ups," *Frontiers of Entrepreneurship Research,* 1994, pp. 88-103, Center for Entrepreneurial Studies, Babson College.

[56] Reynolds, Paul D. and Sammis B. White, "Government Programs and High Growth New Firms," *Frontiers of Entrepreneurship Research,* 1996, Center for Entrepreneurial Studies, Babson College, pp. 621-632.

[57] See Rebello, Kathy (editor), "Silicon Valley: How It Really Works," *Business Week,* Special Issue, August 18-25, 1997. Also see Case, John, "The Most Entrepreneurial Cities in America," *Inc.,* March 1990, p. 41. Many other communities are creating the same type of environment as Silicon Valley with private and not-for-profit organizations serving as the catalyst for bringing new entrepreneurs and mentors together.

A significant percentage of high-growth companies are in technology fields[58] which pose some distinct and important issues in terms of access to knowledge. First, having access to and comfort with computers makes someone far more likely to see new business opportunities in the world of information technology, computers and telecommunications. And in many other industries a solid knowledge of, and comfort with, information technology is vital. But access here is not universal. Recent announcement of a federal initiative to put computers in every primary school classroom seems to acknowledge this reality. Second, we face a growing shortage of people who are going into certain technology fields including engineering, computer science, mathematics and many of the physical sciences.[59]

Access to Markets. Access to markets means having a fair and open opportunity to gain customers for one's products and services.[60] The marketing equation is one of identifying the right markets to target, getting your offerings in front of prospective customers in those markets and then getting them to buy. To stand out from the pack the most successful companies create a unique value proposition that directly meets the target's needs.

There is reason to believe that most markets in the U.S. are open if a company has sound products and services and is skilled in both selling and providing them. Given that entrepreneurs are in the business of offering new or improved products and services or creating new ways of doing business that enhance the choices customers have, one would expect them to receive an open reception in markets where decision-makers are enlightened about the best options. Certain markets are tougher to break into than others for a number of reasons including the cost of entry, the time involved, the scale of production or service delivery required, the lack of an adequate track record, limited know-how or the lack of necessary connections. Selling to large companies, for example, is often a difficult task for new and growing firms. This is because established companies have the luxury of hiring vendors with a proven track record and often view this as the most prudent course of action. To be successful, new and growing firms must offer better products/services or some other unique value that sets them apart. But this is just part of the process of building a new business, and most entrepreneurs will tell their "war stories" of what it took to win their first big account.

Available research does suggest that there may be real barriers to new and growing firms in their ability to gain access to export and procurement

---

[58] Thirty-six percent of the current Inc. 500 companies are in the computer industry and related services, and many of the rest are in industries that rely heavily on information technology. See "The Inc. 500," *Inc.*, October 1997.

[59] National Science Foundation, Division of Science Resources Studies, *Science and Engineering Degrees: 1966-1995* (NSF 97-326), November 1997.

[60] Taub, Richard P. and Connie Marie Gaglio, "Entrepreneurship and Public Policy: Beyond the Credit Crunch," *Frontiers of Entrepreneurship Research*, 1995, pp. 437-444. Center for Entrepreneurial Studies, Babson College.

opportunities.[61] For many high-growth companies, especially in technology fields, the need to expand into overseas markets quickly is a greater imperative than it was even in the recent past. This is due to a combination of factors that include the desire of investors to speed up growth, the strength of domestic and foreign competition and the pace of technological change. In addition, many firms face significant cost and obstacles in protecting their intellectual property rights.

A critical ingredient in gaining access to markets (as well as access to knowledge, talent and capital) is participation in business "networks" and having the relationships and contacts to open doors. These range from the local chamber of commerce, to professional and trade associations, to CEO clubs and other support groups of business owners, to informal networks of like-minded people and companies, to the "old boy" network of established corporations and business leaders. The right networks/organizations to participate in will vary depending on industry and location, but entrepreneurs often can build their market presence more quickly by having colleagues with a better rolodex and better access to key decision-makers.

Access to Talent. Access to talent means having the ability to attract and retain the key employees and managers needed to build a business. From the outset, new and growing firms need management and staff who possess talent in four critical areas: (1) product or service development, (2) marketing and sales, (3) production or service delivery, and (4) finance and administration. While a company might start with one "key" person, its growth depends upon building a strong team. Early on key team members often work for little pay but with equity and the expectation that as the company grows they will be better compensated and their stake in the firm will grow in value. As the company grows it must attract and retain a growing number of talented people.

In today's economy, in particular, the competition to find, hire and retain good people is great. New and growing companies often have great difficulty competing with larger and more established firms for talent. But this is also part of the test of building a new company and simply suggests that entrepreneurial firms must be more creative in building and supporting their talent base—often by combining flexibility with other short and long-term incentives. This problem is most acute for high-technology firms, in particular information technology, which face a growing shortage of technically trained workers. They must go to great expense hiring and training the right people only to see many of them leave for more pay, better benefits or perceived security. A recent report by the Department of Labor and the Information Technology Association of America suggests that a growing number of jobs in information technology and other sciences are going unfilled.[62] Critics, however, suggest that these estimates are inflated. A number of articles in the business and popular press have also recently discussed the growing shortage of "knowledge" workers in high technology fields.

---

[61] Office of Advocacy, Small Business Administration, *Building the Foundation for a New Century,* First Annual Report on Implementation of the Recommendations of the 1995 White House Conference on Small Business, Washington, D.C.: SBA, 1996.

[62] *America's New Deficit: The Shortage of Information Technology Workers,* U.S. Department of Commerce, Washington, D.C., September 1997.

Finding the right talent is also more difficult for firms in some geographic areas such as the plains states and many smaller cities and towns where a high percentage of the most talented young people have left. In these places a "brain drain" threatens their ability to support growing companies.[63]

Access to Capital. Having adequate access to affordable capital is a critical part of starting and building a company. In fact, most of the literature on entrepreneurship identifies capital (access and availability) as the most significant gap faced by entrepreneurs and growing companies. This includes equity and debt financing at the initial stages of growth.[64] The difficulty of obtaining debt financing, primarily from commercial banks, is widely cited as a real and frustrating problem that limits the ability of many new firms to grow.[65] But the risk-averse nature of banks tends to explain why they are generally unwilling to lend money to most emerging businesses. In their minds "newness" often implies greater risk and the downside of losing their funds far outweighs the potential upside of having their loans repaid in a timely manner.

Equity is the most critical type of capital for new and growing firms who need the flexibility to invest in themselves in order to provide the foundation for growth. Here, as well, real issues and opportunities exist. If we divide the world of equity capital into two parts, "seed" and later stage venture capital, the gaps become clear and not unexpected. While there is plenty of private venture capital in the U.S. (and abroad) it tends to go to those companies that have already developed their products, started up and found an initial base of customers.[66] In addition to formal venture capital firms, an even larger amount of venture capital is provided informally by individual investors known as "angels."[67] They are often more flexible about the types of deals that they will do, including making investments in very early stage ventures. But most angels concentrate primarily on companies with some track record. It is the people with new product/service ideas who need money to test them, invest in key

---

[63] Belluck, Pam. "Please Stay, We'll Pay You, Nebraska Begs Its Brightest," *The New York Times,* February 28, 1998, p. 1.

[64] See Taub, Richard P. and Connie Marie Gaglio, pp. 437-444.

[65] Storey, D.J., "New Firm Growth and Bank Financing," *Small Business Economics,* 6 (2), April 1994, pp. 139-150.

[66] For an overview of the venture capital industry see Timmons, Jeffry A. and William D. Bygrave, "Venture Capital: Reflections and Projections," in Sexton, Donald L. and Raymond W. Smilor, *Entrepreneurship 2000,* Chicago: Upstart Publishing, 1997, pp. 29-46. A more in-depth analysis is presented in their book *Venture Capital at the Cross-roads,* Boston: Harvard Business School Press, 1992. In 1992 there were 674 formal venture capital firms worldwide managing an average of $49.5 million each.

[67] For an overview of informal venture capital see Freer, John, Jeffrey E. Sohl and William E. Wetzel, Jr., "The Informal Venture Capital Market: Milestones Passed and the Road Ahead," in Sexton, Donald L. and Raymond W. Smilor, *Entrepreneurship 2000,* Chicago: Upstart Publishing, 1997, pp. 47-70. While data on informal venture investing are harder to find, it is estimated to be between $30 and $60 billion a year, or 10 to 15 times more than formal venture capital firms invest.

infrastructure including people, and launch them that tend to be ignored by formal and informal venture capital sources.

So at the early stage, most entrepreneurs are left to their own devices or to the resources of family and friends.[68] Increasingly using their own devices means being creative and even using credit cards to finance growth. A recent article in the Wall Street Journal stated that more than a third of small firms use credit cards as a key means of financing.[69] John Ledecky, the founder of U.S. Office Products, noted in a recent presentation that he applied for and received credit card cash advances totaling $250,000 prior to starting his business and used that money to seed the start of his firm and to cover costs as he raised the capital to build what is now a $4.0 billion company.[70]

New and growing firms that require a significant level of investment in R&D to develop their products and bring them to market successfully face even greater challenges in their ability to compete. They must be able to raise sufficient capital to build their talent and knowledge base long before they will ever generate sales. And they must make enough progress on limited resources to come to the attention of venture capitalists and larger companies that are interested in either buying into or partnering with their "window" on technology. These issues are particularly acute for firms in the applied sciences including the biotech and pharmaceutical industries in which the cost of doing basic research and developing new products is great. They seem to be less significant in the software industry where the actual cost (and time) to develop new products is not as great and is typically self-financed by the entrepreneur often through "sweat equity." It is worth noting that many small firms do not have the luxury of doing basic R&D. Most of their research is, by necessity, applied and focused on developing and commercializing products as fast as possible.

Gaps Affecting Specific Groups of Entrepreneurs

While women and minorities are starting their own businesses in record numbers, they are less likely to start and build high-growth companies. Only eight percent of the current Inc. magazine list of the 500 fastest growing private companies in America are headed by women.[71] No analysis has been done of the percentage that are headed by minorities. While there is no simple answer, the likely explanations include a lack of access to traditional business markets, established business networks and sources of capital. The question is whether or not there is an appropriate role for public policy in improving their access. While the process of starting and building a high-

---

[68] Over 75% of new and growing firms get started and fund their initial growth through the entrepreneur's own resources and the support of family and friends.

[69] Ho, Rodney, "To Finance a Dream Many Entrepreneurs Binge on Credit Cards," *Wall Street Journal,* March 9, 1998, p. 1.

[70] Remarks before the MindShare Emerging Business Network, Tysons Corner, Virginia, March 12, 1998.

[71] "The Inc. 500," *Inc.*, October 1997. The National Women's Business Council estimates that 20 percent of women-owned businesses have the potential to grow but don't know how to.

growth business was never meant to be easy, it should be fair and open to those people who have sound ideas and the competence and determination to turn them into successful enterprises. It appears that women and minorities have to be even more skillful in tapping markets, building networks and utilizing personal, as well as private and public sector sources of financing.[72]

Women-Owned Businesses. In 1996 there were more than eight million women-owned businesses in the U.S.[73] During the last three years, in fact, women have been starting new businesses at a faster rate than men. This dramatic growth in the number of women-owned firms has been attributed to a several factors including the corporate "glass ceiling," the desire to balance work and family responsibilities, and the more traditional entrepreneurial drive to be your own boss and run with an idea that you are passionate about.[74]

Women entrepreneurs and potential entrepreneurs often face particular problems in gaining access to capital and networks required to build a high-growth company. In an effort to enhance opportunities to receive capital, a limited number of commercial banks such as Harris Bank in Chicago and the United Bank of Philadelphia have begun to focus more attention on the credit needs of women entrepreneurs. There is also an emerging group of venture capital funds that are being targeted to women-owned businesses.[75] These include the Women's Growth Capital Fund in Washington, D.C., started in June 1997, which has raised an initial $8,000,000 to invest in women-owned and managed businesses and the Three Guineas Fund in San Francisco which tries to assist women launching high-technology businesses. Women entrepreneurs are also forming their own networks to strengthen their ability to tap markets.[76]

Minority-Owned Businesses. While minorities constitute a significant and growing part of the entrepreneurial landscape in the U.S., their share of high-growth firms is relatively small. The success of, and obstacles faced by, African-American entrepreneurs have been documented in the popular and business literature and in a number of recent books.[77] Research and analysis on entrepreneurship in the Hispanic,

---

[72] Card, Emily and Adam Miller, *Business Capital for Women*, New York: Macmillan, 1996.

[73] Figures compiled by the National Foundation for Women Business Owners.

[74] Grimsley, Kirstin Downey, "MBA No Ticket to Top for Women," *The Washington Post*, March 24, 1998, p.1.

[75] Stout, Hilary, "The Front Lines," *Wall Street Journal*, November 28, 1997, p. B1.

[76] Henry, Shannon, "The New Girl's Network: Welcome to the Club," *Tech Capital*, Fall 1997, pp. 38-48.

[77] Woodard, Michael D., *Black Entrepreneurs in America: Stories of Struggle and Success*, New Brunswick, NJ: Rutgers University Press, 1997. And Ballard, Donna, *Doing It for Ourselves: Success Stories of African-American Women in Business*, Berkley Publishing, 1997.

Asian-American and Native American communities is more limited.[78] For each of these groups, the lack of access to many markets, networks and capital—tied often to societal and racial barriers—may have limited full entrepreneurial opportunity in many parts of the economy.

Bates argues, however, that a new generation of African-American business owners is emerging that is more educated and building larger businesses—but still faces real obstacles.[79] For example, he found that even when controlling for differences in education, African-Americans starting businesses received 37% less bank financing than their white counterparts for each dollar of equity invested. Perceived risk and bias may be part of the equation. Research indicates that the percentage of Hispanic-owned businesses that receive bank financing is comparable to that of white-owned firms.[80]

A recent feature in *Fortune* magazine titled "The New Black Power" underscores the fact that a new generation of business leaders is emerging and their stories offer a powerful picture of what is possible.[81] But the article acknowledges that the number of successful African-American entrepreneurs is relatively small in leading industries like information technology. It attributes this to a number of factors including the continuing "burden of race" and a shortage of strong mentors to support new entrepreneurs. It concludes, however, that new industries that lack an established "old boy" network hold greater promise for African-Americans and other minorities. Of the 100 largest African-American owned businesses less than 10 percent are in technology and computer-related products and services compared to over 40 percent of the *Inc.* 500 fastest growing firms.[82] Available data suggest a similar pattern for Hispanic-owned businesses.[83]

One emerging model for strengthening minority entrepreneurship is in forming strategic alliances with major corporations.[84] These partnerships, unlike traditional

---

[78] With respect to Hispanic entrepreneurship some data can be gathered from magazines like *Hispanic Business*. The Hispanic Entrepreneurs Project, sponsored by the Kauffman Foundation, is intended to address this void in systematic research and analysis of this diverse and growing segment of the American business community. There is even less information available on Asian-American and Native American entrepreneurship. One does get the sense in reviewing the popular business publications that Asian-American are well-represented in the ranks of high-growth companies.

[79] Bates, Timothy, *Banking on Black Enterprise,* Washington, D.C.: Joint Center for Political and Economic Studies, 1993.

[80] National Survey of Small Business Finances, Washington, D.C.: Board of Governors, Federal Reserve System and the U.S. Small Business Administration, 1993.

[81] Johnson, Roy S., "The New Black Power," *Fortune,* August 4, 1997, pp. 46-82.

[82] See "The Black Enterprise 100," *Black Enterprise,* September 1997. For comparison see "The Inc. 500," *Inc.*, October 1997.

[83] See "The Hispanic Business 500," *Hispanic Business,* June 1997.

[84] See Nelton, Sharon, "Minority Business: The New Wave," *Nation's Business,*

(continued...)

procurement contracts, are providing minority-owned firms with financial support, real equity, access to know-how and technology, mentoring and an ongoing base of business and market clout. This practice is quite common in the automobile industry and is reflected in the disproportionately high number of car dealerships that are on lists of the largest African-American and Hispanic-owned firms in the U.S. It is now spreading to other industries in which minorities comprise a large share of the market, though not necessarily in technology-based industries.

How Do Existing Public Policies Affect Entrepreneurs?

The previous section identifies a set of market gaps in the functioning of the private sector economy as it relates to entrepreneurship and building high-growth companies. This discussion can be viewed as a framework for looking at existing public policies designed to support entrepreneurs. Whether intended or not federal government policies and programs play a role in the lives of entrepreneurs and their ventures. These range from tax and regulatory policies to direct assistance programs. This section of the report provides a brief overview of existing federal policies and programs that either assist or affect entrepreneurs and growing firms.[85]

Direct Assistance Programs And Policies

Most direct federal programs and initiatives to support business growth are aimed at the needs of small businesses with the Small Business Administration being the focal point of assistance. The *Catalog of Federal Domestic Assistance* identifies 304 programs that are intended to assist small firms.[86] Of this total, a limited number are intended specifically to support the start-up and growth of entrepreneurial firms. These programs include:

Access to Knowledge:

Cooperative R&D Agreements (CRDAs). CRDAs are intended to enable the development and transfer of technology from federal labs and agencies to non-federal organizations and commercial businesses. It is based on the widely-held belief that the federal government has a lot of great technology that can be used and developed

---

[84](...continued)
October 1995, pp. 18-27. She describes the partnership between Frito-Lay and Johnson-Bryce Corporation to manufacture packaging for Frito-Lay's products that enabled this "start-up" firm to become a $20 million enterprise in four years.

[85] Vesper, Karl H., *Entrepreneurship and National Policy,* Heller International Corporation Institute for Small Business, Policy Paper No. 3, 1983. While this report is somewhat dated it does provide a good overview of the types of government programs and policies that have been designed to assist small and entrepreneurial firms. In fact, most of the major assistance programs that are in place today were around (in some form) in 1983.

[86] *Catalog of Federal Domestic Assistance*, General Services Administration. Washington, D.C.: U.S. Government Printing Office, June 1997. This is a compendium of all Federal government assistance programs.

in the private marketplace. Under these agreements the resulting intellectual property is shared jointly between the two parties with companies having the rights to all commercial applications. CRDAs are easy to implement because they do not involve the transfer of funds to the non-federal partner, though the government can commit funds to the effort through the acquisition of facilities and property.

The federal government has acknowledged the value of technology transfer for a long time and there are strong examples of success from agencies like the National Institutes of Health. But there is also recognition of the need for new models in which entrepreneurs and the federal labs/agencies work together to do additional testing of promising ideas utilizing the resources and expertise of the federal facilities.[87]

SBA Technical Assistance Programs. The Small Business Administration provides a range of technical assistance and counseling programs that are geared primarily to small rather than growing firms. The two principal programs are its university-based network of Small Business Development Centers (SBDCs) and the Service Corps of Retired Executives (SCORE) that links small business owners with retired business persons.[88] While these SBA programs have the potential to assist high-growth firms, they are rarely used by them.

Access to Knowledge and Capital:

The Small Business Innovative Research (SBIR) Program. Started in 1983, the SBIR program is an interagency initiative that provides equity capital to small businesses to evaluate the scientific and technical merit of ideas, develop them into commercially-viable products and bring them to market. Under the program federal agencies with large research and development budgets designate amounts of their R&D contracts on a competitive basis to small firms. The impetus for the program was recognition that while small firms are major contributors to innovation they received relatively little federal research funding compared to large companies and universities. SBIR was also intended to fund extremely high-risk but potentially high-payoff R&D, including possible breakthrough efforts in emerging fields aimed at technology leadership, where the risk at the idea stage was often too high for private investors.[89] If SBIR research is successful, the risk might be reduced to levels that were acceptable to follow-on investors. As discussed above, while risk capital has never been so abundant, it is still very difficult to obtain private funding in the idea or prototype development stages.

The SBIR program is probably the largest single source of seed capital in the U.S. for young and growing firms.[90] Between 1983 and 1996 a total of $6.0 billion

---

[87] Bibben, Terry, "Industry Maximizing Innovation Through Government and Industrial Partnerships," The Scientific Research Society, November 1997.

[88] A limited number of analyzes and evaluations have been done of these programs.

[89] The standard definition of "small" firms by the SBA and SBIR is 500 employees or less.

[90] For a review of how the SBIR program has worked in one department see *Federal* (continued...)

in SBIR awards was made to roughly 8,000 small high technology companies.[91] While this is a significant number, the small business share of federally-funded R&D is less than four percent.[92]

Advanced Technology Program (ATP). The ATP, under the National Institutes of Science and Technology, Department of Commerce, supports precompetitive generic R&D that is intended to speed up the development of "high-risk" technologies that promise significant commercial payoffs. The focus of the ATP is on enabling technologies rather than the development of specific products, services and business processes. Thus far more than half of the awards have gone to individual small businesses or joint ventures led by small firms.[93]

Access to Markets:

The Section 8(a) Program. The Section 8(a) program provides "set asides" for minority-owned firms in federal procurements.[94] While its primary purpose is to increase participation in government contracting, the program is also viewed as a vehicle for building a stronger base of firms that can compete on equal terms with majority-owned firms in the public and private sector markets, and there is an expectation that participating firms will "graduate." In fact, the 8(a) program has helped to launch the growth of some of the largest minority-owned companies in the U.S. But many firms that have participated in the program have had difficulty making the transition out.

The Federal government has also made efforts, using other "set-asides" programs, to increase procurement opportunities for small, disadvantaged and minority-owned businesses.[95] And, there are special initiatives to support women and

---

[90](...continued)
Research: DOD's Small Business Innovation Research Program, Washington, D.C.: U.S. General Accounting Office, April 30, 1997. This report reviews the SBIR program and the steps that have been taken by DOD to make sure that it results in high quality research.

[91] Tibbetts, Roland, *An Analysis of the Distribution of SBIR Awards by States, 1983-1996,* U.S. Small Business Administration, Office of Advocacy, January 1998.

[92] Bibbens, Terry, "Industry Maximizing Innovation Through Government and Industrial Partnerships," paper presented at the *Trends in Industrial Innovation Conference,* Sigma Xi, The Scientific Research Society, Crystal City, Virginia, November 20, 1997, pp. 3-4.

[93] Advanced Technology Program, "Small Business Participation in the Advanced Technology Program," 1998. Additional information on the ATP can be found on the worldwide web at www.atp.nist.gov.

[94] The "Small Business Act of 1958," Public Law 85-536, as amended. Section 8(a) was originally used to target government contracts to small businesses.

[95] *Efforts to Provide Federal Procurement Dollars to Women-Owned Businesses,* Washington, D.C.: U.S. General Accounting Office, July 1992.

minority-owned businesses in building their companies. In addition, initiatives have also been attempted to increase export development.[96]

Access to Capital:

ACE-NET. The ACE-NET system at the Small Business Administration is a new service that is designed to create a network of informal investors for entrepreneurs and companies seeking equity capital using the Internet as a mechanism for exchanging information. It is intended to facilitate investments by "angel" or "accredited" individual investors in companies with high-growth potential. The program is based on the realization that while informal venture capital is a large source of potential equity funds for small growing firms, there is no systematic means of matching companies and investors.[97]

Small Business Investment Companies (SBICs). The SBIC program, under the Small Business Administration, is intended to make capital available to small businesses with growth potential. SBICs are privately organized and managed investment firms that combine their own funds with monies borrowed from the federal government to provide debt and venture capital including some seed capital. There are also "Specialized Small Business Investment Companies" that target capital to socially and economically disadvantaged firms.

The Section 7(a) Lending Programs. SBA's 7(a) Loan Guarantee Program which is designed to provide loans, through commercial banks, to qualified small businesses that are unable to receive credit on reasonable terms through normal lending channels. According to SBA, 25 percent of the programs' funds are now being used for start-up companies. Related programs can also be used for lines of credit, export working capital and to fund pollution control.

Regulatory and Tax Policy

Federal regulatory and tax policies also have significant effects on new and growing businesses and receive a great deal of attention in forums on the barriers faced by small and growing firms.[98]

---

[96] *Export Promotion: Improving Small Businesses' Access to Federal Programs,* Washington, D.C.: U.S. General Accounting Office, April 28, 1993. This report describes barriers that limit the ability of small and medium-sized firms to export and identifies ways to improve federal assistance programs.

[97] Additional information on ACE-NET and the other SBA programs described briefly in this section of this report can be found on the worldwide web at www.sba.gov.

[98] *The Annual Report of the Chief Counsel for Advocacy on Implementation of the Regulatory Flexibility Act,* Office of Advocacy, Small Business Administration, Washington, D.C., 1997, summarizes the key actions being taken by federal agencies to reduce the regulatory burden on small businesses.

Compliance With Federal Regulations

Many believe that a broad range of government regulations imposes a great and disproportionate cost on small and growing companies. In fact, regulatory relief is the one area in which businesses uniformly agree that reforms are needed to make it easier and less costly to do business. There are some 60 Federal agencies that develop and enforce regulations affecting business.

The areas of the greatest concern to many entrepreneurs and high-growth firms include health, safety and workplace rules, environmental regulations, the protection of intellectual property, healthcare reform, "social" policies that are designed to protect or support people in disadvantaged groups, and dealing with the judicial system to resolve business matters. New and growing manufacturing firms often find environmental regulations particularly burdensome.[99] Entrepreneurial firms in the pharmaceutical industry face obstacles in their ability to finance the development of potential breakthrough drugs. Hopkins contends that the overall cost of compliance with government regulations as a percentage of sales is almost twice as high for smaller firms (i.e., under 20 employees) than for larger firms (i.e., with more than 500 employees).[100] A big part of this is the complexity of regulatory and tax policies. But suggestions to lessen many types of regulations are quite controversial. Many believe that forms of regulation including environmental, health and safety, etc., reflect societal standards or market failures, and that these costs should be part of doing business.

The "Small Business Regulatory Enforcement Fairness Act of 1996," Public Law 104-121, reduced the burden of regulation on small and growing companies by providing for judicial review. (It should be noted that businesses must also deal with state and local regulations.)[101] And some recent articles and analysis suggest that it is beginning to make a positive difference.[102]

---

[99] Dean, Thomas J., Robert L. Brown and Mark De La Mater, "Pollution Regulation and Small Establishment Formations in U.S. Manufacturing Industries: A Longitudinal Analysis of Deterrent Effects," *Frontiers of Entrepreneurship Research,* 1994, pp. 499-513, Center for Entrepreneurial Studies, Babson College. See also Carlson, Eugene, "Small Firms Spend Much Time, Money Complying With Environmental Rules," *Wall Street Journal,* June 15, 1992, p. B1.

[100] Hopkins, Thomas D., *Regulatory Costs in Profile,* Center for the Study of American Business, Washington University, Policy Study Number 132, August 1992.

[101] Osborne, David and Ted Gaebler, *Reinventing Government*, Reading, MA: Addison-Wesley, 1992. In this book the authors profile government agencies that have rethought their mandate and practices in order to be more responsive to their constituents. See also Richman, Tom, "Public Policy: What Does Business Really Want from Government?," *Inc.*, State of Small Business, 1995, pp. 92-103.

[102] An initial look at the emerging benefits of the Small Business Regulatory Enforcement Act is presented in *The Annual Report of the Chief Counsel for Advocacy on Implementation of the Regulatory Flexibility Act*, Office of Advocacy, Small Business Administration, Washington, D.C., 1997.

Tax Policy and Capital Formation

There is considerable debate about the effect of the tax system on small and growing businesses. Holtz-Eakin argues that there is no clear economic basis for providing special tax treatment to small businesses.[103] Poterba's analysis suggests that a reduction in the capital gains tax rate for investment in entrepreneurial firms would likely spur a modest increase in the flow of venture capital.[104] Gravelle's analysis indicates that there is no sound economic basis for targeting tax investment incentives — an investment tax credit or targeted capital gains tax — to small businesses.[105] But laws and regulations that restrict investment in venture capital by pension funds or foreign investors, as well as a wide range of specific tax provisions that deal with issues such as the treatment of capital gains, accounting treatment of stock options, costs when businesses change hands, and the deductibility of investment expenses are important issues for entrepreneurial high-growth companies.[106]

Young and growing firms often have little or no income tax obligation. Instead they view the tax code as a potential ally in increasing capital investment (i.e., equity) in their businesses and encouraging the R&D investments needed to support growth. The available analysis suggests that small businesses—especially firms that are organized as S-corporations—receive significant tax advantages over larger firms.[107]

What Should the Public Policy Framework Be?

We examined earlier whether or not there are gaps in the private market that might warrant federal policy consideration. This section of the paper summarizes the key public policy issues that relate to these gaps as they affect entrepreneurial high-growth companies.

Areas for Federal Policy Consideration

At its best, public policy can help to create an environment that supports innovation and the start-up of businesses with growth potential. And it can help to assure that the playing field is level for every talented person who has a strong business idea and the ability and commitment to try to build a growing enterprise. To accomplish these objectives in today's and tomorrow's global economy there are a

---

[103] Holtz-Eakin, Douglas, "Should Small Business Be Tax-Favored?," *National Tax Journal,* Volume XLVIII, No. 3 (September 1995), pp. 387-395.

[104] Poterba, James M., "Venture Capital and Capital Gains Taxation," in *Tax Policy and the Economy 3*, Lawrence H. Summers (editor), National Bureau of Economic Research. Cambridge: MA: The MIT Press, 1989.

[105] Gravelle, Jane G., *Small Business Tax Subsidy Proposals,* Congressional Research Service, Report 93-316 S, March 15, 1993.

[106] Issues raised at a high-growth company forum sponsored by the Kauffman Foundation, 1997.

[107] See Gravelle, Jane G., *Small Business Tax Subsidy Proposals.*

several areas in which federal policy could play a role in making sure that the environment is right and the barriers are eliminated. These are:

1. Understanding the Unique Needs and Nature of High-Growth Companies. As noted at the outset, many federal policies and programs focus on the needs of small businesses and large established corporations. The situation of entrepreneurial high-growth firms is unique. To the extent their attributes and contributions are better understood in the policy-making arena, better grounded policies are likely to emerge.[108] *(See the discussion on pages 1-8.)*

2. Facilitating New Business Formation. As discussed above, entrepreneurial firms have been significant contributors to economic growth, innovation, jobs and training. Yet a relatively small share of new firms ever become high-growth companies. Research on the dynamics of business formation and the "churning" process suggests that the way to increase the number of growing companies is to maintain a high level of new business start-ups. Improving access to knowledge about entrepreneurship, business opportunities and the entrepreneurial skills required to succeed should help people to self-select and increase the likelihood of business success. Improving the availability of and access to high-quality entrepreneurial education and training—even beginning at an early age with school children—can support the success of potential and existing entrepreneurs. There might be a role for the federal government, partnering with the private sector and states, in creating a high-level database on business and educational opportunities and in testing alternative models for entrepreneurial education. *(Access to knowledge: see the discussion on pages 14-17.)*

3. Facilitating Mentoring. Mentors have served as influential role models and sounding boards for many successful entrepreneurs. Improving the availability of, and access to, skilled and dedicated mentors may improve the number of people—especially women and minorities—who successfully build growing companies. There might be a role for the federal government, partnering with the private sector and states, in testing models for expanding mentoring opportunities. *(Access to knowledge: see the discussion on pages 14-15 and 19-21.)*

4. Facilitating Networking. Participation in the right business networks serves as an important support structure for growing a successful business. Improving access to strong business networks can strengthen the ability of existing and potential entrepreneurs to start and build their companies. It is especially important for women and minorities who often have limited options for building these critical relationships. Again, there might be a role for the federal government, partnering with the private sector and states, in supporting the development of effective networking models. *(Access to knowledge, markets and capital: see the discussion on pages 14-15 and 19-21.)*

---

[108] It is also worth noting that most high-growth firms, especially in high-technology areas, rarely do business with the federal government. This is due in large part to the difficulty of doing government work and the limited potential for profit. As a result, many observers believe the federal government misses out on much of the best talent and expertise available in meeting its objectives.

5. Facilitating Knowledge of, and Access to, Foreign Markets. Today many entrepreneurial companies, especially high-tech firms, are looking to foreign markets as a vital area for growth but they lack the knowledge or resources to make a successful entry. A specific aspect of this involves helping them to protect their intellectual property rights overseas.[109] There might be a role for the federal government in bridging the knowledge gap with respect to both markets and the protection of intellectual capital. *(Access to markets: see the discussion on page 16.)*

6. Building a Pool of Skilled Workers. Strengthening the educational system and workforce development is essential to providing a pipeline of skilled workers who can meet the requirements of growing businesses. In addition to basic work and business skills, there appears to be an emerging "gap" in the availability of trained workers in information technology. There might be a role for federal policy in creating an environment that helps focus resources on educating the workforce of the future.[110] *(Access to talent: see the discussion on page 14-17.)*

7. Facilitating Continued Investment in R&D and Commercialization. While the private sector is by far the largest funder of research and development in the U.S., small and growing firms often have limited access to the resources required to do basic and applied research. Facilitating continued investment in R&D and technology commercialization through partnerships with universities, established corporations and the public sector can enable entrepreneurial firms to develop important innovations and commercialize existing technology. There is a potential opportunity for the federal government, partnering with high-tech companies, established corporations and states to create more effective approaches to commercializing federally-developed and sponsored research and technology. *(Access to knowledge and capital: see the discussion on pages 14-16 and 21-24.)*

8. Facilitating Capital Formation. Access to capital, especially seed capital, is critical to building a growing company. When entrepreneurs and their firms are excluded from traditional sources of private sector capital there might be public policies that can enhance availability and access. *(Access to capital: see the discussion on pages 17-19 and 21-24.)*

Areas for Further Study

There are a number of areas in which further study would be beneficial in order to strengthen our knowledge of the most appropriate federal policy role in assuring the proper environment for entrepreneurial high-growth companies. These include understanding:

---

[109] Protection of intellectual property rights is an important element of U.S. international trade policy. See CRS Report 96-847, *Intellectual Property Rights and U.S. Foreign Trade,* by Glennon J. Harrision.

[110] Recently there have been calls for increasing the number of foreign workers under H-1B visas. For a detailed discussion see CRS Report 98-531, *Immigration: Nonimmigrant H-1B Specialty Worker Issues and* Legislation, by Ruth Ellen Wasem.

1. The role and dynamics of mentoring and networking in supporting the start-up and development of high-growth firms. *(Access to knowledge)*

2. The keys to effective education and training programs for entrepreneurs and potential entrepreneurs, and the value of starting educational programs at an early age. *(Access to knowledge)*

3. The importance of computer and information-technology literacy as an ingredient for entrepreneurship and business building. How do we make access as universal as possible to support potential entrepreneurs and knowledge workers? *(Access to knowledge and talent)*

4. The reasons why women are under represented as the founders and owners of high-growth companies? What are the critical barriers and can public policy enhance private initiative to bridge the gap? *(Access to knowledge, markets and capital)*

5. The reasons why minorities—in particular African-Americans and Hispanics—are under represented as the founders and owners of high-growth companies? What are the critical barriers and can public policy enhance private initiative to bridge the gap? Are new models such as strategic alliances with established companies a viable and replicable model for increasing entrepreneurship? *(Access to knowledge, markets and capital)*

6. How to maintain our strong position in research and development as a vital catalyst for entrepreneurial initiative. Are opportunities for new and growing companies to invest in and commercialize new technologies being missed? And can we improve the process of transferring technologies from Federal R&D to entrepreneurial firms and successful products? Also, what is the return on investment of R&D dollars placed with entrepreneurial firms vs. established corporations and university-based research centers. *(Access to knowledge and capital)*

7. The nature of the "capital gap." Is there a real gap in the availability of seed capital, as well as debt financing, for new and growing companies that warrants public policy attention? *(Access to capital)*

8. The nature of the "human capital gap." Is there a real shortage of skilled technology-trained workers today? Will there be one in the future? If so, does it pose a serious threat to new and growing firms? And, what are the right options for addressing it? *(Access to talent)*

9. How existing and potential tax policies affect entrepreneurship and the start-up and development of high-growth firms. Are there sound policies that would facilitate increased investment in new ventures with great potential?

10. The effect of government regulations on new and growing firms. Do they place an unfair burden on entrepreneurs at a critical time in the growth of their companies? If so, which ones and what can be done about them? Also, are

there industries in which regulations make it very difficult to start and build a new high-growth venture?

11. The dynamics of the most entrepreneurial communities in America and what it takes to build the right support networks of entrepreneurs, investors and the other professionals. Are there lessons that can be learned and applied to encourage appropriate local initiative?

A Note on the Role of State and Local Policy

While many of the issues discussed here fall into the federal policy sphere, many are to a large extent local, state or regional concerns. As such, states and localities have important roles to play in building environments that nurture entrepreneurs and growing companies. Much of the public policy innovation that is taking place in support of entrepreneurial high-growth firms is occurring at the state level. Initiatives being pursued or discussed include the use of a (small) percentage of state pension funds for venture capital, public seed capital funds, programs to build partnerships between growing companies, universities and research centers (such as the Ben Franklin Partnership in Pennsylvania and the Thomas Edison Program in Ohio), and innovative workforce development and entrepreneurial education efforts.[111]

Entrepreneurs start and grow their businesses in localities, and some localities appear to be more conducive places for this effort.[112] As noted above, many communities have become strong generators of new and growing companies with limited public sector involvement. In other communities, often in the battle for economic development and the growth, jobs and taxes that it brings, localities and states have taken an active role in trying to spur entrepreneurship. The key issue here is what states and localities can and should do in partnership with the private sector to create a culture and climate that facilitates innovation and entrepreneurial initiative.

---

[111] For a discussion of some of these initiatives see Bingham, Richard D., Edward W. Hill and Sammis B. White (eds.), *Financing Economic Development: An Institutional Response,* Newbury Park, CA: Sage Publications, 1990.

[112] Gregerman, Alan S., *Competitive Advantage: Framing a Strategy to Support High Growth Firms,* Council for Urban Economic Development, 1985. See also Gregerman, Alan S., "Rekindling the Future: Creating a More Entrepreneurial Environment," *Economic Development Commentary,* Winter 1991, pp. 7-14.

ENERGY SYSTEMS CRITICAL TECHNOLOGY

P.B. Smucher
(Editor)

Highlights

- Energy system technologies provide for power generation and energy conversion, energy storage, and power conditioning.
- The Joint Chiefs have recognized high-energy density power converters as a technology priority.
- The military's goal is to increase system power densities by two to five times, depending upon the specific system.
- Power converters typically take up approximately 50 percent of system volume.
- Stringent military requirements demand reliable, intelligent/survivable, and affordable energy systems that can perform in battlefield threat environments.
- Budget pressures and the military's need for reduced volume and weight, increased performance, and graceful degradation make it difficult for military and commercial compatibility.
- Advances in power electronics, packaging, and thermal management will drive the performance advances in the state-of-the-art.

OVERVIEW

This section discusses the developing critical technologies for energy systems embedded in the following subsections:

- Energy Conversion and Power Generation
- Energy Storage
- Power Conditioning
- Biological Energy.

Energy and power system technologies are among the least visible, and their functions are usually taken for granted because we assume that reliable power will always be available. However, many weapon system platforms, because of different energy and power requirements, integrate a host of diverse technologies for power generation, energy storage, and power conditioning. Because of the weapon system platforms' variance in application and mission criteria, a wide range of energy requirements exist. These platforms need not only pulsed or continuous power but also need many niches and diverse technologies to satisfy low-, medium-, and high-power requirements.

Factors that influence the development of military energy systems include the critical need to occupy smaller volume, have low weight, provide long life-cycle performance, and be highly reliable and survivable in threat and extreme environments. Because of the military's stringent performance requirements, the advances in power electronics packaging and integration and in thermal management will be the key drivers in the successful realization of the military's future vision. The Joint Chiefs have recognized high-density power converter systems as a technology priority, and these systems will require new technology design and packaging approaches that are expected to revolutionize the power electronics industry.

Specific load requirements (e.g., power level, duration, pulsed, or continuous) drive the system design and the component technologies. To a large extent, the power level required determines the primary energy source and often severely limits the options available to choose from. The mission duration is also a critical factor since certain power-system technologies have inherently useful lifetimes, while others are capable of continuous and extended missions. Finally, the distinctions between single-shot, repetitive, and continuous power systems are critical since the duty cycle places different stresses and performance requirements (e.g., voltage, current, frequency, and temperature) upon the components used. . . . Power levels range over many orders of magnitude—from milliwatts of continuous power in handheld electronic devices to terawatts in certain directed energy weapons. Durations range from small fractions of a second for single-shot, pulsed-power applications to months or even years for remote power and/or space systems (ATAR, p. I-3).

Figure 7.0-1 addresses the complex taxonomy of energy systems. As a real-world example, take a mobile platform (e.g., an air vehicle) as the system. Integrated mission requirements include the ability to hunt and kill (mobility, navigation, and lethality), to communicate, and to ensure the safety of the aircraft and its crew (environment and survivability). Each mission requirement results in the need for distinct subsystems driven by either continuous or pulsed power. The power supply system is, therefore, a conglomeration of energy conversion and power-generation systems driving energy storage, power conditioning, and pulse-forming networks. The energy systems are subject to strict constraints of thermal management, scalability, and integrated functionality. Beneath this layer lie the supporting technologies and components that make everything possible. Stepping from one function to the next reduces output and capability by less than perfect efficiencies.

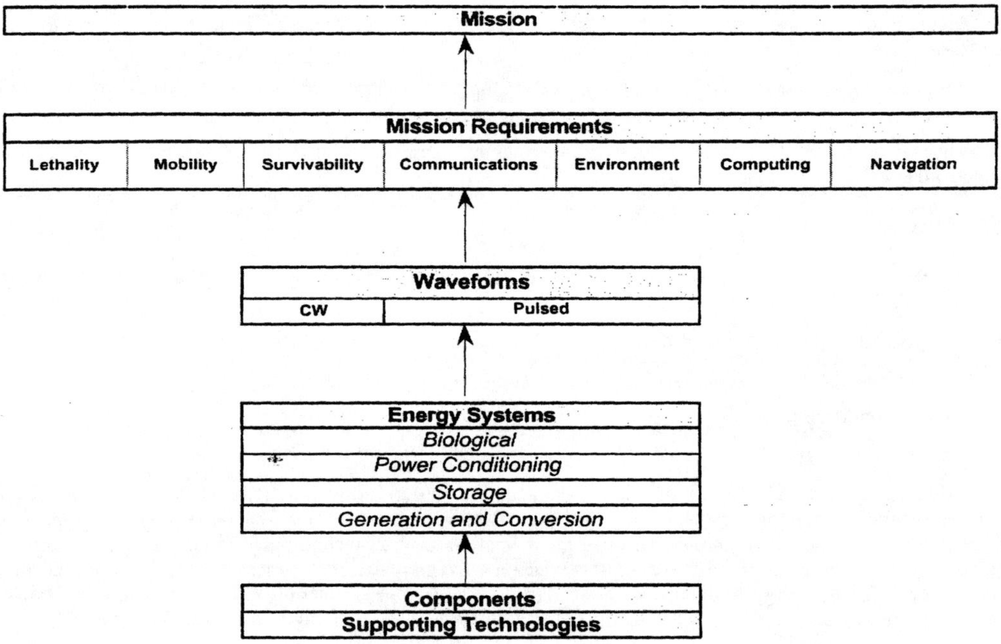

Figure 7.0-1. Energy Systems Taxonomy

RATIONALE

Energy is generally collected or available from a prime source at low-power levels and power densities. To meet common military load requirements, such as high power (density), the energy is released from storage over an extremely short duration and converted into pulsed-power form. Power is energy per unit time. Therefore, by compressing the duration of time over which energy is supplied, one achieves higher powers. After additional compression of the pulsed power, it is delivered to a load with great amplification of power and power density (see Figure 7.0-2).

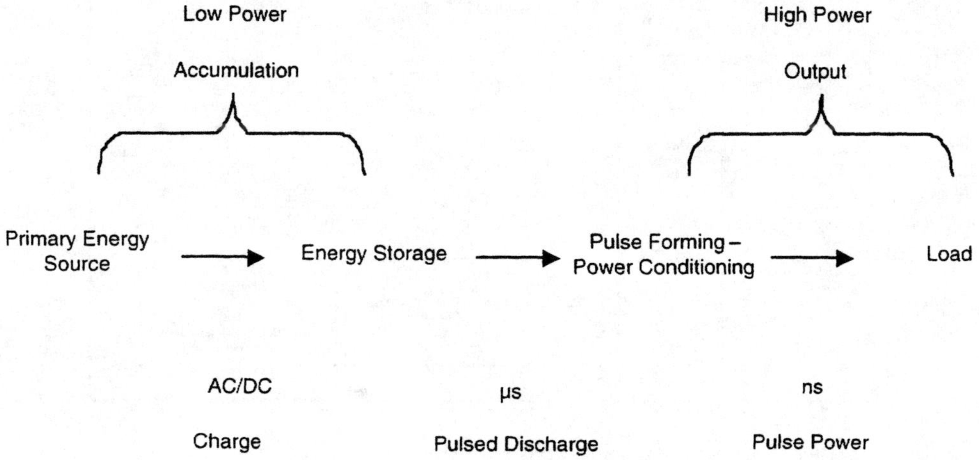

Figure 7.0-2. Typical High Pulsed-power System (Adapted From Pai, 1995)

In many cases, military energy requirements have a much broader range of demands than the commercial segment. Figure 7.0-3 shows the large variation in peak power output and duty cycle for several types of missions. With the exception of a few commercial radars, navigation systems, communication systems, and industrial processes, typical commercial energy systems do not have the high-output requirements of the military systems shown in Figure 7.0-4. Because of this disparity, only a portion of advanced technologies developed to serve as commercial energy system components can be directly transferred into military systems. Commercial-off-the-shelf (COTS) technologies are competitive in size and cost and are enabling when components can meet requirements of military systems.

There is an enduring military vision for reducing weight and volume while increasing performance, reliability, and, most importantly, survivability. This is true for most sizes of applications—from the individual soldier to the largest platforms. Figure 7.0-5 highlights current Soldier Power System and Land Warrior power requirements, and Figure 7.0-6 highlights the electrical power applications for a sample mobile platform (repetition rates are measured in pulses per second). In the latter case, system design is a significant challenge because of the need for both continuous wave (CW) and pulsed subsystems and dense packaging. In one sense, Soldier Power Systems are a mobile platform; however, we mention them specifically because of their unique requirements.

> Equipment for the dismounted soldier must be both compact and rugged. Availability of fuel, specific energy, specific power, minimal signature (electronic, thermal and acoustic), simplicity of operation, and environmental impact are also major considerations. The mass a soldier can carry on a mission is already approaching its limits forcing a trade-off of bullets or food for batteries to power electronics (NRC, p. 28).

Mobile Platform Energy Systems combine the propulsion and electric energy generation functions. In the future, mobile platforms may benefit from integrating the power system to include power for weapons and for propulsion (mobility) and other vehicle electric functions. With careful planning and a well-thought-out control strategy, a central power system may feed the continuous, transient, and pulsed loads while reducing the platform size and increasing the efficiency. Developing technologies are going to enable the military to move toward replacing heavy, inefficient systems with more efficient "more electric" platforms. Critically developing technologies are essentially converging towards all-electric vehicles. Currently, some commercial hybrid vehicles, primarily transit buses, are being made in quantity. In the United States and Europe, several programs have made prototypes or electric or hybrid-electric military vehicles, and some are even being tested.

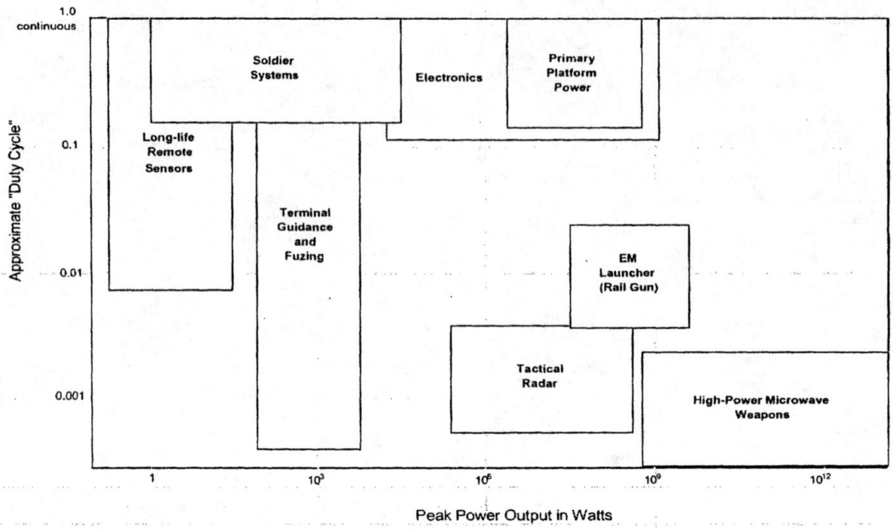

Figure 7.0-3. Peak Power and Duty Cycle for Pulsed and High-Energy Systems and CW Energy Systems for Several Applications or Missions (Adapted From ATAR)

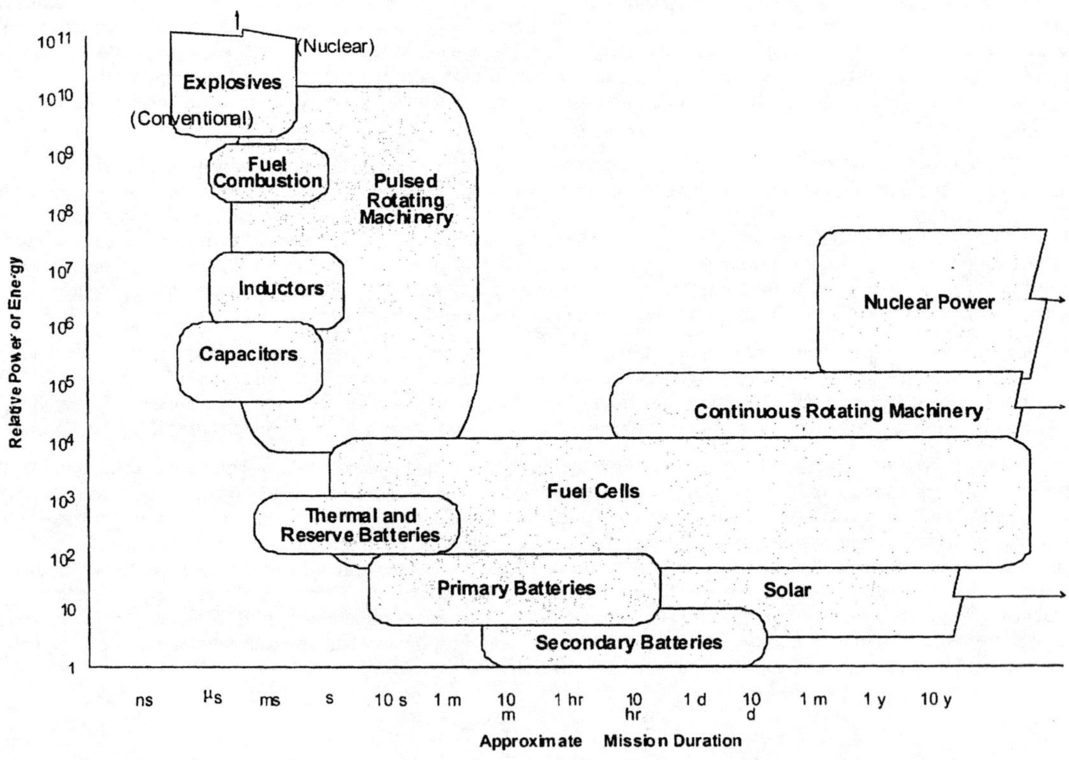

Figure 7.0-4. Trends in Discharge Times of Power System Technologies (Adapted From ATAR)

Figure 7.0-5. Sample Power Requirements of Soldier Power Systems (Source: NRC, 1997)

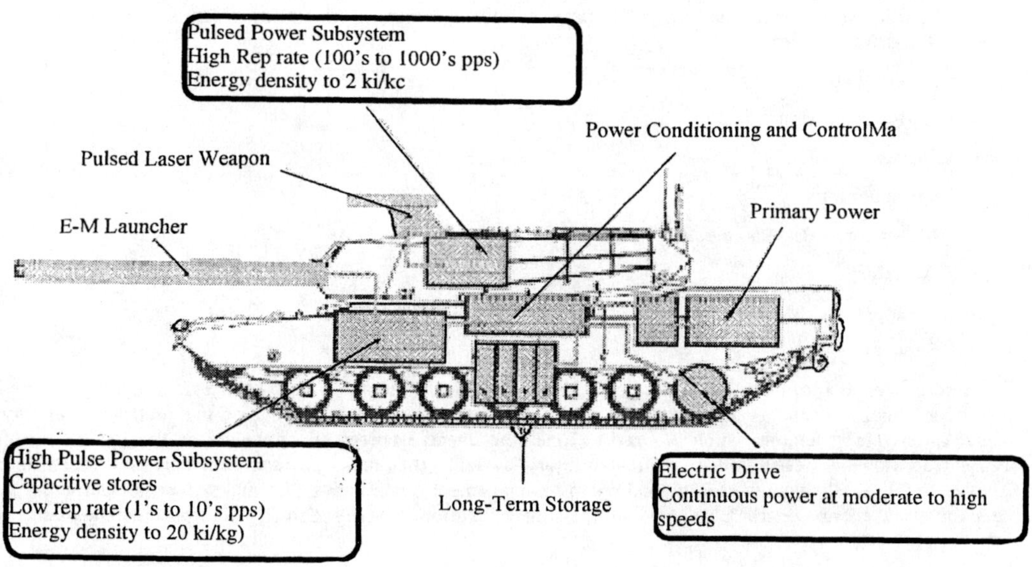

Figure 7.0-6. Advanced Hybrid Ground Vehicle (Source: ATAR)

BACKGROUND

To create useful power, energy is converted from potential to kinetic and either stored or conditioned and then delivered to a load. This power can be provided continuously or in pulses. Varying military applications require power to be delivered in time periods ranging from extremely short durations (on the order of nanoseconds, 10^{-9} sec) to continuous duty. Specifically, many specialized advanced weapons require pulsed power. Power electronics and other components used in continuous systems have very different characteristics from those used in pulsed systems.

Not only are the materials and component advancements in energy systems key, but design philosophies are essential for delivering a product in the most efficient and affordable manner. The approach of Synchronizing Enabling Technologies (SET) is the assessment and the development or implementation of enabling technologies to support a thrust area. It is analogous to knowing the ingredients for a recipe and timing their preparation and introduction into the amalgam. To be successful, a proper assessment of needed enablers should be completed during the early phases of a technology's development. Enablers can range from parallel developments to components or materials, such as composites, that are in the research and testing phase. Synchronizing technologies is an important management issue that will be reiterated further in relevant technology areas.

Soldier Systems

Soldier Power Systems (see Figure 7.0-5) will provide the 21st century warrior with power sources that are enabling for a host of man-portable electronic devices ranging from communications and sensors to weapons. The warfighter will be the keystone of the future digitized Army. Power requirements have been estimated as high as 500 W, with 22.9 W as the current power assessment for the (Army) approved Land Warrior system (NRC, 1997).

In the far term, nuclear power and rotating machinery (microturbines) offer choices for soldier power. Nuclear power has not yet come to practical design because of significant system-implementation issues associated with radiation near biological tissue and social feelings towards ionizing radiation. The specific energy of radioisotopes, however, which can be greater than 100 MW hrs/kg, combined with more efficient energy converters, could easily solve power-generation issues for the unmanned future soldier. Continued efforts may also see other power technologies come into use for the infantry. Such technologies include biological energy sources (for milliwatt applications) and advanced capacitors and converters.

Mobile Platform Energy Systems

Mobile military systems combine the prime propulsion and onboard electrical power-generation functions. The net effect is an energy system that has operational advantages of reduced signature, increased energy density, lower weight and volume, greater flexibility in configuration, and greater economy/reliability. Electrical energy components are characterized by a significant reduction in moving parts, the elimination of rigid connections, and an improved ability to use small and irregular spaces within a vehicle. Potential application areas for migrating to a more-electric drive include:

- Active suspension
- Active protection
- Drives
- Computations
- Motor controllers and electric filters
- Actuators
- Weapons
- Sensors.

Energy systems for all types of combat vehicles are broadly characterized by the level of energy required and the time over which the energy is demanded (pulsed or continuous). Components of the platform's energy system include an engine-generator, which is sized to meet the average energy requirements of the vehicle. (This is not always true, however, especially in critical military systems that must be redundant and survive battle damage.) Coupled to this are the energy-storage and energy-management components that make it possible for the system to meet the peak energy demands required in mission execution. Energy density, thermal management, cost, and packaging are critical parameters for military energy systems. Aggressive management, combined with deliberate restraint of system designs to those that satisfy required parameters, will be essential to producing economical, reliable, and high-performance systems. Thermal and electromagnetic capability (EMC) properties will remain essential design and performance parameters.

Electronic Packaging And Thermal Management

Some of the major considerations in the Soldier Power and Mobile Platform Energy Systems are electronic packaging and thermal management. The method by which components and assemblies are integrated play a vital role in the performance and reliability of an end product. Some of the major advances in power technology—both in performance and in cost—will be the result of electro-mechanical-magnetic packaging. Packaging occurs at all levels of product development. Because it is ubiquitous, the following definition indicates how it is used in our discussions: Each level in the electronics hierarchy has within itself the combining of parts to make an assembly. For example, a semiconductor has a die, wire, housing, and leads to form a component. The manner in which these parts are put together is called packaging. Packaging mainly involves thermal management, structural integrity, interconnections, and electromagnetic compatibility. We can apply this same example to each level of assembly in an electronic system (e.g., begin with components and then move to subassemblies, power supplies, DC-DC converters, and functional systems). Finally, when all these parts are integrated together (system packaging), the result is an end product.

One of the revolutionary new packaging approaches is to integrate individual components into each other to form a new component that has increased functionality, lower cost, and higher reliability because of reduced interconnections. An example of this is integrating resistors and capacitors into the internal layers of a printed circuit board. This results in greater packaging density, lower cost and reduced interconnection (compared with discrete components) and, thus, higher reliability. Another example would be integrating semiconductor dies directly into magnetic assemblies, such as a transformer. This is a new approach and will have a profound impact on discrete component manufacturers as the technology matures. For example, if capacitors were directly embedded in a printed circuit board, the demand for discrete capacitors would be less. Understanding the physics and manufacturing processes needed in this functional integration is critical for making significant advances in packaging.

Electronic packaging is a rapidly advancing technology that has been fairly dormant for many years. While advanced electronic packaging techniques have been developed in microelectronics, little significant work has been done in the higher levels of the interconnect hierarchy (subsystems and systems packaging). Packaging, with respect to power systems, must be viewed from an end-use perspective in order for the system to be efficient and effective. This is becoming more important with the introduction of integrated functionality, where multiple discrete components are being integrated and built as a new discrete component. Advanced packaging is critical for meeting the demands of efficiency, thermal management, and power density in future power systems. Integration at all levels will be needed, including interconnections such as direct die attach technology.

To illustrate, capacitors and magnetics are critical components that require improvements to meet the new packaging demands. Further, high-temperature and high-efficiency power semiconductors are also important component needs. In all cases, necessary technological advances include the materials and processes used in building such components. Advances in thermal management will come from heat spreaders and selective cooling approaches.

In cases where integrated functionality is used, hybrid test equipment will be needed to evaluate the parameters and characteristics of multiple components in one (component). High-technology processing equipment will also be needed for areas previously satisfied by lower technology processes.

Education and establishing a systems-level packaging mentality are the main issues in evolving to advanced packaging capabilities. Change is required to overcome the headset of single-level packaging. Effecting this change will be difficult because of the subsystem mentality that exists and the way parts are designed and manufactured in today's environment.

Power technology has moved forward in the last several years at a fairly rapid pace, although, in general, it has not moved nearly as fast as it has within the semiconductor industry. This difference in sector performance is evident across the military and the industrial communities. Many of the packaging demands for higher density, higher performance, lower weight, and lower cost power systems are common to both areas, especially in the area of more cost-effective semiconductors. The comparatively large physical size of power system components, compared with the rest of the electronic package, has inhibited system density improvements. Reactive components tend to be large, heavy, and inefficient. Spatial compression of these components produces increased power density (W/cm^3), without decrease of thermal impedance, resulting in localized heat build up.

Semiconductor design philosophies can be extended to high-density power converters and to regulators, where system packaging is critical. Proper partitioning, with special attention given to electromagnetic interference and thermal management, will be needed. Thermal management will enlist a combination of active and passive cooling technologies to control "hot spots." Thermally conductive materials will play a bigger role in thermal management, with "heat spreading" being used to a larger degree. Systems packaging will require the integrator to understand clearly each subassembly and how it will affect system performance. The application of high-density power regulators into the systems architecture will be key to enhanced system performance.

Integrated Functionality

An area that is emerging from the packaging discipline is Integrated Functionality. The role of the packaging engineer is to integrate components into a functional assembly using processes and assembly techniques made available to the manufacturer. This in itself may be viewed as integrating functionality; however, it is also starting to emerge at a lower level. This function actually involves the discrete components, such as resistors, capacitors, and semiconductors used in assemblies. We call this integrating functionality at the component level. The result of this activity is the creation of new components that contain the collective features of multiple components (see Figure 7.0-7). This change is profound since it will change the supplier base to the power supply industry. The rationale behind the development of integrated functional components is increased reliability, better performance, space reduction, and cost benefits—features that are desired in power product development programs at all levels.

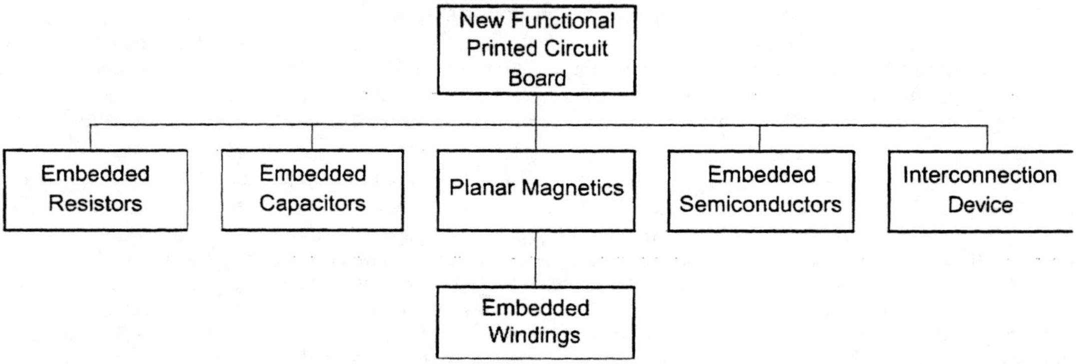

Figure 7.0-7. Example of Integrated Functionality

The challenges in integrating the functionality of components are significant but can result in major advances in end-product cost and reliability. For example, to be able to combine the desired features of multiple components, a good understanding of the individual components is necessary. This includes understanding the way in which the components are manufactured and knowing their constraints. With this knowledge, combining features of multiple components into one part can, in many cases, become a reality.

An example of integrated functionality can be seen in planar magnetics. This is where the windings for the magnetics are incorporated in the printed circuit board using etched circuit traces. The core material is inserted in holes or slots in the printed circuit board and bonded, to complete the magnetic assembly. This operation, in essence, replaces a discrete component. It also eliminates the core winder and creates a more functional printed circuit board. It takes manual labor out of the winding process, which, in turn, will improve cost and quality. The printed circuit board manufacturer is developing new skills in producing planar magnetics, but these new skills do not significantly affect the manufacturing process. A result of planar magnetics is the elimination of a discrete component in the overall assembly, which, in turn, increases reliability. The business ramification is a reduction in purchased magnetic assemblies. Eliminating the need for particular discrete magnetic components takes a certain amount of business from the transformer companies but also provides some packaging flexibility by core geometry selection, especially in low-profile applications.

While this change had little affect on the printed circuit industry, other component integrations will have a profound impact. An example is the emerging technology of embedded capacitors. Significant reductions to the size of a power supply can be realized by replacing discrete capacitors mounted on the printed circuit board by embedding capacitors within the layers of a printed circuit board. As new dielectric material technology with high-energy storage features becomes available, a breakthrough in power density could result. This integration would have a large impact on printed circuit board manufacturers. They would be building a capacitor(s) into their boards, thus becoming a manufacturer of an electronic component and an interconnection component. This would involve enhancing the knowledge base of printed-circuit people in many areas. They would be required to learn new manufacturing processes and deal with new materials and would have to be able to test their embedded capacitors. They would require much of the knowledge of capacitor manufacturers. This would be like building a capacitor without a package (as in a discrete component). Terminations of the capacitor would be included in circuit traces and plated through holes. The new processes that will be needed to build this product presently exist in the semiconductor and other industries and can be easily transitioned. The equipment required would include vacuum deposition equipment. Although the printed circuit business will change, the financial rewards could be huge. The power supply industry desperately needs higher power density and the replacement of discrete capacitors. Capacitors currently consume about 20 to 30 percent of the volume of a power supply, depending on the type of power supply. Capacitors also account for about 20 percent of the cost of a power supply. This could be a very lucrative new business. However, again, there are business ramifications. New capital equipment purchases would be needed and the labor force would have to be trained. This "new component" will affect the discrete capacitor manufacturing companies by decreasing the need for discrete components.

Other examples of integrated functionality have also had an impact on power electronics [e.g., application-specific integrated circuits (ASICs)]. ASICs have replaced many discrete components with a single package. Before ASICs, many electronic circuits required several discrete components to provide a certain function, such as housekeeping circuitry for sensing power supply functions. Some standardization was developed and ASICs became very practical devices. Replacing several discrete components with a single component that has the reliability of an integrated circuit has had a large effect on power supply reliability. A mean time between failure (MTBF) of over 1 million hours for a DC-DC converter is not unusual today. Much of this improvement can be attributed to integrated functionality and improved designs.

Direct die attach technology is another area being used more frequently. This eliminates a level of interconnection and can increase power density and improve reliability.

Other designs that demonstrate additional ways of integrating functionality (by combining parts and reducing the amount of interconnections) will soon be introduced to the market. This interesting trend will have a significant impact on today's discrete component supplier. Suppliers that today sell discrete components must become involved in this activity if they want to preserve their markets or open new markets.

Improvements from military mobile systems with integrated functionality include:

- Simplicity
- Permanent magnetic materials
- Reliability
- Efficiency
- Operating temperature
- Power density
- Thermal management
- Energy extraction.

Electromagnetic Compatibility (EMC)

When dealing with electromagnetic emissions, two important considerations are conducted emissions and radiated emissions. Conducted emissions are generally managed with in-line electromagnetic interference (EMI) filters. Radiated emissions usually require containment with ferrous metal materials and shielding design approaches to eliminate radiation. Shielding materials would include products such as fingerstock and braided mesh for seams and doors or ports to the electronic equipment.

High-Energy Density Conventional Systems

High-density conventional systems will be used for many applications in mobile platforms. High-density conventional systems are those comprising components/subsystems rated at less than 500 kW. Such systems are found in almost every military operating system, including multiple components of major weapons. Critical technologies in this section cover both energy sources and power-conditioning activities. Energy systems are fundamental components, which are often specifically tailored for each of the other major technology areas within the list of Military Critical Technologies (MCTs).

Technologies for components or systems rated at less than 500 kW are migrating toward revolutionary changes in energy density and precision. This is critical to the military because of the needed improvements in system performance: new capabilities are becoming possible. Power system miniaturization is an important element and has led to micropackaging on a chip and distributed packaging to perform a function such as track motor or flight control. In the near future, many weapons systems that now commonly use mechanical, hydraulic, or pneumatic components will rely on electrical systems. Major initiatives include the More-Electric Aircraft, All-Electric Ship, and the Future Combat Vehicle (Army). Military requirements for energy systems are distinguished primarily by the need for ruggedized packaging, and high-energy and high-power densities. Thermal stress is the primary concern, with shock stress an important secondary regard. Thermal management or components capable of functioning at high temperatures are keys to reduced failure rates and improved system survivability. Military demand for low failure (total life-cycle reliability) drives new design architectures. As a result, approximately 40 percent of the total demand for military energy systems results in custom-designed applications to fit military systems requirements. Such requirements create pervasive pressures to reduce production and product costs.

High-Density Converters

An example of high-density conventional systems would include high-energy density converters. DC-DC power regulators in the military and the commercial sectors have become smaller and more powerful. One of the driving factors in the military sectors is weight as it applies to a payload. In the commercial arena, this trend has been driven by the fast-moving semiconductor industry. Hardware size in telecommunications, networking, and computer systems has been greatly reduced by the high-performance semiconductors now in the marketplace. Slow to follow this technology trend are the DC-DC power regulators. However, in industry, significant gains have been achieved in reducing of the size and increasing the reliability of DC-DC power regulators. In industry, regulators designed today can be packaged in volumes of 70 to 100 W/in^3 and have a MTBF of over 1 million hours—numbers that would look very much like a military requirement just 5 to 10 years ago. The resulting designs of these high-power density regulators are accomplished by designing new topologies, developing new components, increasing the functionality of components, and mastering electro-mechanical packaging. This is the way to get to high-density regulators. However, as the semiconductors progress to lower and lower voltages and the chip power levels increase, the issues of high current distribution and the associated power losses become significant, putting more difficult constraints on the design engineers.

The National Technology Roadmap for Semiconductors developed by the Semiconductor Industry Association has projected the power supply voltage to be 1.8 V by the year 2001 and down to 0.9 V by the year 2010. The maximum power associated with these numbers is also projected to increase significantly. In 2001, the power at 1.8 V is expected to be 120 watts, and, in 2010, the power at 0.9 V is expected to be 180 W. With this level of chip power density, power regulator designers must consider several strategies.

Power system architecture ranks as one of the most important considerations. With the very-high-power requirements of the semiconductors, one design option is point-of-load power architecture, which means placing the regulator as close to the actual load as possible. This would minimize the power distribution losses. This would also require the designer to view the power supply at the system level, integrating it into the system efficiently. This effort requires the use of advanced systems-packaging designs with careful considerations for thermal management. Thermal cooling technologies become an important factor in this overall package, and chilled air becomes an important choice, especially in chip cooling.

Another option is distributed power. This approach would locate the power regulators close to the load but not at the point of load. This approach would use larger regulators that provide power to more than a single chip. The problem of power losses caused by the high current levels is addressed with power bussing techniques. The interconnection design must receive careful consideration. To minimize power losses, hard connections, such as bolted on bussbars, would be preferable to removable connectors. An advantage of this power architecture is that the approach will easily accommodate *n+1* redundancy. This configuration provides an extra regulator paralleled with the primary regulators. If a regulator fails, the extra regulator will automatically be activated. This will afford the overall system higher availability and will appear to the user as increased reliability. If current sharing is being used with the paralleled regulators, actual reliability is increased because of the reduced loading on the individual regulators. This approach is possible but not nearly as suitable as the point-of-load power architecture, since it would take up real estate that is not readily available and would not be as cost effective.

TECHNOLOGY ASSESSMENT

Table 7.0-1 highlights technology groups, their function and relevant waveforms, and some applications where they may be found.

Table 7.0-1. Technology Groups and Their Functions (Adapted From ATAR)

	FUNCTIONALITY			MILITARY APPLICATION													
ENDURING TECHNOLOGY GROUP	GENERATION & CONVERSION	STORAGE	POWER CONDITIONING	SOLDIER SYSTEMS	COMMUNICATION	COMPUTERS	ELECTRIC VEHICLES	STEALTH	HPM	DIRECTED ENERGY WEAPONS	ELECTRONIC WARFARE	RADAR	ACTIVE PROTECTION	NUCLEAR EFFECTS SIMULATION	LASER WEAPONS	RAIL GUNS
Batteries	CW, P	CW, P		X	X	X	X	X		X	X	X				
Fuel Cells	CW, P			X	X	X	X	X								
Capacitors		CW, P	CW, P		X			X	X	X	X	X	X	X	X	X
Switches	P		CW, P		X		X		X	X	X	X	X	X	X	X
Rotating Machinery	CW, P	CW, P	P	X			X									X
Inductors		CW, P	P						X	X			X		X	X

CW = Continuous
P = Pulsed

Batteries

A battery is a device that converts chemical energy to electrical energy. The two types of batteries are primary and secondary (rechargeable). Batteries are desirable because of their reduced thermal and acoustical signature, ability to be molded into many configurations (mission specific), portability, and ability to store energy at high densities. Table 7.0-2 displays the status of the North American military battery industry as of 1994.

Battery design and fabrication is a proprietary, art-intensive activity in which details of know-how are closely held and protected. Factors of particular interest include high-temperature materials; high-conductivity electrolytes; the selection, refinement, and use of specific catalysts; electrode design and fabrication; packaging (especially corrosion and safety seals); additives for electrodes and electrolytes; and flexible small-lot manufacturing technology. In addition, the controlled (graceful) degradation in performance over time—as opposed to a sudden catastrophic failure—is a very desirable military trait. Figure 7.0-9 displays the inherent tradeoffs in battery technology. Increasing stored energy comes at the expense of power performance—with one caveat: For traditional battery construction and electrochemistries, an apparent trend indicates that a single battery is typically not able to provide both high power and high energy efficiently.

Table 7.0-2. Status of North American Military Battery Industry by Chemistry (Source: NATIBO)

Chemistry	Military/ Commercial Demand	Present Supplier Health	Base Health	Projected Business Climate	Convertibility
Thermal	Medium military demand. No commercial demand. Potential for emergency power systems in commercial.	One healthy supplier. Others considering entry.	Healthy.	Stable and flat.	N/A.
Lithium	Military demand for lithium sulfur dioxide decreasing. Increased military interest in other lithium chemistries. Small commercial demand. Commercial demand for lithium ion and lithium polymer increasing.	Five companies with poor-to-good health.	Diversification from military to commercial market is a concern because of a rapid military decrease and a slow commercial increase. Military transition to lithium maganese dioxide should support at least two suppliers. Suppliers looking for commercial market to sustain business.	Need product diversification based on increased commercial demand.	Lithium sulfur dioxide to be replaced by lithium maganese dioxide. Lithium thionyl chloride dual use today (bobbin configuration). Lithium ion dual use for future.
Mercury	Military demand decreasing. Small, niche commercial demand.	Presently stable. Eventually supplier will disappear.	Not required.	Market going away.	N/A.

Table 7.0-2. Status of North American Military Battery Industry by Chemistry (Source: NATIBO) (Continued)

Chemistry	Military/ Commercial Demand	Present Supplier Health	Base Health	Projected Business Climate	Convertibility
Silver	Mostly military demand. Some commercial demand.	Four suppliers with fair-to-good health.	Healthy. Silver cadium going away.	Stable, but overall demand decreasing.	N/A.
Lead Acid	High commercial demand. Piggyback military applications. New commercial applications for the future.	Four healthy suppliers.	Healthy.	Excellent and expanding.	Full convertibility and dual use.
Nickel Cadium	High commercial demand. Piggyback military applications.	Three healthy suppliers.	Healthy.	Excellent and expanding.	Full convertibility and dual use.
Magnesium	Small military demand. Small commercial demand.	Two healthy suppliers.	Healthy.	Stable and flat.	Army uses dual-use batteries. Navy does not use dual-use batteries.

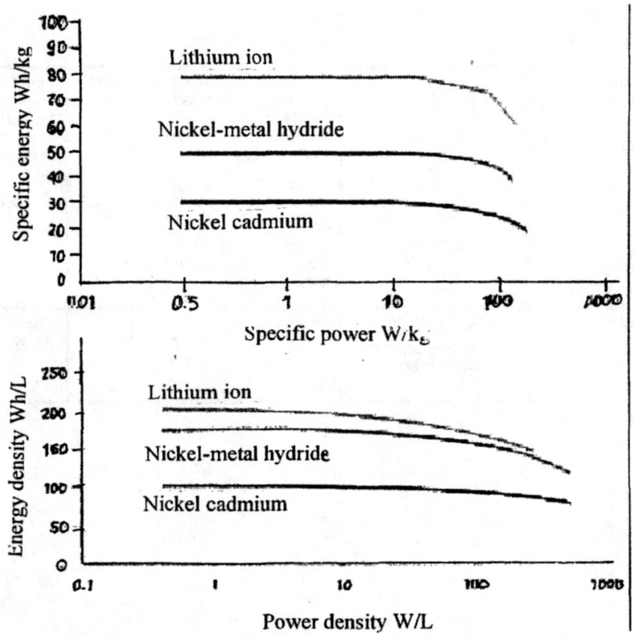

Figure 7.0-9. Tradeoffs in Battery Technology (Source: IEEE, Nov. 1998)

Fuel Cells

"A fuel cell is a continuous-feed electrochemical device of limited pulse power capability" (ATAR, p. II-14). It produces electricity by oxidizing hydrogen. Fuel cells are categorized by the electrolyte that is contained within them and by their operating temperature. Air or oxygen is used as the oxidant, and hydrogen or hydrogen derived from a carbonaceous fuel is used. (Hydrogen is the simplest fuel from an operational viewpoint, with increasing difficulty in consuming more complex fuels from methanol through diesel.) However, storing hydrogen on the battlefield is difficult because it needs to be maintained at near absolute zero temperature and is a formidable fire hazard. Present systems require that such hydrocarbon fuels be reformed or partially oxidized to yield hydrogen. Current smaller (200–300 W) systems are anticipated to use hydrogen and, subsequently, methanol as a fuel. For small power source applications, direct methanol fuel cells will emerge early in the next century (about 2005). Although hydrogen is a difficult logistics issue, it can supply more kilowatt-hours per kilogram than other fuels (see Table 7.0-3).

Table 7.0-3. Specific Energies of Various Fuels (Source: NRC, 1997)

Fuel	Specific Energy (kW-hrs/kg)
Hydrogen	33.3
Gasoline	12.2
Diesel	11.9
Methanol	5.5
Propane	12.8

Fuel cells that can use diesel JP8 fuel are of special interest to the military since these fuel cells would reduce the logistics burden of supplying multiple fuel types to the battlefield. This requires a reformer, however, to process the fuel before it is used.

As electric vehicles (EVs) and high-technology, low-weight, electrically powered devices expand in number and importance, the market for fuel cells is likely to expand rapidly. All fuel cells have tradeoffs. The advantages: they are lightweight; have high energy output, significantly reduced thermal and acoustic signatures [e.g., proton exchange membrane fuel cells (PEMFCs)], cogeneration abilities, and clean operation (primary byproduct being water; and are very durable. The disadvantages: they are expensive, occasionally inconvenient, and sensitive to air-borne contaminates. Most scientists believe that these problems will be solved relatively quickly as research and development (R&D) proceeds in this area. Early in the next century, fuel cell costs should approach U.S. $50/kW and be more competitive with internal combustion engines.

Among the vehicular applications for fuel cells are spacecraft, aircraft, industrial, automotive, ship, and submersible vehicles. Of these, only spacecraft and submarine applications have reached the commercial stage. Industrial and automotive applications are in the near-commercial and R&D stages, respectively. Ships and aircraft powered by fuel cells are the applications that are furthest from the market because they are still in the conceptual stage. Ballard of Canada has developed small fuel cells for possible use in manpack radio sets and laptop computers. Fuel cells are still in the early phases for battlefield usage. Currently, only PEMFCs demonstrate practical applicability for soldier power. Figure 7.0-10 forecasts the average power for fuel cells. These data represent the maximum demonstrated power of the largest fuel cells.

Fuel cells are also being tested for electrical power for buses, cars, and other vehicles. Many prototypes currently exist. Among the military applications for the fuel cell are EVs and the Soldier Integrated Protective Ensemble (SIPE), both of which would provide more power and less weight for the vehicles and man-portable units.

Another application for large-scale fuel cells is prime power generation. Fuel cell power plants have been developed and are being used in several remote areas to generate power for small communities. Moreover, fuel cells have been used in large-scale cogeneration applications, where they produce heat and electricity for buildings. This is possible because fuel cells generate a great deal of heat during operation. As fuel cells become more commercially available, they are expected to take a significant market share from batteries because of their greater power and

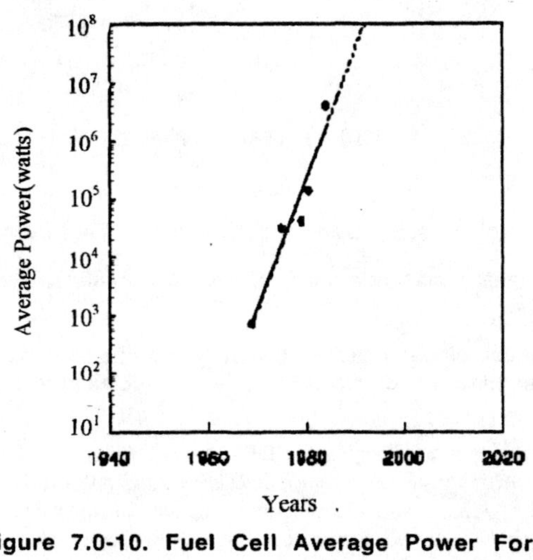

Figure 7.0-10. Fuel Cell Average Power Forecast

lighter weight. The final size of the market for vehicular fuel cells will be determined by the developers' ability to decrease the cost-per-watt of energy. As the cost for the fuel cells is reduced, the market will grow. Commercial production of automotive fuel cells is expected by 2005, and production over 100,000 units per year by 2010.

Capacitors

Capacitors (see Figure 7.0-11) are a basic component for energy storage and power conditioning in a variety of pulsed and continuous applications. Areas of application include filters in alternating current (AC) and direct current (DC) systems, AC resonant-charging power supplies, switched-mode power supplies, energy storage system, and high-frequency bypass circuitry. Major types of capacitors are usually described by their insulators, such as polymer, ceramic, chemical double layer (CDL), electrolytic, and mica (see Figure 7.0-12). Performance depends upon the usage (burst vs. continuous), the duty cycle, and the amount of energy or charge transferred. The major technical problem involves developing new dielectric materials and the related manufacturing processes to produce very-high-energy density capacitors that will operate at very-high-power levels.

Capacitors have wide applications in consumer electronics, telecommunications, computers, medical instruments, aircraft, spacecraft, robotics, and automobiles in the defense and the commercial sectors. These applications determine the performance of the capacitor market. For example, if a country has a strong consumer electronics sector in its economy (such as Japan), the capacitors found in those types of applications are going to perform particularly well. Most of the capacitors being manufactured are for small electronic devices, primarily consumer oriented. The economic forces affecting these devices are well known, and there is a wealth of market data available.

The economic forces affecting the consumer market are quite different from those affecting the capacitors of interest to this analysis. Capacitors capable of very-high-energy density at very high power are a special subcategory, and little economic information is available since this is such a specialized market. However, certain comparisons can be made. Capacitors, in general, are following the trend in semiconductor technologies: components are getting smaller, lighter, and more powerful. Greater amounts of capacitance are being fit into the same or smaller sized packages. These developments have compelled manufacturers to keep up with industry demands, and these demands have created pressure on the manufacturers to make huge capital expenditures on manufacturing equipment so they can produce state-of-the-art technology. This causes a market consolidation. Firms who cannot maintain these

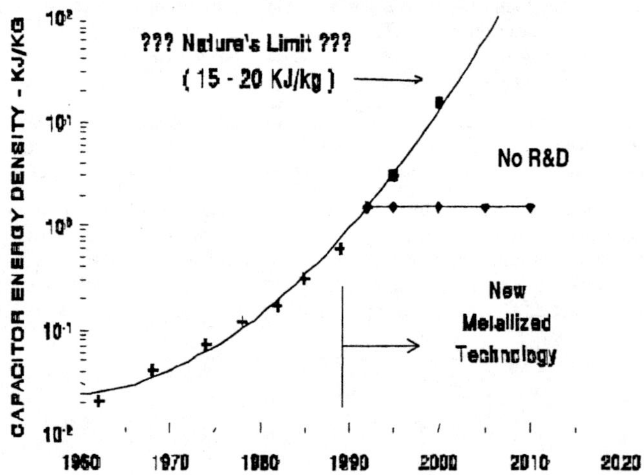

Figure 7.0-11. (High-Energy Density) Capacitor Technology Forecast

Note for Figure 7.0-11: *The diamonds represent a future without R&D in capacitors, while the squares identify an extrapolation with continued R&D.*

expenditures either fold, get bought out, or find a niche where they can remain solvent. Consolidation has occurred mostly in the capacitor market, and those firms who are present and functioning in the market will most likely continue be a factor in the future.

Ultracapacitors (also known as supercapacitors and CDL) are emerging as viable energy storage devices for hybrid-electric vehicles (HEVs), uninterruptible power supplies (UPS), and automotive and aircraft ignition systems. Presently, France, Germany, Italy, Japan, and Russia have active ultracapacitor research, development, and engineering (RD&E) programs. HEVs with ultracapacitors have been demonstrated in France, Italy, Japan, and Russia.

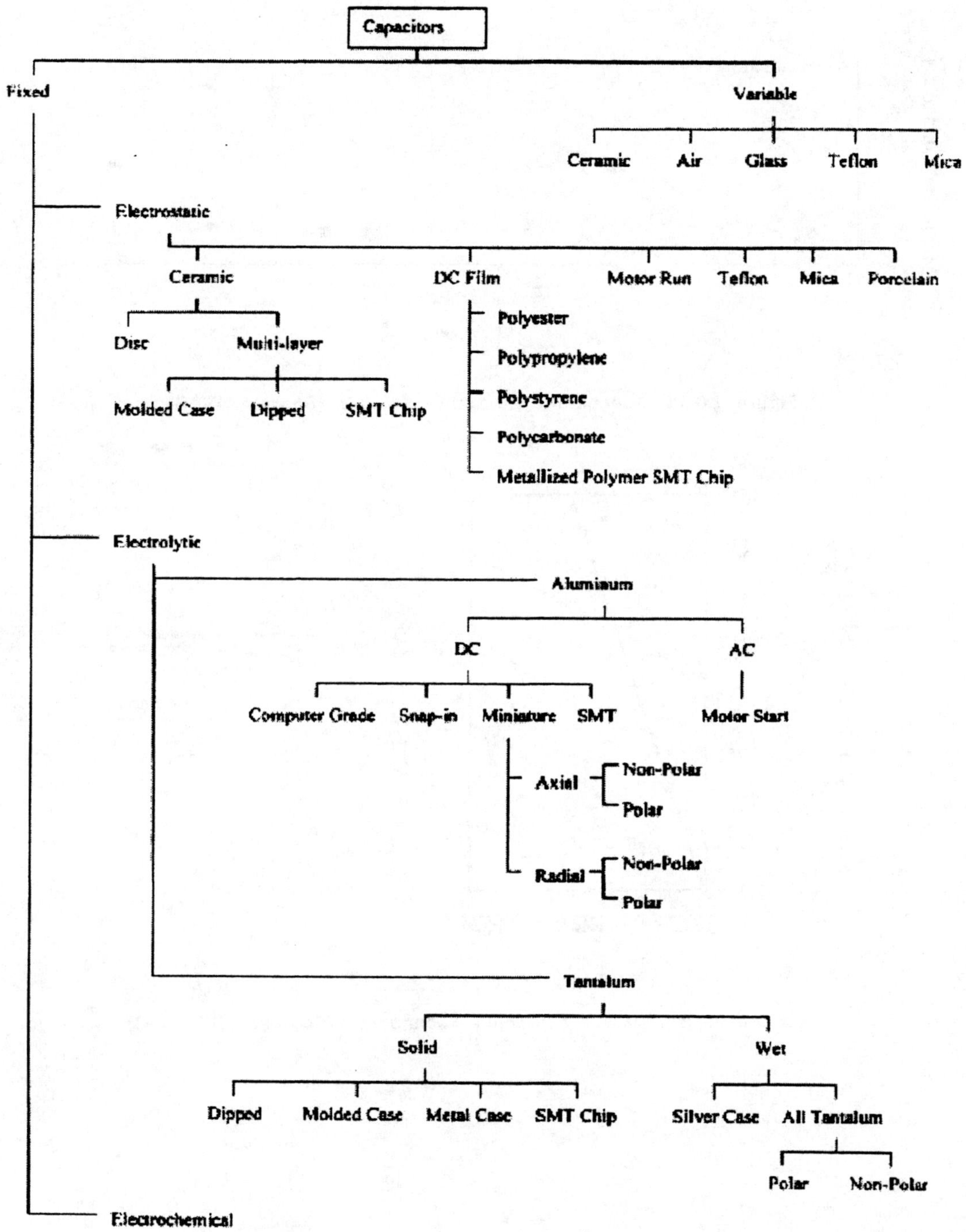

Figure 7.0-12. Types of Capacitors (Source: Mallory Capacitors)

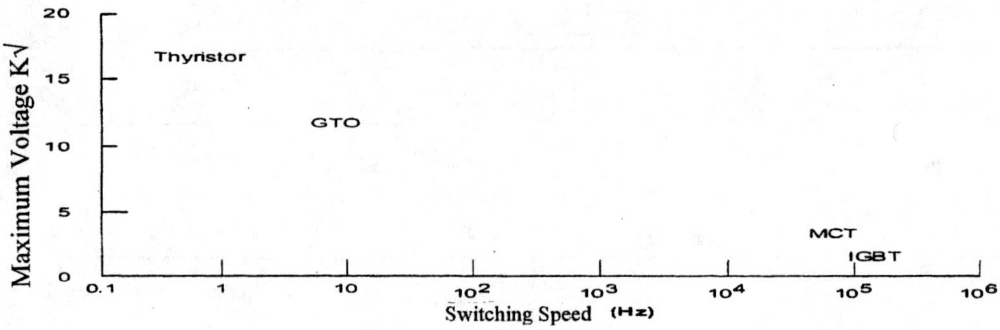

Figure 7.0-13. Speeds for Switching Devices (Source: ATAR)

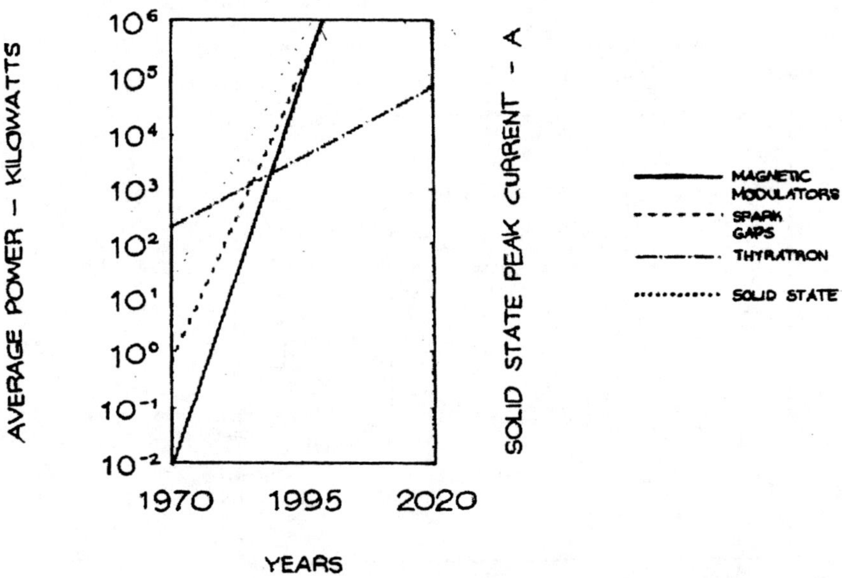

Figure 7.0-14. Trends for Switching Technology (Source: NRC, 1993)

Several foreign countries produce high-quality capacitors for the electronics industry (France, Germany, Japan, Russia, and the United Kingdom). The United States is the world leader (based on energy density, W hrs/kg) in high-power capacitors, especially those suited for the very-high-pulsed-power applications needed by the military. Maxwell Labs and Aerovox are the two leading manufacturers. In addition, Martin Marietta, General electric (GE), and some smaller companies (Power-One, Giner, Pinnacle Research, Evans) are involved in certain types of special-purpose high-power or high-density capacitors.

Filtering applies to mainly AC ripple suppression in DC bus and power transmission systems. The specific capacitor technology employed is application specific and dominated by ripple frequency. In some applications, large fractions of the stored energy are removed from the capacitor during some portions of the mission, making the capacitor internal losses a dominant consideration in applicability and efficiency.

Switches

Switches are *critical* components for high-power conditioning and pulse formation. They are mainly used in networks designed to convert power from one form to another. A switch's internal environment can be solid state, gas, magnetic, or liquid. The driving requirements in switch technology are high efficiency (greater than 98 percent), negligible internal inductance (<< 1 nH/A), and increased operating temperatures (–55 through +500 °C desirable). The latter is being realized through Silicon Carbide (SiC) switches, which are an emerging approach to the temperature problem. Applications are motor control systems, UPSs, power conditioning, high-voltage DC transmission, inductive heating, and many other commercial applications. Figure 7.0-13 compares voltages and speeds for various switches, and Figure 7.0-14 highlights the trends for several switches.

> In establishing operating parameters of high-energy switches, care must be taken to avoid subjecting the device to conditions that are potentially capable of causing a fault. For the SCR [Silicon Controlled Rectifier] switch, a fault is a catastrophic event. The spark gap and the triggered vacuum switch, however, regain most, if not all, of their capabilities after a fault. A fault can also cause damage to the overall system as well as the loss of the effectiveness of the shot or the shot itself (H. Singh et al. 1999).

Rotating Machinery

Rotating machines offer the potential for storage and generation of electrical energy. The key requirement is a reduction in size. The Department of Defense (DoD) is interested in several types of rotating machines various pulsed and continuous applications. These applications range from electromechanical energy storage systems using flywheels to high-power density alternators and generators for power generation. Measuring the size of the rotating machine market depends upon what type of device is being explored. Many different devices are referred to as rotating machines, including various motors, generators, and flywheels.

The market for the EV application of a rotating machine is likely to expand quickly. California has enacted a series of strict emission laws that require all car manufacturers to produce zero-emission vehicles by the early 21st century. Among the technologies being introduced to meet these deadlines, the most viable seems to be the EV.

WORLDWIDE TECHNOLOGY ASSESSMENT (See Figure 7.0-15)

Foreign capability will not necessarily follow the same progression as the United States. Culture and diverse needs may be factors that affect developing technologies in different nations.

An extensive list of global fuel cell developers can be found at http://216.51.18.233/fcdevel.html .

Country	Section 7.1 Energy Conversion and Power Generation	Section 7.2 Energy Storage	Section 7.3 Power Conditioning	Section 7.4 Biological Energy Systems
Australia	•	•	••	
Canada	••	•••		•••
China	••••	••	•••	••
Finland				•
France	•••	••••	•••	•••
Germany	••••	••••	••••	••••
India	•	•	••	•
Ireland	•••		•••	
Israel	•••	•••	••	
Italy	••	••	••	•
Japan	••••	••••	••••	••••
Norway				
Russia	••••	••••	••••	•
South Africa	•	•	••	
South Korea	•••		••	
Spain	•			
Sweden			•••	•
Switzerland			•••	
Ukraine	•	•	••	
UK	•••	•••	•••	••••
United States	•••	••••	••••	••••

Legend:
Capability in technology elements — Extensive R&D •••• Significant R&D ••• Moderate R&D •• Limited R&D •

Figure 7.0-15. Energy Systems Worldwide Technology Assessment (WTA) Summary

SECTION 7.1—ENERGY CONVERSION AND POWER GENERATION

Highlights

- The future battlefield will require systems supplying power at levels ranging from milliwatts to gigawatts.
- Advanced precision weapons, weapons of mass destruction (WMD), radar, and electronics countermeasures and communications systems are enabled through next-generation high-energy electronics systems, including the prime energy sources.

OVERVIEW

Energy conversion and power generation encompass the transformation of biological, chemical, electromagnetic, nuclear, mechanical, and thermal energy or reactions into electrical power. The output may pulsed, burst, or continuous. The future battlefield will require systems supplying power at levels ranging from milliwatts to gigawatts.

Pulsed power is generated by the inherently pulsed nature of the source, such as explosive technologies, or derived through the use of pulse-forming networks. Further, pulsed sources can be used in conjunction with fast switches to provide pulses of shorter duration, higher frequency, and greater peak power. Burst is a form of pulsed power with the duration between shots ("off") being much greater than the "on" duration. Continuous power, by definition, is supplied on a continual basis by sources ranging in size from batteries to large turbines.

RATIONALE

All military platforms are subject to volume and weight constraints. In addition, since the ability to initiate actions without awareness by any adversary is becoming more significant, the need for stealthy land, sea, and air platforms is more important. With the exception of hydrocarbon-fuelled air vehicles, all others can be optimized for limited missions using electrical energy.

HEVs and all-electric vehicles will integrate traditional power-generation devices (e.g., combustion engines and generators) with power sources (e.g., batteries and fuel cells). Further, ultracapacitors and batteries have been used with regenerative braking—the process of reclaiming energy through deceleration. All-electric vehicles rely strictly on batteries for power generation, while HEVs combine batteries with other sources of power. Series HEVs use combustion engines or fuel cells to generate electric power and use electric motors to drive the wheels. Parallel HEVs take some of the power from the combustion engine as shaft power and deliver this power directly to the wheels in parallel with the electric motors. These robust platforms eliminate the use of hydraulic, pneumatic, and mechanical power, and the benefits are numerous.

Mobile platforms are important in this discussion because their subsystems require pulsed or continuous power, which is provided by the larger power-generation system. Mobility itself has its own power requirements, which vary with the motion of the vehicle. For example, acceleration requires more power than deceleration. In battle conditions, an HEV would need to provide continuous power to computer, environmental, and low-observable subsystems, radar and mobility. Pulsed power would be required for weaponization [including high-power microwave (HPM), lasers, electronic warfare (EW), and directed energy (DE)], for active protection, and for augmenting propulsion. Mobile platforms include attack vehicles and support vehicles.

Pulsed-Power Generation

Pulsed-power systems are necessary to parts of current and future military systems. Current use ranges from laser range-finders and tactical communication to radars, jammers, and electronic countermeasures. In the future, pulsed-power systems will be critical enabling subsystems for military weapons and other systems ranging from kinetic energy (KE), microwave, and laser weapons to pulsed jammers, mine destruction, and active protective systems. Such particular applications may be powered by a unique combination of the prime power source, energy storage, and power-conditioning and pulse-forming networks.

Continuous Power Generation

Continuous power-generation systems are those ranging from several watts for soldier systems to many gigawatts for EVs. Fuel cells, one source of continuous power, are of interest to all branches of the military because of their low signature, high-energy density, and efficiency characteristics. A significant problem in current fuel cell design is weight because of the requisite pressurized vessels that are heavier than the fuel. While this is not as great an issue for ship platforms, non-electric ground mobility will find more supportability and functionality in internal combustion turboshaft engines in the near term and probably the mid-term.

WORLDWIDE TECHNOLOGY ASSESSMENT

China, Germany, Japan, and Russia are the world leaders in this category, whereas the United States is the world leader in most facets of mobile electric platform energy systems. Commercial counterparts are unlikely to be developed because of performance, volume, and environmental demands, combined with low production volumes. Most foreign endeavors to develop total mobile electric platform energy systems will be regarded as highly competitive military activities and would require monitoring in the absence of a strong outlook for cooperation

Japan has very advanced commercial fuel cells, and Canada has strength in the niche area of fuel cells.
In establishing operating parameters of high-energy switches, care must be taken to avoid subjecting the device to conditions that are potentially capable of causing a fault. For the SCR [Silicon Controlled Rectifier] switch, a fault is a catastrophic event. The spark gap and the triggered vacuum switch, however, regain most, if not all, of their capabilities after a fault. A fault can also cause damage to the overall system as well as the loss of the effectiveness of the shot or the shot itself (H. Singh et al. 1999).

Rotating Machinery

Rotating machines offer the potential for storage and generation of electrical energy. The key requirement is a reduction in size. The Department of Defense (DoD) is interested in several types of rotating machines various pulsed and continuous applications. These applications range from electromechanical energy storage systems using flywheels to high-power density alternators and generators for power generation. Measuring the size of the rotating machine market depends upon what type of device is being explored. Many different devices are referred to as rotating machines, including various motors, generators, and flywheels.

The market for the EV application of a rotating machine is likely to expand quickly. California has enacted a series of strict emission laws that require all car manufacturers to produce zero-emission vehicles by the early 21st century. Among the technologies being introduced to meet these deadlines, the most viable seems to be the EV.

WORLDWIDE TECHNOLOGY ASSESSMENT (See Figure 7.0-15)

Foreign capability will not necessarily follow the same progression as the United States. Culture and diverse needs may be factors that affect developing technologies in different nations.

An extensive list of global fuel cell developers can be found at http://216.51.18.233/fcdevel.html .

Country	Section 7.1 Energy Conversion and Power Generation	Section 7.2 Energy Storage	Section 7.3 Power Conditioning	Section 7.4 Biological Energy Systems
Australia	•	•	••	
Canada	••	•••		•••
China	••••	••	•••	••
Finland				•
France	•••	••••	•••	•••
Germany	••••	••••	••••	••••
India	•	•	••	•
Ireland	•••		•••	
Israel	•••	•••	••	
Italy	••	••	••	•
Japan	••••	••••	••••	••••
Norway				
Russia	••••	••••	••••	•
South Africa	•	•	••	
South Korea	•••		••	
Spain	•			
Sweden			•••	•
Switzerland			•••	
Ukraine	•	•	••	
UK	•••	•••	•••	••••
United States	•••	••••	••••	••••

Legend:
Capability in technology elements Extensive R&D •••• Significant R&D ••• Moderate R&D •• Limited R&D •

Figure 7.0-15. Energy Systems Worldwide Technology Assessment (WTA) Summary

SECTION 7.1—ENERGY CONVERSION AND POWER GENERATION

Highlights

- The future battlefield will require systems supplying power at levels ranging from milliwatts to gigawatts.
- Advanced precision weapons, weapons of mass destruction (WMD), radar, and electronics countermeasures and communications systems are enabled through next-generation high-energy electronics systems, including the prime energy sources.

OVERVIEW

Energy conversion and power generation encompass the transformation of biological, chemical, electromagnetic, nuclear, mechanical, and thermal energy or reactions into electrical power. The output may pulsed, burst, or continuous. The future battlefield will require systems supplying power at levels ranging from milliwatts to gigawatts.

Pulsed power is generated by the inherently pulsed nature of the source, such as explosive technologies, or derived through the use of pulse-forming networks. Further, pulsed sources can be used in conjunction with fast switches to provide pulses of shorter duration, higher frequency, and greater peak power. Burst is a form of pulsed power with the duration between shots ("off") being much greater than the "on" duration. Continuous power, by definition, is supplied on a continual basis by sources ranging in size from batteries to large turbines.

RATIONALE

All military platforms are subject to volume and weight constraints. In addition, since the ability to initiate actions without awareness by any adversary is becoming more significant, the need for stealthy land, sea, and air platforms is more important. With the exception of hydrocarbon-fuelled air vehicles, all others can be optimized for limited missions using electrical energy.

HEVs and all-electric vehicles will integrate traditional power-generation devices (e.g., combustion engines and generators) with power sources (e.g., batteries and fuel cells). Further, ultracapacitors and batteries have been used with regenerative braking—the process of reclaiming energy through deceleration. All-electric vehicles rely strictly on batteries for power generation, while HEVs combine batteries with other sources of power. Series HEVs use combustion engines or fuel cells to generate electric power and use electric motors to drive the wheels. Parallel HEVs take some of the power from the combustion engine as shaft power and deliver this power directly to the wheels in parallel with the electric motors. These robust platforms eliminate the use of hydraulic, pneumatic, and mechanical power, and the benefits are numerous.

Mobile platforms are important in this discussion because their subsystems require pulsed or continuous power, which is provided by the larger power-generation system. Mobility itself has its own power requirements, which vary with the motion of the vehicle. For example, acceleration requires more power than deceleration. In battle conditions, an HEV would need to provide continuous power to computer, environmental, and low-observable subsystems, radar and mobility. Pulsed power would be required for weaponization [including high-power microwave (HPM), lasers, electronic warfare (EW), and directed energy (DE)], for active protection, and for augmenting propulsion. Mobile platforms include attack vehicles and support vehicles.

Pulsed-Power Generation

Pulsed-power systems are necessary to parts of current and future military systems. Current use ranges from laser range-finders and tactical communication to radars, jammers, and electronic countermeasures. In the future, pulsed-power systems will be critical enabling subsystems for military weapons and other systems ranging from kinetic energy (KE), microwave, and laser weapons to pulsed jammers, mine destruction, and active protective systems. Such particular applications may be powered by a unique combination of the prime power source, energy storage, and power-conditioning and pulse-forming networks.

Continuous Power Generation

Continuous power-generation systems are those ranging from several watts for soldier systems to many gigawatts for EVs. Fuel cells, one source of continuous power, are of interest to all branches of the military because of their low signature, high-energy density, and efficiency characteristics. A significant problem in current fuel cell design is weight because of the requisite pressurized vessels that are heavier than the fuel. While this is not as great an issue for ship platforms, non-electric ground mobility will find more supportability and functionality in internal combustion turboshaft engines in the near term and probably the mid-term.

WORLDWIDE TECHNOLOGY ASSESSMENT

China, Germany, Japan, and Russia are the world leaders in this category, whereas the United States is the world leader in most facets of mobile electric platform energy systems. Commercial counterparts are unlikely to be developed because of performance, volume, and environmental demands, combined with low production volumes. Most foreign endeavors to develop total mobile electric platform energy systems will be regarded as highly competitive military activities and would require monitoring in the absence of a strong outlook for cooperation

Japan has very advanced commercial fuel cells, and Canada has strength in the niche area of fuel cells.

SECTION 7.2—ENERGY STORAGE

Highlights

- Energy is stored either chemically, mechanically, or electrically, with durations of microseconds to years.
- Fast-storage devices are militarily critical for weaponry.

OVERVIEW

Energy storage devices are intended for pulsed applications requiring short-term storage (microsecond-to-second) through long-term storage (several years). These devices can be charged and discharged in a pulsed or near-continuous manner. Examples of technologies used as energy storage include batteries, capacitors, inductors, and flywheels or rotating machinery.

Energy can be stored chemically, mechanically, or electrically. For mechanical storage, the energy is stored in the rotary motion of machinery, such as a flywheel. For electrical storage, the energy is stored in the electric field of the dielectric medium, such as a capacitor, or in the magnetic field of the dielectric medium, such as an inductor. For chemical storage, the energy is stored in the reactants (as in batteries).

For pulsed-power applications, once energy is stored, it can then be extracted completely or in small portions. In the mechanical case, the flywheel storage device can be coupled to a switch or a switching network that, depending on the speed and other characteristics, can deliver pulses of power to some load. Some devices integrate the switching into the storage device and feed its pulsed output directly to the load. Once charging is complete, the switch is closed and a pulse is delivered to the load. High-frequency inverters (DC-to-AC) can be used as a power supply to charge the storage device in a few milliseconds. Once charging is complete, a switch is closed, and pulsed power is delivered to some load.

In the high-energy case, mainly for weapons applications, systems of average powers in excess of 100 MW, for times of seconds to minutes, are integrated with pulsed-energy conditioning to create gigawatt-class repetitive pulses of energy from milliseconds down through submicroseconds, at voltages from several kilovolts up to megavolts. There may emerge selected military applications, such as high-power radars, electronic countermeasures, and directed energy weapons (DEWs), that demand a class of precisely conditioned electric energy that has no direct foreseeable commercial or industrial application. However, industry may be able to use for different applications the components used to produce these levels of output. High-energy electronics have no current commercial applications. Assured development, therefore, will require government sponsorship.

Peak power, pulse shape, pulse duration, repetition rates, firing rates, silent watch, and system energy storage recharging times represent militarily critical performance parameters that, in many cases, transcend known commercial, industrial, or consumer applications. In addition, high-energy electronics packaging currently requires parallel/series combinations of components in the power train to achieve reliability, fault tolerance, and graceful aging at performance levels greater than 10 times today's commercial standards.

RATIONALE

Fast-storage devices are militarily critical for weapons. However, energy is generally collected or available from the prime source at low-power levels and power densities. To meet common load requirements, such as high power, the energy is released from storage over an extremely short duration and converted into a pulsed-power form.

Electric power systems are critical enabling technology subsystems in all nuclear devices and nuclear weapons effects simulators. Some of these applications require single-shot devices. Many of the specific technologies involved are dual use.

WORLDWIDE TECHNOLOGY ASSESSMENT

The United States is a world leader in many areas of battery technology, but Japan is ahead in specific areas. In addition, the Europeans have a strong capability in small-lot production of high-reliability batteries that could be of great benefit to the United States. The U.S. Advanced Battery Consortium (ABC,) consisting of the Electric Power Research Institute (EPRI), Argonne National Laboratory, Sandia National Laboratories, and the Nonlinear electronics Laboratory (NOEL) are major government players working with the big three auto makers [Chrysler, Ford, and General Motors(GM)] to develop batteries for EVs. In addition, the following private firms have ongoing R&D in advanced batteries:

- Eagle Picher (nickel-metal hydride, lithium thermal)
- Yardney (lithium ion)
- Electro Energy (nickel-metal hydride)
- Valence Technology (lithium polymer)
- SAFT America (lithium ion)
- Ovonics (nickel-metal hydride)
- Duracell (nickel-metal hydride)
- Eveready (lithium ion)
- Rayovac (lithium ion)
- Wilson Greatbatch (lithium ion).

Lithium battery research elicits particular interest. Table 7.2-1 lists the main North American R&D contributors in industry. Table 7.2-2 lists the North American contributors from laboratories and universities. Table 7.2-3 lists the lithium battery R&D contributors from foreign industry. Table 7.2-4 lists the lithium battery contributors from foreign universities. Table 7.2-5 lists the lithium battery R&D contributors from foreign laboratories.

Table 7.2-1. North American Industry Lithium R&D (Source: NATIBO)

Companies		
Advanced Energy Technologies, Inc.	ECO Energy Conversion	Mine Safety Appliances Company
Advanced Technology and Research, Inc.	EIC Laboratories	Moli Energy
Alliant Techsystems Inc.	Electrochemica Corporation	Panasonic
Applied Power International	Electrofuel	PolyPlus Battery Company
ARCO Medical	Energy Conversion	Power Conversion, Inc.
Arthur D. Little, Inc.	Eveready	Ray-O-Vac
ATT Bell Laboratories	GE Neutron Devices	SAFT - America
Aurora Flight Sciences Corporation	General Dynamics Space Systems	Sony
Ballard	General Motors	Sharp
Battery Engineering, Inc.	Gould, Inc.	SRI International
Bellcore	HED Battery Corporation	Technochem
Bell Labs	HEDB Corporation	The Aerospace Corporation
Catalyst Research	Hoppecke Battery Systems, Inc.	Ultralife Technologies
Coastal Systems Station	Johnson Controls	U.S. Advanced Battery Consortium
Covalent Associates	Lithium Energy Associates	UCAR Caron Company
Dowty Batteries	Martin Marietta Aerospace	Valence Technology
Delco Remy	Maxell Corporation of America	Westinghouse Electric Corporation
Duracell, Inc.	Medtronics, Inc.	Wilson Greatbach Limited
Eagle-Picher Industries	Microchip Technology, Inc.	W.R. Grace

Table 7.2-2. North American Laboratory and University Lithium R&D (Adapted From NATIBO)

Laboratories	Universities
AFRL	California Institute of Technology
ARLs	Harvard University
Argonne National Laboratories	Polytechnic University
Bell Labs	Texas A&M University
Marshall Space Flight Center	University of Dayton Research Institute
Naval Warfare Center	University Kentucky
Hydro Quebec	University of Minnesota
Jet Propulsion Laboratory	University of Pennsylvania
Lawrence Berkeley Laboratory	University of Texas
National Research Council, Canada	University of Texas at Austin
Naval Surface Warfare Center, Crane	University of Waterloo
Naval Surface Warfare Center, White Oak	University of Ottawa
Naval weapons Support Center	
Oak Ridge National Laboratory	
Sandia National Laboratories	
U.S. Air Force Space and Missile Center	
U.S. Army EDTL	
Wright Patterson Air Force Base	

Table 7.2-3. Foreign Industry Lithium R&D (Source: NATIBO)

Country	Company	Chemistry
United Kingdom	AEA	Rechargable lithium batteries
	SAFT NIFE	Lithium ion secondary and lithium thionyl chloride
	Dowty Batteries	Lithium manganese dioxide batteries
	Dorutcy	Lithium manganese dioxide batteries
	British ARE	Lithium iron sulfides
	Oakdale Batteries	Lithium aluminum iron sulfide secondary batteries for submarines
	Vickers	Lithium iron sulfides
	VSEL	Lithium aluminum iron sulfide secondary batteries for submarines
Germany	Varta AG	Lithium polymer secondary and lithium thermal
	BAFF	Lithium polymer secondary and lithium thermal
	Bayer Fulrik	Lithium polymer secondary and lithium thermal
	Hoppecke Betterien	Lithium primary batteries
	Siemens	Lithium primary batteries
Israel	Tadiran	Lithium ion
	Israel Defense Forces Power Sources Division	Lithium ion
Belgium	SEDEMA	Rechargable lithium cells
Switzerland	RENATA	Secondary lithium batteries

Table 7.2-3. Foreign Industry Lithium R&D (Source: NATIBO) (Continued)

Country	Company	Chemistry
Japan	Central Glass Company Ltd.	Lithium ion secondary, lithium polymer, lithium iron sulfide, and lithium ion cells
	Sony	Lithium ion secondary, lithium polymer, lithium iron sulfide, and lithium ion cells
	Toshiba Asahi Chemical Joint Venture	Lithium ion secondary, lithium polymer, lithium iron sulfide, and lithium ion cells
	Hitachi	Lithium ion secondary, lithium polymer, lithium iron sulfide, and lithium ion cells
	Sanyo	Lithium ion secondary, lithium polymer, lithium iron sulfide, and lithium ion cells
	Asahi	Lithium ion secondary, lithium polymer, lithium iron sulfide, and lithium ion cells
	Yuasa	Lithium ion secondary, lithium polymer, lithium iron sulfide, and lithium ion cells
	Matsushita	Lithium ion secondary, lithium polymer, lithium iron sulfide, and lithium ion cells
	Mitsubishi Petrochemical Company	Lithium ion secondary, lithium polymer, lithium iron sulfide, and lithium ion cells
	Sharp	Lithium ion secondary, lithium polymer, lithium iron sulfide, and lithium ion cells
	Honda Research and Development Corporation	Lithium ion technology
France	SAFT	Lithium ion secondary, lithium polymer, lithium thionyl chloride, lithium thermal, and lithium vanadium pentoxide
	Alcatel	Rechargable lithium cells
	INSA	Rechargable lithium cells

Table 7.2-4. Foreign Universities Lithium R&D (Source: NATIBO)

Country	University
Bulgaria	Bulgarian Academy of Sciences
China	Bejing Institute of Spacecraft System Engineering
	Changchun Institute of Applied Chemistry
	Chinese Academy of Sciences
	Tianjin Institute of Power Sources
	Institute of Physics, Academia Sinica
	Tianjin University, Department of Applied Chemistry
Czech Republic	Wuhan University
Denmark	Technical University of Denmark
France	CNRS - FranceCzech Academy of Sciences
	SORAPEC
	Universite Bordeaux
Germany	Dresden University of Technology, Institute of Physical Chemistry and Electrochemistry
	Ernst-Moritz-Arndt University of Greifswald
	Fraunhofer-Institute for Chemical Technology
	Merseburg University, Institute of Macromolecular Chemistry
	University of Munster

Table 7.2-4. Foreign Universities Lithium R&D (Source: NATIBO) (Continued)

Country	University
Greece	Aristotelian University University of Ulm
Israel	Bar-Ilan University Tel Aviv University
Italy	Universita di Bologna
Japan	Chuba University Iwate University Keio University Kyoto University Mie University Rikkyo University NTT Interdisciplinary Research Laboratories Shinshu University Tokyo Institute of Technology Yamaguchi University
Netherlands	Delft University of Technology
Poland	Central Laboratory of Batteries and Cells Technical University of Poznan, Institute of Chemistry and Applied Electrochemistry University of Warsaw
Russia	A.N. Frumkin Institute of Electrochemistry of the Russian Academy of Sciences Frumkin Institute of Electrochemistry Institute of Chemical Engineering Institute of Transportation Engineering Russian Academy of Sciences Saratov State University
Scotland	University of St. Andrews
Slovenia	National Institute of Chemistry
Taiwan	National Central University Chung Shan Institute of Science and Technology

Table 7.2-5. Foreign Laboratories Lithium R&D (Source: NATIBO)

Country	Laboratory
Australia	Australian Army Engineering Development Establishment
France	Laboratoire d'Ionique et d'Electrochimie du Solide, Institute National Polytechnique de Grenoble Laboratoire de Chimie du Solide Mineral

SECTION 7.3—POWER CONDITIONING

Highlights

- Power conditioning is the process involved in modifying the source output to meet load characteristics.
- Voltage, current, time (pulse length), pulse shape, and frequency characterize power conditioning.
- Loads require power to be delivered in specifically shaped continuous or pulsed waveforms.
- Advanced weapons of precision and wide-area mass destruction, radar, countermeasures, and communications systems are enabled through next-generation-plus-high-energy electronics systems, including the prime energy sources.

OVERVIEW

Power conditioning is the process involved in modifying the source output to meet load characteristics. Such loads can include military and commercial applications. Military applications range from charging a battery or forming pulses for a weapon subsystem to electric motor drives. Two types of conditioning are defined by the amount of time over which power is supplied: continuous duty and pulsed or burst mode. To revisit, continuous duty is defined as power supplied to a load continuously, which, in this discussion, implies bursts lasting longer than 1 second (since semiconductor devices reach their thermal equilibrium in less than 1 second). Bursts lasting less than several seconds are considered pulsed power and generally have higher average powers. Pulsed power conditioning discharges a high-energy storage bank in a low repetition-rated burst mode (see Figure 7.3-1). For some applications, long pulses that last from several microseconds to seconds may be fed into a fast conditioning stage, which creates bursts with high power lasting from microseconds and femtoseconds.

Pulse conditioning subsystems are comprised of both discrete and transmission-line-type components. The circuit functions can be described by lumped element equivalent circuitry provided that the characteristic times of energy discharge are far longer than parasitic IRC (inductor, resistor, capacitor) time constants of any system component. These subsystems are intimately tied to the load, with efficiency and reliability often constraining the types of near-term solutions available. In the future, advanced component development and enabling new energy control topologies will permit considerable reduction in system size and mass for next-generation systems.

A considerable aspect of power conditioning is the pulse-forming functions. Pulse formation is dominated by the characteristics of the switches used. Switch performance is summarized and defined in terms of the following parameters:

- **Peak voltage** is the maximum voltage that can be applied while the switch is open without breakdown.
- **Peak current** is the maximum current reached by the pulse.
- **Charge** is the amount of current a switch can pass while conducting over a period of time.
- **Voltage drop** is the maximum voltage across the device when the current has reached its maximum value.
- **Current density** is the current per unit area of conduction.
- **Action** in a switch is the impact that occurs in a short interval of time because of a force associated with the high-energy pulse. The force introduces stresses and strains in the switch. These stresses can induce microcracks and fractures, which ultimately lead to device failure. If the value of the action exceeds the fusing capability of a solid-state device, failure is catastrophic. As a result, solid-state switches have to be operated at action levels well below their fusing limit. When devices are connected in parallel, including a fuse in series with each switch may be necessary (H. Singh, et al. 1999).

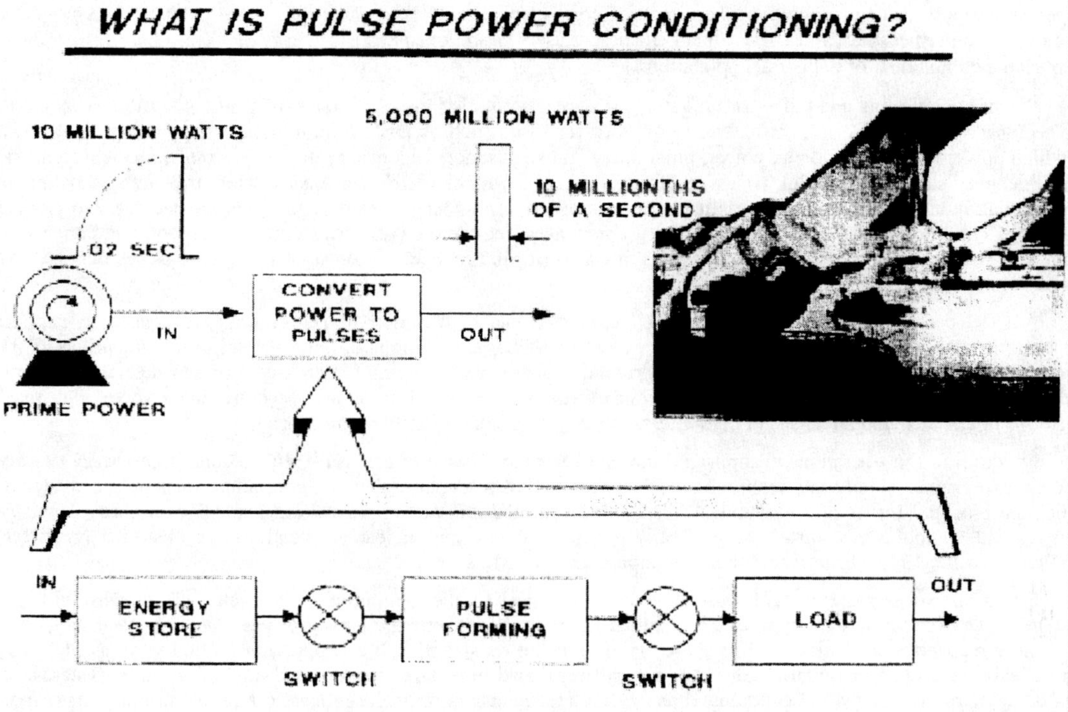

Figure 7.3-1. Pulsed Power Conditioning

- **di/dt** is the maximum rate of current rise that the switch can tolerate.
- **dv/dt** is the maximum rate of voltage rise that will not turn the switch on.
- **Thermal management** is the ability to remove heat energy generated by device.
- **Reverse blocking** is the peak voltage that can be applied across the device in reverse direction immediately after a high current in the forward direction.
- **System volume and mass** include the componentry: gate drives, protective drives, and so forth.
- **Life** is the predicted number of shots or pulses the device may deliver before failure.
- **Cost** is measured in dollars per switch.

Additional parameters include series and parallel operation, reliability, commercial availability, duration, and repetition rate.

RATIONALE

In the high-energy case, mainly for weapons applications, systems of average powers in excess of 100 MW for times of seconds to minutes are integrated with pulsed conditioning to create gigawatt class repetitive pulses of energy from milliseconds through submicroseconds. Voltages range from several kilovolts in the first case up through nearly megavolts in the latter case. This technology is used in commercial radar, smelting plants, and so forth. In addition, all aspects of the energy system demand significant technology base development at high risk. Thus, if needed, industry would possibly develop the capability but only if fully funded by the government. In addition, with the government technology investment base continuing to erode in militarily unique aspects of modern military systems, many industry senior management personnel feel that they are working as rapidly as possible to develop other customers, and, when these are in place, will be unwilling to redirect valuable internal expertise into government programs. There may emerge selected military applications, such as super energy radars, electronic countermeasures, and DEWs, that demand precisely conditioned electric energy of a class that has no direct foreseeable commercial or industrial application.

Power conditioning and pulse-forming networks are systems for the conversion of prime electrical energy into the necessary short pulses of electrical energy needed for loads such as DEWs and kinetic energy weapons (KEWs) and high power microwaves. Peak power, pulse shape, pulse duration, repetition rates, firing rates, silent watch, and system energy storage recharging times represent militarily critical performance parameters that transcend known commercial, industrial, or consumer applications. In addition, high-energy electronics, which are necessary to provide the power conditioning for a variety of military consumers, require parallel/series combinations of components to achieve reliability, fault tolerance, and graceful aging at performance levels an order of magnitude better than today's commercial standards.

Advances in technology for conditioning, regulation and distribution of power will be required to support advanced sensors, avionics, vehicle electronics (VETRONICS), and command, control, and communications (C3). These advances will address EMI, reliability, thermal management, packaging for high-energy densities, and so forth and will be equally critical to future commercial developments. Fielding smart weapons and remote unmanned sensors will also demand advances in high-energy density, high-energy military batteries.

Essential to the operation of almost all advanced weapons systems are well-regulated and high-energy density electronics systems supporting technology. This also includes technologies for continuous and pulsed systems, which are essential for a variety of industrial, commercial, and military applications. Typical examples of energy systems and components controlled specifically for nuclear and non-nuclear applications are identified in Part I, Section 15 of the 1996 Militarily Critical Technologies List (MCTL).

Electrical power systems, although absolutely essential to the operation of virtually all modern military systems, have been viewed as supportive rather than enabling. Where energy density drives enabling feasibility, they are clearly recognized as critical. While these are among the most noticeable applications, reliability, availability, maintainability, and energy quality constitute a significant portion of size, weight, and cost of military systems. In this context, prime energy and conditioned energy are recognized as critical elements for future military battlefield systems. Mission-compliant energy forms critical elements, with the application areas in systems fielded at present and in systems that will be fielded in the future.

Critical to future mission areas are affordable energy systems that enable stealthy mission compliance. This includes operating regimes that range from long life, continuous low drain (AA penlight cell energy levels for many weeks (submilliwatt energy levels at unattended locations) through terawatt peak power (single or limited burst duration) systems for DEW or nuclear electromagnetic pulse (NEMP) special effects weapons. Single-shot pulsed power at multi-megajoule energy-per-pulse levels for nuclear effect simulators is quite specific to that application and, being of little general technology applicability, is covered separately within the MCT Part III.

In general, conditioning can be divided into three areas: bus power conditioning, slow power conditioning, and fast power conditioning (NRC, 1993).

Bus Power Conditioning

Bus power conditioning takes the feed from one of the prime power sources into a conditioning stage where the voltage or frequency level is altered for optimum, reliable power distribution to other power conditioning stages.

Slow Power Conditioning

The slow power conditioning stage produces repetitive electrical pulses in the millisecond time frame. Switches, capacitors and transformers are the major components used.

Fast Power Conditioning

If required, particularly for weaponization, a final stage of power conditioning can be used to compress or transfer the millisecond-duration input pulses into microsecond, nanosecond, or picosecond pulses. This stage uses transmission lines/capacitors, switches, and magnetic devices as the primary components.

Figure 7.3-2 and Figure 7.3-3 display the trends for repetitive and microsecond power conditioning, respectively. In both figures, burst duty implies 100 sec "on" with several hours "off," while continuous duty indicates 24-hour operation. The barrier in Figure 7.3-3 is a performance barrier, which can be overcome with significant R&D.

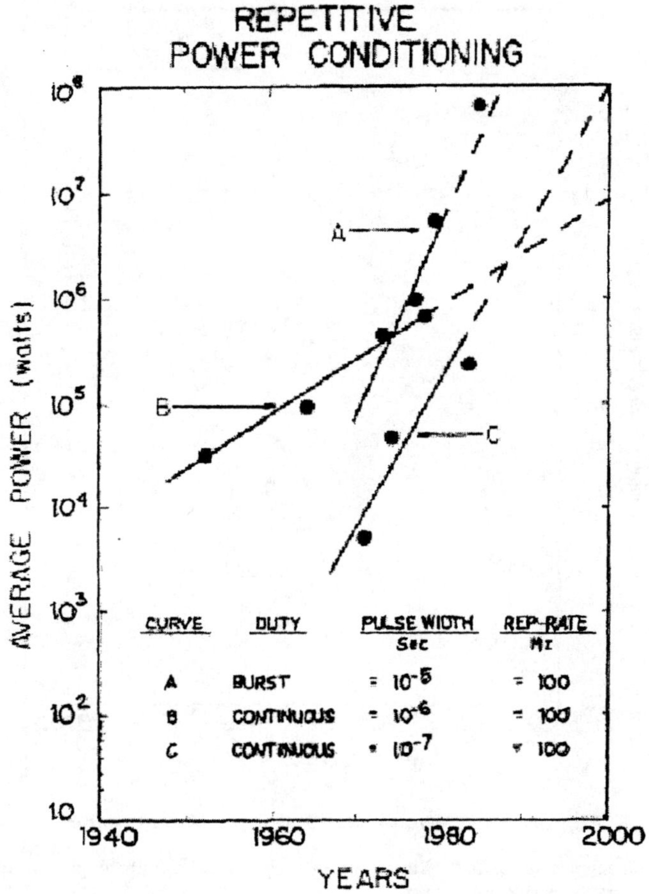

Figure 7.3-2. Repetitive Power Conditioning

To reiterate, power conditioning is the delivery of reliable and quality power, specifically pulsed power, where energy is provided in the necessary short pulses required by the load, such as weapons or other systems. For pulses of extreme short duration, special storage, switching devices, and packaging are required. Technologies for conditioning include inductors, resonant transformers, inverters, semiconductor diodes and switches.

In contrast to high-density conventional systems, pulsed and high-energy systems operate at average powers well above the 100-kW level. In this power regime, almost all applications are military, and many of these militarily unique applications project small-quantity procurements for the foreseeable future.

Since the ability to initiate actions without awareness by any adversary is becoming more significant, the need for stealthy land, sea, and air platforms is more important. With the exception of hydrocarbon-fuelled air vehicles, all others can be optimized for limited missions using electrical energy. Integration of hydrocarbon/nuclear fuels into future scenarios remains a possibility if signature management can be incorporated into initial designs.

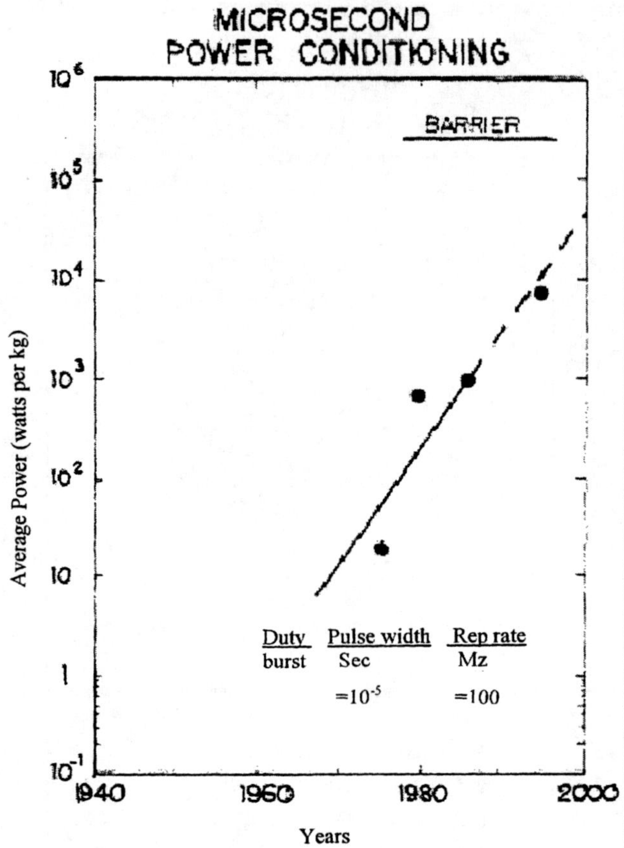

Figure 7.3-3. Microsecond Power Conditioning

For high-energy electronics systems for military systems, operating at multimegawatt average powers becomes viable only if available in very compact volumes.

Successful implementation will only proceed through direct DoD investment in high-energy switching, enhanced through international cooperative activities with our allies who have similar interests.

WORLDWIDE TECHNOLOGY ASSESSMENT

U.S. manufacturers are capable of meeting approximately 30 percent of the actual U.S. demand for high-power semiconductors. The remaining balance is imported from France, Germany, Japan, Sweden, Switzerland, and the United Kingdom. Several other countries, therefore, dominate the current power electronics market. Several U.S. manufacturers and the government are making headway towards expanding the U.S. market share in the high-power, high-temperature electronics area. Specifically, they are looking toward the potential future market in the Silicon-Carbide- and Nitride-based systems.

SECTION 7.4—BIOLOGICAL ENERGY SYSTEMS

Highlights

- Results in biotechnology research indicate that low-power, energy-generating elements will be produced commercially.
- Proton pumps and electron transfer systems have been created from lipids that are capable of self-assembly (and thereby forming membranes) and from proteins that perform pump functions.
- Proteins functioning as molecular motors have diameters of 20–50 Å.
- Energy is derived either from the oxidation of organic molecules to carbon dioxide and water through the process of metabolism or from the absorption of energy from light sources.

OVERVIEW

Electrical power systems are intrinsic elements of offensive and defensive operation in the modern military. Examples of these systems include communications, weapons detection, weapons guidance, and sensors. Associated with new technologies is the increasing burden borne by the infantry soldier; approximately 25–35 percent of the weight carried is related to power generation. Because of the high weight burden devoted to power supply, alternative power sources are in development. One source of energy may be based on biological system models or mimetics of biological power generators.

The new era of biotechnology has created opportunities for using biological systems or biomimetics as power-generating systems and as nanoscale motors. These electrical power systems are based on the ability of biological membranes to generate electro-osmotic gradients and proton pumps.

RATIONALE

During the past 2 decades, much has been learned about the mechanisms involved in the living systems' production of energy from nutrients and light. Living organisms generate electrical potentials, synthesize large organic molecules, and perform other tasks that require energy. Energy is derived either from the oxidation of organic molecules to carbon dioxide and water through the process of metabolism or from the absorption of energy from light sources.

The mechanisms involved in the oxidative phosphorylation metabolism of organic compounds and in the transduction of light energy to chemical energy require the generation of an electro-osmotic/electrochemical gradient across a membrane. The gradient is associated with transfer of electrons and/or protons, which results in the synthesis of high-energy compounds in the cell. The most ubiquitous of these high-energy compounds is adenosine triphosphate (ATP), which is recognized as a common currency of energy needs in a living cell.

In the metabolic process, the oxidation of the major nutrients results in the removal of electrons from the fat, carbohydrate, or amino acid substrate compounds and the stepwise transfer of the electrons in the mitochondrion to molecular oxygen, which then reacts with protons to form water. The removal of protons from the substrate and the creation of a proton pump across the cell membrane is an integral aspect of substrate oxidation. The production of ATP from adenosine diphosphate (ADP) by oxidative phosphorylation operates at an efficiency of approximately 40 percent and is, therefore, relatively efficient. The energy released by oxidation, not coupled to ATP production, is emitted as heat. Hibernating mammals use the oxidation of brown fat in a relatively uncoupled manner to generate heat needed to maintain body temperature.

In the photosynthetic process, larger organic compounds are generated from water and carbon dioxide using the energy of light. In this process, light interacts with a porphyrin-like tetrapyrrole component of chlorophyll in the thylakoid membrane of plant chloroplasts to cause the generation of molecular oxygen from water. The chlorophyll obtains an electron from the water as the molecular oxygen is generated. In this process, the electrons are passed from an antennae-like system of clustered chlorophyll molecules to a photochemical reaction center (PCRC). The PCRC has a particular arrangement of chlorophyll molecules that permit transfer of excited electrons from the antennae to electron receptor molecules. A proton pump, which generates ATP in a manner similar to that seen in the mitochondria during oxidative phosphorylation, is generated as part of the process. The time required for the processes is approximately 100 μs. In a related system, halophilic bacteria-like organisms, called halobacterium halobium, use a tetrapyrrole bacterial rhodopsin to generate a proton pump. For these systems to operate, an intact membrane is required, and the proteins acting as electron transfer agents are required to be organized in two dimensional (2-D) space. The proton pump is generated across the membrane, and the electron transfer process is generated in the plane of the membrane.

Because lipids are capable of self-assembling and thereby forming membranes, constructing proton pump and electron transfer systems is technically feasible. Such systems have been constructed. The generation of one molecule of molecular oxygen requires the transfer of four electrons. The production of 6 molecules of oxygen therefore requires 48 photons. Approximately 10,000 kJ are required to make 1 mole (mol) of glucose possessing 2,840 kJ.

From these considerations low power, energy generating elements will probably be produced commercially. These components can be used for generation of power for changing opaqueness of a visor, for driving sensors, or for read-write information storage components. In the Biological Technology section of Part III of the MCTL, the utility of bacterial rhodopsin as an information storage material is described.

Insertion of bioreceptor-ion channel molecules into lipid bilayers permits production of electrical switches at the nanometer scale. The lipid bilayer of electrically excitable tissues, such as muscle and nerve cells, contain molecular pumps that generate an ion gradient (hence a voltage drop of 70 mV) across the membrane. Sodium and potassium are the ions involved. Nerves and muscles also contain protein molecules that serve both as receptors and ion channels. The acetylcholine receptor (AChR) is an example of such a protein. When incorporated into a lipid bilayer 70-Å thick, these AChR molecules will bind to acetyl choline on one side of the membrane and cause an opening of the relevant ion channel.. The ions will transiently cross the membrane resulting in a reduced gradient and lowering of the electrical potential.

MANUFACTURING AND FABRICATION CRITICAL TECHNOLOGY

S.T. Paul
(Editor)

Highlights

- The continued development of rapid prototyping and near net-shape manufacturing will result in reduced costs, faster prototyping, and the ability to form a product closer to design (final product tailored to need).
- Higher speed machining capability means less time for the machining operation and, thus, reduced costs.
- The development and implementation of nanotechnology should result in improved military hardware.
- Quieter, longer lasting bearing assemblies will be used in submarines and helicopters.
- As manufactured dimensions become smaller, improved metrology equipment is necessary to maintain quality control and keep costs down.
- As manufactured dimensions become smaller and hardware becomes more expensive, more effective methods of in-process evaluation and non-destructive evaluation (NDE) are necessary to determine the quality of final product.
- Advanced land, sea, and air military robots will extend military capabilities in several areas, including reconnaissance and mine detection.

OVERVIEW

This section describes selected developing critical technologies for the production of U.S. military hardware. Such technology is important if the U.S. military is either to produce increasingly superior military performance or reduce the costs of existing hardware. In most cases, the technologies, the equipment, and the know-how are dual use and impact civil applications, where considerations of costs, agility, flexibility, competitiveness, and so forth have also become major concerns. All technologically advanced countries are pursuing similar programs if only to maintain a commercial advantage.

Several technologies included in subsection 12.1 are rather mature and are included as affordability issues. Affordability usually is considered in the context of a life-cycle allocation of resources. Major considerations for affordability include the following: (1) meets the consumer's performance parameters; (2) has resources available; (3) is available when the consumer requires it; (4) can be maintained through its life cycle with or without hostilities and without undue shortages at projected budget levels; (5) has importance established compared with other requirements; and (6) reduces cost of operation; and (7) increases reliability, effectiveness, or efficiency.

The technologies addressed in this section either allow prototypes to be produced in a fraction of the time required by conventional methods or, in some cases, actually produce one-of-a kind parts, in a first-time-right concept, without the need of expensive stocks of spare parts. Other technologies (e.g., nanotechnology and several coating technologies) are aimed at effecting significant improvement in military hardware. Bearings (see subsection 12.2) addresses the development of three new types of bearings. In two of the technologies, the new bearings would be critical parts of the improved (high-speed) spindles used in machine tools. There is also the possibility that larger types might find application in quiet-running machinery (e.g., submarines). The remaining bearing development has potential application in electric power generation.

Developing technologies (see subsections 12.3 and 12.4) address techniques either to inspect dimensionally (metrology) the final product or test the final product for latent failures. Continued development of these technologies will result in more accurately produced and reliable products. Subsection 12.5 addresses production equipment. Technologies listed include both affordability issues (high-speed spindles, machine tool monitoring, and so forth) and technologies that might improve the overall capability of military hardware [hexapods, precision grinders, micromachines/microelectromechanical systems (MEMS), and so forth]. Subsection 12.6 discusses robots that might find increased use in actual battlefield applications.

In short, the items addressed in this section address two critical points: Can we make it better and can we make it more affordable?

RATIONALE

The U.S. military presently faces two challenges: One is well known, but the other is rather new. The former addresses the requirement to improve existing hardware and to develop newer, more advanced hardware. The latter addresses the need to accomplish both of the former requirements at reduced costs.

This section includes the various technologies used to manufacture not only military hardware but also a wide range of commercial hardware. The United States is not the world leader in all these technologies, and less technologically advanced countries are slowly improving their capabilities, by either purchasing advanced technologies or developing indigenous capabilities. As these advanced manufacturing technologies are proliferating throughout the world, the United States must continue to improve the technologies used to manufacture hardware, both from a military viewpoint and a commercial, competitive standpoint. At the same time, in a period of reduced military budgets, technologies must be developed to produce the hardware faster and at less cost.

BACKGROUND

For years, manufacturing and fabrication equipment has been a mainstay of industrial societies. This equipment was instrumental in bringing about the Industrial Revolution in the 18th century, the continued development of a wider range of machines in the 19th century, and the development of the concept of automation in the early 20th century. The mid-to-late 20th century witnessed a rapid expansion, with the introduction of automatic control of the machining axes and the incorporation of additional axes of motion. Indeed, one can trace the development of our present industrial society, as well as the sophistication of military hardware, to the development of manufacturing and fabrication equipment.

While rudimentary machines have been used throughout history, machines, as we know them today, were first developed in England and the United States. In England in 1775, J. Wilkinson invented a precision horizontal-boring machine to bore out cylinders for the newly invented steam engine. In the United States in 1798, Eli Whitney, invented machinery to produce interchangeable parts for the assembly of army muskets. The 19th century saw the development of milling machines and turning machines (for rifle stocks), gear-cutting machines, sewing machines, harvesters, grinding machines, and automatic screw machines.

The early 20th century witnessed the development of automation. This, coupled with the existing machines, opened the world to mass production. Products could now be manufactured in higher volume and at much lower cost, and the world experienced the mass-market appeal of automobiles and numerous other consumer products. Consumer products became affordable to a much wider range of the populations. At the same time, the military used this capability during World War II to produce tanks, planes, and ships in unprecedented numbers and at costs previously unheard of. Such machines were critical for the manufacture of engine parts and nuclear weapons.

However, automation alone was not sufficient to meet some of the post-World War II military needs, as more sophisticated weapons were developed. Production of these weapons required not only high accuracy, but also high repeatability. From this need came the development, in 1952, of a three-axis machine with the rudiments of numerical control (tape instructions). Development continued, with the introduction of automatic tool changing, four- and five-axes machines, and computer numerical controllers. Most subsequent improvements involved materials, better cutting tools, more accurate raceways, and faster and more stable spindle assemblies.

At the same time that these later developments were being introduced, composite materials were developed. To make best use of these new materials, new machines—tape-laying and filament-winding machines—were developed. This revolutionized the production of a wide range of commercial and military products (e.g., strong and lighter aircraft assemblies and automobile and tank parts).

Along with these continued improvements in manufacturing technology came continued requirements to perform both dimensional inspection and non-destructive inspection (NDI) of the final product.

The technology for coating various substrates has also experienced rapid growth and development during the 20th century. In earlier centuries, coatings were mainly applied for surface protection, and the most common media were paints or similar coatings. The perfection of equipment to produce vacuum environments increased rapidly the range of coating materials and technologies. Technologies moved from simple vacuum evaporation to chemical vapor deposition, plasma spraying, sputter deposition, ion implanting, and so forth. The refinement of these various technologies resulted in faster, more reliable jet aircraft (improved gas turbine engines); improved canopies for aircraft; longer-life bearings for applications in jet engines, machine tools, drive trains of automobiles, trucks, and tanks; specially designed dielectric layers and wear coatings for optical systems and sensors; and coatings to reduce low observability of weapon systems.

Other products addressed in this section, bearings and robotics, have experienced similar developments. While bearings, in their simplest concept, have been used for many years, it was not until the 19th century, with the introduction of machine tools, that they were recognized as individual, important components. In the mid 20th century, the development of tapered bearings, high-speed bearings, and miniature bearings were instrumental in improving automotive drive trains; high-speed machine tool spindles; and navigation systems and gyroscopes, respectively. Robots, as they are known today, have developed as a direct result of the invention of computers. They have matured from interesting playthings to important production tools in most state-of-the-art factories, whether as robotic welders or material delivery tools. In more sophisticated applications, they are used in nuclear facilities and are being developed as a battlefield replacement for soldiers in some dangerous environments. In addition, unmanned flying vehicles (robots) are receiving increasing attention as reconnaissance vehicles (including normal reconnaissance and as sensors for chemical or biological weapons).

WORLDWIDE TECHNOLOGY ASSESSMENT (See Figure 12.0-1)

For most of the technologies included in this section, the United States is not the world leader but is a major player among other advanced technological countries. This is particularly true in manufacturing equipment, where Germany, Japan, and Switzerland are world leaders, either in the number of machines sold or the technical level of their machines. Advances in these technical areas (including bearings) are driven more by commercial needs than by military needs. Metrology and non-destructive testing have received considerable support from the European Union (EU) and, as a result, several European countries have active research programs in these areas. Japan is also very active.

The United States appears to have the unquestioned lead in battlefield robots. Other countries have developed robots more for industrial purposes [e.g., robotic welders and robots for space application (Japan)] or for use in the nuclear industry.

Country	Sec. 12.1 Advanced Fabrication and Processing	Coating	Sec. 12.2 Bearings	Sec. 12.3 Metrology	Sec. 12.4 NDI and NDE	Sec. 12.5 Production Equipment	Sec. 12.6 Robotics
Argentina						●●	
Austria				●		●	
Australia	●			●		●	●●
Brazil	●					●●	
Canada	●●	●●		●●●	●●	●●	
Croatia			●				
Czech Republic	●●			●	●	●●	
China	●●	●		●●	●●	●●	
Denmark				●	●		
Finland		●	●	●			
France	●●●●	●●●	●●●		●●●	●●●	●●●
Germany	●●●	●●●●	●●●●	●●●●	●●●●	●●●●	●●●
Greece						●	
Hungary	●			●		●	
India						●	
Indonesia	●			●		●	
Israel	●	●		●●	●●	●●	●●
Italy	●●	●●	●	●●	●●	●●●	●●
Japan	●●●●	●●●●	●●●	●●●●	●●●	●●●●	●●
Korea			●		●	●	
Netherlands	●	●●●	●	●●●	●●●	●●	
Norway						●	
Romania	●			●		●	
Russia	●●	●	●	●●	●●●	●●●	●●
Singapore					●	●	●
Spain				●●	●●	●●	
Sweden	●●	●		●	●●	●●	
Switzerland	●●	●●●	●●●	●●●	●●●	●●●	
Taiwan	●			●●		●●	
Ukraine		●●				●	
UK	●●●●	●●●	●	●●●	●●●●	●●●	●●●
United States	●●●●	●●●●	●●●●	●●●●	●●●●	●●●●	●●●●

Legend:
Capability in technology elements ●●●● most ●●● many ●● some ● at least one

Figure 12.0-1. Manufacturing and Fabrication Technology WTA Summary

12.1—ADVANCED FABRICATION AND PROCESSING

Highlights

- Rapid prototyping and near net-shape manufacturing will result in reduced cost of producing prototypes and end products.
- More accurate grinding machines will result in reduced cost and improved engine performance.
- Breakthroughs in nanotechnology will result in significant improvements in electronics, chemistry, and materials, with widespread military applications.
- Continued advances in coating technologies will result in improved hardness capability of various military hardware surfaces and improved performance in multiple domains.

OVERVIEW

This subsection covers two groups of technologies. The first includes equipment for fabricating structures using a wide range of different equipment, ranging from various thermal furnaces and equipment for bending, stretching, or rolling material to form-desired shapes to scanning tunneling microscopes (STMs). The second group includes the development, refinement, and production of nonorganic coatings on non-electronic substrates. Such substrates include ceramics, low-thermal-expansion glass, metals, polymers, and so forth. Coating materials include ceramics, metals, dielectric layers, abradable materials, and so forth. The coating procedures include chemical vapor deposition (CVD), various techniques of physical vapor deposition (PVD), sputter deposition, plasma spraying, and ion implantation.

Many of the developing technologies associated with the fabrication equipment involves new and continuing procedures to produce what is needed accurately and affordably. While manufacturing most products, prototypes have to be produced. An ongoing program, rapid (virtual) prototyping allows for a quick three-dimensional (3-D) fabrication of the prototype so that a design can be evaluated before production is initiated. This technology consists of:

- A computer-aided design (CAD)-generated virtual-reality model of the design, which can be analyzed and altered, as necessary
- The ability to perform virtual machining (manufacturing) of the virtual part, followed by a computer comparison of the virtual part with the original digital design
- The capability to produce a model from the pattern in a fraction of the time required to produce a model using conventional model-shop procedures.

Additional effort is being aimed toward extending this procedure to produce not only prototypes, but also final products, thus significantly reducing the time/cost of production.

Rapid prototyping is an integral part of the product realization cycle. It permits the fabrication of complex parts in a fraction of the time required by traditional prototyping techniques and, thus, will play a significant role in making products more affordable. Rapid prototyping can be categorized into two areas: virtual and physical. In the former, the computer system (CAD) is used to generate a 3-D image that can be analyzed before final manufacture. In the latter, the CAD program is fed directly to the manufacturing tool that produces either a final product (most often made of ceramic, wax, or plastic) or a mold to produce a metal object. Much effort is being expended to produce high-quality metal objects without the need of a mold.

Another ongoing developing technology included in this section is near net-shape. Near net-shape is an ongoing program with the goal of lowering the costs of manufactured products by reducing the amount of raw material used (less waste) and by producing a product so close to the design shape that only a minimum amount of subsequent machining is required to achieve the desired final shape. Casting is an example of a technology in which much effort is being expended in near net-shape. The traditional trial-and-error approach of selecting an optimum combination of design and process parameters results in long lead times and excessive material waste. To circumvent this approach, computer simulation of the casting process is being used and evaluated as a means to ensure that defect-free castings are produced in the first attempt.

Generally, items manufactured using techniques such as casting, injection molding, hot isostatic pressing, and so forth result in a final shape different from the shape of the container. As a result, historically, many products were manufactured using several steps until the proper shape of the container could be determined. Near net-shape technology uses CAD programs to calculate the resultant changes in shape that occur during cooling of the product so that the final product is "near the designed shape." This is of particular importance when fabricating ceramic parts because such items are very hard and subsequent finishing steps can be quite time consuming and expensive. Near net-shape is an approach similar to the philosophy of "first-time-right" manufacturing.

Ausforming is another developing near net-shape technology to manufacture superior spur and helical gears at less cost. It involves contour austenization of case-hardened gear teeth and quenching to metastable austenite, followed by plastic deformation of the gear tooth surface layers to final dimensions and then quenching to marstenite. The potential advantage of this procedure is that it forms the final shape of the gear without damaging the hard surface. It eliminates the need for gear tooth grinding while improving surface durability and fatigue behavior and reducing overall processing time and costs. The objective of the research program is to produce gears of better than 12 American Gear Manufacturing Association (AGMA) quality.

Semi-solid metalworking (SSM) is another example of near net-shape manufacturing, incorporating elements of casting and forming. In this process, the raw material is melted and allowed to cool to form a "mush" of liquid/solid material. This slush is then forced into a die, forming a final product that has higher structural integrity than castings but can be produced at lower cost than forgings.

Parts produced with SSM enjoy several advantages:

- Have higher structural integrity than castings, yet can be produced at lower cost than forgings
- Are less porous than parts produced with conventional high-pressure die casting
- Have equivalent properties to parts produced by either forging or conventional machining but require fewer steps.

To date, SSM has been demonstrated with aluminum, titanium, and copper; however, its implementation in either civil or military applications has been hindered by lack of process specifications, process models, training, and experience.

The most radical technology in this section deals with nanotechnology, defined as an anticipated manufacturing technology giving thorough, inexpensive control of the structure of matter (i.e., the ability to design and manufacture devices that are only tens or hundreds of atoms across). Nanotechnology is still an emerging technology. Significant advances have been made (e.g., buckyballs and fullerenes), but there has been little commercial activity. Sandia National Laboratories has reported the development of "supersensitive coating to improve detection of dangerous materials," and some use has been made of conducting fullerton as a filler material in plastics to make the plastic conductive. However, nanotechnology holds promise of an industrial revolution, as witnessed by the organizations—both government and industry—that are active in research. At a recent government nanotechnology workshop, agencies in attendance included the National Science Foundation (NSF), Department of Defense (DoD) (many organizations), DOC, National Institute of Standards and Technology (NIST), National Aeronautics and Space Administration (NASA), National Institutes of Health (NIH), and DOE.

Most of the research technology areas include materials, electronics, and medical devices. In materials, the bulk behavior of materials can be dramatically altered when constituted from nanoscale building blocks, the hardness and strength of nanophase metals can be greatly increased, and nanophase fillers in composite materials yield unique properties. Buckyballs and fullerenes (sometimes called buckytubes) are examples of nanophase materials. They are specific forms of carbon that are one-fourth the weight of steel and greater than 100 times stronger. Such capabilities hold great promise not only for use as improved cables, but as filler material for composite materials. Other potential applications include metal-doped buckytubes that theoretically would be 50 times more conductive than copper. Such tubes would bring about a revolution in power transmission.

Such a capability holds great promise in biology, electronics, chemical catalysis, and materials. This technology requires the capability to manipulate microscopic atoms, molecules, and so forth to form the desired structures. The procedures used in nanobiotechnology might be significantly different from those used in nonbiological applications. In the former, researchers are studying the molecular behavior of nucleic acids and proteins. In the latter, researchers are actively studying ways to use "buckyballs" and "buckytubes" (nanotubes) to produce usable end-items. Some military applications of this technology include biosensors, stronger ceramics and composites, superior coatings, abrasion-resistant materials, and so forth.

Buckyball structures are extremely malleable, can be compressed to less than half their original volume, and can be joined readily to other atoms, creating new capabilities. Although there have been no practical applications, as yet, for buckyballs, possible applications include superior armor, superconductor applications, ferromagnetic applications, superior lubricants, and medical applications.

Another fallout from buckyballs, and possibly more promising, are buckytubes (or nanotubes). These nanotubes are similar to buckyballs, but, instead of a sphere, the carbon atoms are linked together in a chicken-wire pattern. In certain cases (controllable), the carbon sheets roll together, and the edges join seamlessly, forming a nanotube. In some arrangements, the tube is a conductor. In other arrangements, the tube is a semiconductor. If two such tubes are joined to form a single molecule, the junction acts like a diode. Other possibilities for nanotubes include use as ultrastrong, thin cable and as a means of delivering medicine internally. Presently, conductive nanotubes are used in plastics to make the plastic conductive.

Coating technologies include several programs of developing technologies. Research has been conducted for several years on the technology for the deposition of diamond coatings. While some success has been achieved, the widespread application of the technology is still 5 to 10 years away. Nanophase coatings and cubic boron nitride coatings are more recent research programs, and both offer potential in the same technology area as diamond coatings—extremely hard, durable coatings. All these technologies have wide potential applications in military and civil applications. Of the three, nanophase coatings (related to nanophase materials) is probably the most esoteric.

Items fabricated from nanophase materials have superior characteristics. For example, nanophase copper and palladium have hardness and yield strengths 500 percent greater than conventionally produced metal, and nanophase ceramic material can be manufactured with much greater ductility than conventionally manufactured ceramics. These same bulk-nanophase characteristics result in similar improvements in the characteristics of nanophase coatings, particularly in applications requiring improved wear-resistant surfaces and increased thermal protection.

More recent studies have focused on the use of multilaser sources to heat the surface of a substrate selectively—resulting in extremely hard, graded surfaces. The technology can be modified by adding specific constituents to the plasma around the surface, altering the surface layer of the substrate and resulting in an extremely hard surface alloy.

RATIONALE

The fabrication equipment and technologies have significant military application. Along with machine tools (see subsection 12.5), these technologies form the cornerstone for producing mechanical hardware; composite sections for military aircraft, ships, and land vehicles; and various engine and transmission parts. The coating technologies included here address new technologies to improve the high-temperature operation or corrosion resistance of the base materials, or to effect changes in the optical characteristics of the substrate, and so forth.

WORLDWIDE TECHNOLOGY ASSESSMENT (See Figure 12.1-1)

Advanced manufacturing techniques are present, in varying degrees, throughout the industrial world. However, the main countries using these techniques are France, Germany, Japan, the United Kingdom, and the United States. In the United States, these efforts are supported by DoD and NSF, with considerable effort by industry and universities. In Europe, the European Commission has funded studies in several companies, and other companies are quite active. These advanced manufacturing techniques include several technologies that are primarily aimed at shortening turn-around time and reducing costs of final product.

Coating technologies research is primarily aimed at improving some aspect of hardware operation, whether it be improved hardness, resistance to corrosion, or improved thermal protection.

Country	Advanced Fabrication and Processing	Coating
Brazil	●	
China	●●	●
France	●●●●	●●●
Germany	●●●	●●●●
Hungary	●	
Indonesia	●	
Israel	●	●
Italy	●●	●●
Japan	●●●●	●●●●
Netherlands	●	●●●
Romania	●	
Russia	●●	●
Sweden	●●	●
Switzerland	●●	●●●
Taiwan	●	
Ukraine		●●
UK	●●●●	●●●
United States	●●●●	●●●●

Legend:
Capability in technology elements ●●●● most ●●● many ●● some ● at least one

Figure 12.1-1. Advanced Fabrication and Processing WTA Summary

12.3—METROLOGY

Highlights

- As critical dimensions of hardware become smaller, the positioning of the machining tool and the ability to measure that position become more important.
- The accuracy to which gears can be machined has a direct bearing on the noise and heat generated by the gears. Efforts to extend the metrology capability of gears to the 100-nm range should result in quieter and more efficient gear trains.
- The ability to operate high-quality coordinate measuring machines (CMMs) in the shop environment would result in a cost saving because the necessity of taking the part to a centrally located measuring room for high-precision dimensional measurements is very time consuming.

OVERVIEW

Metrology, in some form, has been used since the beginning of civilization. It took on additional importance during the Industrial Revolution when parts were no longer manufactured as one-of-a-kind. During the intervening years, metrology has advanced from measuring angles and lengths in inches and degrees to measurements in fractions of microns and arc-seconds. Modern metrology equipment includes gauge blocks, surface profilers, angular measuring equipment, laser-based measuring systems, coordinate measuring machines, and so forth.

Present developing technology can be separated into two parts: increased measurement capability and the use of advanced equipment in the shop working area (as opposed to operation only in controlled environments). This subsection addresses two technologies in the first part and one in the second. The following programs to increase measurement capability are listed: (1) to produce gear-measuring equipment that is not only faster than conventional equipment but will have measurement accuracies less than 1 μm and (2) to improve the capability of more accurately measuring the position of the spindle of a machine tool. Work is also being carried out to produce high-quality coordinate measuring machines that can withstand the environment of the shop floor and, thus, carry out inspections next to the production machines.

RATIONALE

Accurate dimensional inspection (metrology) is essential for designing, developing, manufacturing, and using most military and commercial hardware. In addition, meaningful interchangeability of parts requires that dimensional accuracy be maintained when replacement parts are manufactured. Until recently, it was satisfactory if these instruments had resolutions and/or accuracies in the 0.5- to 1.0-μm range. However, demands for ever-decreasing dimensions in products—whether for use in micromachines, MEMS, or semiconductor devices—have placed increasing demands on the metrology discipline. Considerable effort is being expended in what is sometimes called "micro-metrology." Linear measurements in the sub-100-nm range will assume greater importance as the capability of lithography is extended to that range.

Improvements in metrology should result not only in more reliable military hardware, but also in reduced costs. A universal agreement concerning the best techniques could make cost reductions significant.

WORLDWIDE TECHNOLOGY ASSESSMENT (See Figure 12.3-1)

Metrology is a critical technology in any industrialized country, and, as a result, most such countries have ongoing programs to improve metrology. The United States, Germany, and Japan appear to have the most extensive programs. However, Denmark, France, Italy, and the United Kingdom have substantial efforts, with Australia, the Netherlands, and Sweden having niche efforts.

Country	Machine Tool Metrology	Gear Metrology	Factory Floor CMMs
Austria			●
Denmark	●●	●●	
France	●●	●●	●●
Germany	●	●●	●●●
Italy	●●		
Japan	●●●●		●●●
Netherlands		●	
Sweden			●
Switzerland	●●		●
Taiwan	●		
UK	●●	●●	●
United States	●●●	●●●	●●●

Legend:

Capability in technology elements ●●●● most ●●● many ●● some ● at least one

Figure 12.3-1. Advanced Fabrication and Processing WTA Summary

12.4—NON-DESTRUCTIVE INSPECTION AND EVALUATION

Highlights

- The use of shearographic techniques should increase the technical capabilities for the detection of flaws.
- Smart materials could revolutionize the inspection and evaluation of hardware by incorporating within the hardware sensors that could detect flaws.
- The application of data fusion technology to NDE should reduce the time/cost to perform more advanced analyses.

OVERVIEW

NDE encompasses a wide range of disciplines, including eddy current, magnetic testing, penetrant testing, radiographic testing, and ultrasonic testing. Each of these techniques is used in several different applications; however, they all have one thing in common: They are used for the detection of defects after an item or structure has been manufactured and requires periodic inspection. All these techniques have been used for quite a few years, and only minor changes have been incorporated into the procedures. Most changes have resulted from improvements in detection equipment. In this subsection, we describe three new NDE techniques: digital shearography, the use of smart materials, and data fusion.

Shearography is more traditional. It is a technique that is used to analyze finished product. In its simpler sense, shearography is an existing technology; however, it has several drawbacks. Digital shearography is an approach aimed at improving the technique. In short, shearography is a nonscanning, laser-based interferometry system used to detect areas of stress concentration caused by anomalies in materials. Digital shearography combines these techniques with charged-coupled device (CCD) cameras and computers to improve the image and allow analyses of the results. Digital shearography can examine substances as large as 1 ft square and is portable so inspection need not to be place in a laboratory.

Shearography has a major advantage over conventional non-destructive test (NDT) techniques (e.g., dye penetrant, magnetic particles, and radiography) used to detect flaws. The conventional techniques detect all visible flaws, but shearography, by examining not the flaws but the flaw-induced strain, provides more information on the criticality of the flaw. Potential military applications include the detection of subsurface flaws in aircraft panels and the detection of voids in composite materials.

The use of smart materials is a more radical approach because the sensors used to detect flaws in the structure are built into the structure and can be used to monitor quality/reliability either during manufacture or after the product is completed. The concept of smart materials is also an emerging technology in materials technology. In that application, the sensors detect some "designed-for" parameter and respond with a counteraction (e.g., damping engine vibrations, silencing refrigerators and aircraft cabin noise, and so forth).

Smart materials include technologies such as piezoelectric materials, shape memory alloys, and magnetostrictive fluids. The field of piezoelectric materials is the most advanced, and some commercial applications have been developed. However, widespread use of these technologies is still in the future. Potential civil and military applications abound, with civil applications including sensor determination of aging in building, bridges, and so forth; determination of maintenance requirements for engines, aircraft, and so forth; and chromogenic applications (e.g., self-dimming rear-view mirrors, architectural windows, and so forth). Potential military applications include smart skins for improved submarine signature control (cancel incoming acoustic radiation), improved helicopter rotor-blade trailing edge flaps, and aircraft wings that can change shape to minimize drag.

Data fusion is a technique that uses data from a group of NDE sensors rather than using data from a single sensor. It has potential application in a wide range of disciplines. In essence, data fusion is a technique in which data from multiple—and perhaps diverse—sensors are correlated into digitally formatted products. These products provide a user with complex information in a user-friendly format so that decisions can be made quickly and accurately.

RATIONALE

The optimal use of structural materials, as well as the introduction of high-performance systems for military applications, is dependent on the ability to detect and characterize strength-limiting flaws and defects. As military systems become more complex and, at the same time, military budgets decline, existing NDE techniques must be refined and new, more sensitive and less expensive techniques must be developed.

The technologies addressed in this section meet both of these requirements. They result not only in increased technical capabilities of the final product, but also in cost savings.

WORLDWIDE TECHNOLOGY ASSESSMENT (See Figure 12.4-1)

Countries with aerospace industries have shown a particular interest in digital shearography because of its potential in detecting flaws in the composite structures commonly used in aircraft structures. (To date, shearography has not detected disbonds in some composite materials.) The United States, France, Germany, and the United Kingdom appear to be the leading countries in shearography development. China, Italy, Portugal, Singapore, and Switzerland have lesser programs, although ETH-Zurich is a world-class facility.

Research in smart materials is one of the most intense research efforts worldwide. The United States, France, Germany, Japan, and the United Kingdom are the world leaders. Austria, China, Italy, and the Netherlands are also pursuing research in this area.

Country	Shearography	NDE with Smart Materials	NDE Data Fusion
Austria		●	
Belgium			●
Denmark			●
France	●●	●●●●	●●●
Germany	●●	●●●●	●●
Hong Kong (PRC)	●		
Italy	●●●	●	
Japan		●●●●	●●
Korea		●	
Netherlands	●	●	●●
Singapore	●		
Switzerland	●		●
UK	●●	●●●	●●●
United States	●●●●	●●●●	●●●●

Legend:
Capability in technology elements ●●●● most ●●● many ●● some ● at least one

Figure 12.4-1. Non-destructive Inspection and Evaluation WTA Summary

12.5—PRODUCTION EQUIPMENT

Highlights

- The development of spindles capable of operating at higher speeds and power would reduce the cost of machining.
- Improved grinding machines would result in more powerful jet and diesel engines and more accurate guidance systems.
- The continued development of parallel kinematic machine tools may result in improved machining capabilities and reduced cost of the machining operation.
- The use of MEMS technologies to improve military hardware and to reduce costs may be one of the most significant advances in developing superior and less expensive military hardware.
- Improved monitoring of machines and tools should reduce the cost of military hardware.

OVERVIEW

This subsection addresses various developing production technologies that either will provide superior final product (military and/or commercial product) or will reduce the costs of manufacturing existing product. Included in the subsection are several technological advances that may improve the final product and several modifications that may result in decreased costs. The former includes the following technologies:

- Improved grinding machines that could result in product manufactured to closer tolerance
- Machine tools based on the parallel kinematic concept (e.g., Stewart platform or hexapod)
- Extremely small machine tools (micromachines) that can be used to manufacture miniature hardware
- MEMS, a revolutionary technology or manufacturing miniature hardware that combines both electronic and mechanical on a monolithic substrate.

The Stewart platform is the basis of the hexapod milling machine design. The design also has application in flight simulators, telescope mounts, and space-borne laser reflectors. In the machine tool arena, particular attention is being given to its development as a tool for manufacturing nanometer-size items (nanotechnology).

Hexapods are based on the Stewart platform. Companies in numerous countries are investigating the use of hexapods as machine tools. Indeed, initial designs have been manufactured and are in the process of being evaluated. While some limited use has been found for the equipment, sufficient problems still limit their use at the present time. The main problem is that the positioning accuracies are on the order of five times worse than that of the best milling machines (although comparable to many existing machines). Such equipment might find a niche in the manufacture of micro- or nano-scale products.

Tetrahedral tripods are a third class of parallel kinematic machines. Efforts are still in the research stage, but such equipment may avoid some of the problems inherent in Stewart-platform-based equipment.

Micromachines are being pursued primarily in Japan, coordinated by the Micromachines Center and supported by MITI. The goals of this program are to manufacture minute items, such as invasive medical micromachines and maintenance systems for power plants. The technology, although different from MEMs, has the same general goals as the MEMS concept.

MEMS uses the techniques developed for the manufacture of integrated circuits to fabricate devices that incorporate, on a semiconductor substrate, not only integrated circuits, but also mechanical structures that can be used for a range of applications. MEMS technology includes three important characteristics: miniaturization, multiplicity, and microelectronics. By using semiconductor device manufacturing technology, these packaged

MEMS-based systems are very small (ranging from 1 mm to 2 cm in size), can be manufactured in quantity using conventional pholithographic techniques, and incorporate complex electronic circuitry.

Until the advent of MEMS, miniaturization was a concept most often associated with electronics. The continual miniaturization of electronic systems has been one of the pillars of the advanced technologies used in military and civil hardware. The decrease in size, coupled with the increase in processing power, of computers is a prime example of the advances in electronic technology. In MEMS technology, actuators, sensors, and medical devices incorporating mechanical structures (e.g., levers, springs, motion sensors, and so forth) and integrated circuits are manufactured on semiconductor substrates, thus using the miniaturization capabilities of integrated circuit technology. The use of semiconductor manufacturing technology also results in the batch processing inherent in photolithographic-based semiconductor processing and makes it possible to fabricate thousands of components as easily as a single component. As a result, the cost of the components is reduced significantly, thus extending the use of the products to a wider range of applications. The incorporation of these mechanical structures either into microcircuit chips or on associated substrates (hybrid device) provides the intelligence to the devices that allow their use without the cumbersome external components that were often required to interconnect discrete components.

Technologies that may result in decreased costs include:

- Tool monitoring and sensing, which may result in more cost effective technique for determining tool life
- Machine degradation monitoring, which may result in a more cost-effective technique for determining machine tool maintenance
- High-speed spindles, which will decrease the time required to machine many parts, thus reducing the final cost
- First-part-correct programs, which would reduce the cost of having to make two or three items before the design/manufacturing process was corrected to produce a product that met design specifications.

RATIONALE

Numerically controlled (NC) machine tools find wide application in the defense and aerospace industries. Their ability to produce complex shapes with minimum setup time and fixturing makes them critical to the batch processing of components for modern weapons systems. While it is difficult to determine the results of developing technologies (some may never reach fruition), any manufacturing technologies that improve the characteristics of military hardware and/or reduce the cost of that hardware are of significant military importance. For example, improved grinding machines could result in more powerful and efficient jet engines and more accurate guidance systems. While the future of hexapods and micromachines is not clear, significant effort is being expended in these technologies. The use of improved hexapods may result in superior product and/or less expensive product or it may find use in complex parts. The significant interest in the use of the Stewart platform as the "motion system" for sensitive telescopes should also be noted. Micromachines (primarily pursued by Japan) is envisioned as a technology that may find use in miniature medical and industrial products. Success in these applications would certainly lead to applications in military hardware. MEMS, while included here as a developing technology, is, at the same time, a technology that has found present applications in several commercial applications (e.g., accelerometers for automobile airbag deployment). The future applications of this technology appear to be extensive, and few doubt that the military applications will be significant.

WORLDWIDE TECHNOLOGY ASSESSMENT (See Figure 12.5-1)

Several countries possess advanced machine tool technology, whether in grinders, milling machines, or turning machines, and all are pursuing continued improvements. Efforts to achieve high-speed spindles with large force and feed-rate capabilities, as well as the capability to monitor tool and machine tool condition, has been a continuing effort by many universities and machine tool companies. A large international consortium, IMS-SIMON, has been formed to address the problem of sensor monitoring of machine tools and cutting tools. The consortium includes companies from the United States, Canada, Germany, Italy, Japan, Spain, and Switzerland. Research in parallel kinematic machine tools is extensive in most of the countries with advanced machine tool industries. The United States, Germany, Japan, Sweden, and Switzerland appear to be the most involved.

Country	Manufacturing	Parallel Kinematic Machines	MEMS
Argentina	●●		
Brazil	●●		
Canada	●●		●●
Czech Republic	●●		
China	●●		
France	●●●		
Germany	●●●●	●●●	●●●
India	●		
Israel	●●		
Italy	●●●		
Japan	●●●●	●	●●●●
Netherlands			●
Norway		●	●
Romania	●	●	
Russia	●●	●	
Spain	●●●		
Sweden	●●	●	●
Switzerland	●●●●		
Taiwan	●●	●●●	
UK	●●	●	●●●
United States	●●●●	●●●	●●●●

Legend:
Capability in technology elements ●●●● most ●●● many ●● some ● at least one

Figure 12.5-1. Production Equipment WTA Summary

MEMS research is extensive throughout the industrialized world. While the United States, Germany, and Japan appear to have the most extensive programs, France, the Netherlands, Norway, Sweden and the United Kingdom also have advanced programs.

12.6—ROBOTICS

Highlights

- Advanced sea, air, and land robots will carry out a wide range of military operations, ranging from chemical and biological warfare (CBW) detection and reconnaissance to mine detection.
- The development of fractal, self-duplicating, self-repairing robots would be a significant breakthrough in the design of military hardware.

OVERVIEW

This section covers the developing technology for advanced battlefield robots, encompassing robots for land, sea, and air. Possible robots include:

- **Ground robots.** Although these robots could be designed for many different applications, early planning has addressed the fields of cameras, sensors, or equipment for gathering and disseminating information.
- **Seaborne robots that resemble jet skis.** These robots can be equipped with a various types of equipment, including underwater cameras, laser scanners, night vision equipment, radar, and so forth.
- **Micro and mini air vehicles (MAVs).** Proposed designs of these pilotless machines vary from micro (smaller than 6 in.) to extremely small (8- to 10-in. size) to moderate (on the order of 3- to 5-ft size). Initial plans would be for the craft to carry some type of sensor, although they could also carry small amounts of explosives.
- **Fractal shape-changing robots.** These robots can, theoretically, change shape, replicate, self-repair, and perform a wide variety of military tasks.

Considerable effort is being expended to develop MAVs that can serve as observation planes or sensors for WMD. Such a concept will most probably make use of MEMS technology. To be effective, the MAVs would require sensors and transmitters. Classical sensor development has been a major part of MEMS research, and engineers at MIT have developed a TV camera on a chip. Some of the major obstacles are the aerodynamics of such a small plane and the weight requirements of sensors, transmitters, and propellant systems.

The fractal shape-changing robots are a form of nanotechnology and are still in the conceptual stage. However, interest has been increasing since the inventor was awarded the 1996 European Invention Competition of Monaco. In addition, statements by personnel at British Telecomm predict, in their "1997 Technology Calendar," that early forms of these robots are most likely to exist by 2005. The concept of self-replication is a basic tenet of nanotechnology, whether one considers classic nanotechnology or molecular nanotechnology. The goals of nanotechnology are low cost, molecular precision, and flexibility. Some authors believe these can only be reached with the development of self-replicating objects.

RATIONALE

Robots are indispensable in many hazardous military operations, including the handling of munitions, operating in highly radioactive or electromagnetic pulse (EMP) environments, and performing tasks in space. The use of robots in these applications extends the military capability much further than that which could be accomplished with "protected" humans. The robots addressed in this "developing technology" volume are extensions of the existing robotic technologies. The more typical sea, air, and land robots addressed are significant advances over what the military presently fields. They should extend the observation/reconnaissance capabilities of existing technologies not only because of their additional capabilities, but because their size should limit their "detectability" and extend their ability to gain access.

The eventual applications of the less conventional fractal shape-changing robots are more difficult to quantify. Theory says that they will be very small, almost indestructible (since they are self-repairing); can change shape, if necessary, to gain access to small or odd-shaped openings; and can carry out most operations presently performed by soldiers in the field.

WORLDWIDE TECHNOLOGY ASSESSMENT (See Figure 12.6-1)

Robotic research is a major international endeavor. Although most research is carried out by commercial/ university laboratories and most is directed toward civil applications, much of the technology is also applicable to military applications. Military applications include land-based robots (many designed for mine clearing), undersea robots, and unmanned aerial vehicles (UAVs).

Country	Land Robotics	Sea Robotics	Unmanned Air Vehicles	Fractal Robots
Australia		●●		
Belgium	●			
France	●●●●	●●●		
Germany	●●●●	●●●●	●●	
Italy		●		
Japan	●●●●	●●●●		
Russia			●	
Singapore		●		
South Africa	●			
Sweden		●		
UK		●●●●	●●	●●
United States	●●●●	●●●●	●●●●	●●

Legend:
Capability in technology elements ●●●● most ●●● many ●● some ● at least one

Figure 12.6.-1. Robotics Systems WTA Summary

POSITIONING, NAVIGATION AND TIME CRITICAL TECHNOLOGY

Arnold Gieger
(Editor)

Highlights

- Positioning, navigation, and time (PNT) technology usage has doubled every 5 years, mostly because of the U.S. Global Positioning System (GPS) program and the miniaturization of electromechanical components. Future PNT usage is expected to double every 2 years because of telecommunication and automobile navigation commercial markets.
- The economic engine for PNT is both the nonmilitary commercial community sector and the expanding need for more accurate position and especially precise time.
- Military exploitation and harnessing of a three-dimensional position (latitude, longitude, and altitude) and precise time (POSITIME) common battlespace grid reference and use of hybrid multisensor arrays are in the embryonic stage. The impact on the military in terms of situational awareness—that is, the use of multiple sensor data to reduce fratricide and positively identify friendly forces, foe targets, and neutrals—will be significant.
- Significant advances in PNT technologies should be anticipated from developed nations and less developed nations. This will allow more nontraditional sources in manufacture of PNT products.

OVERVIEW

Recent localized conflicts and the evolving role of the U.S. military in antiterrorism points to (1) the future use of more autonomous unmanned vehicles (AUVs) for precision strike and tactical combat, in addition to surveillance, targeting, and covert operations, and (2) more use of long-range, stand-off, precision and laser-guided weapons and other smart weapons, including artillery shells and munitions. Joint U.S. and allied military land, air, and sea forces will also need new levels of situational awareness to reduce fratricide and provide rapid battlefield tracking of troops, both friend and foe. The need to minimize time over target and maximize kill will require accurate and continuous knowledge and location of targets, especially mobile targets, rapid battlefield damage assessment, and in-flight retargeting of missiles and weapons. These capabilities will translate directly into tactical and strategic military advantage, saved lives, and reduced military cost. To achieve this will require using advanced technologies in computer processing, navigation and precise time, telecommunications, and positive combat identification sensors in a "system of systems." This will provide the capability to determine accurate locations of friendly and enemy forces, as well as to collect, process, and distribute relevant data that is position and time (POSITIME) tagged across the battlespace. This interactive "tactical picture" will gain U.S. forces an enormous dominant battlespace awareness, which will decrease response time and make the battlespace considerably transparent for the warfighter.[1] Included in this section are descriptions of the technologies necessary to achieve dominant battlespace awareness, including autonomous and cooperative positioning, data-based navigation systems, positive combat identification, and nonintrusive detection of military force elements. Figures 16.0-1 and 16.0-2 show battlespace awareness for two warfare scenarios: sophisticated conventional warfare and military operations in urban terrain, respectively. The latter is the more complex and resource-intensive environment; it requires very precise situational awareness.

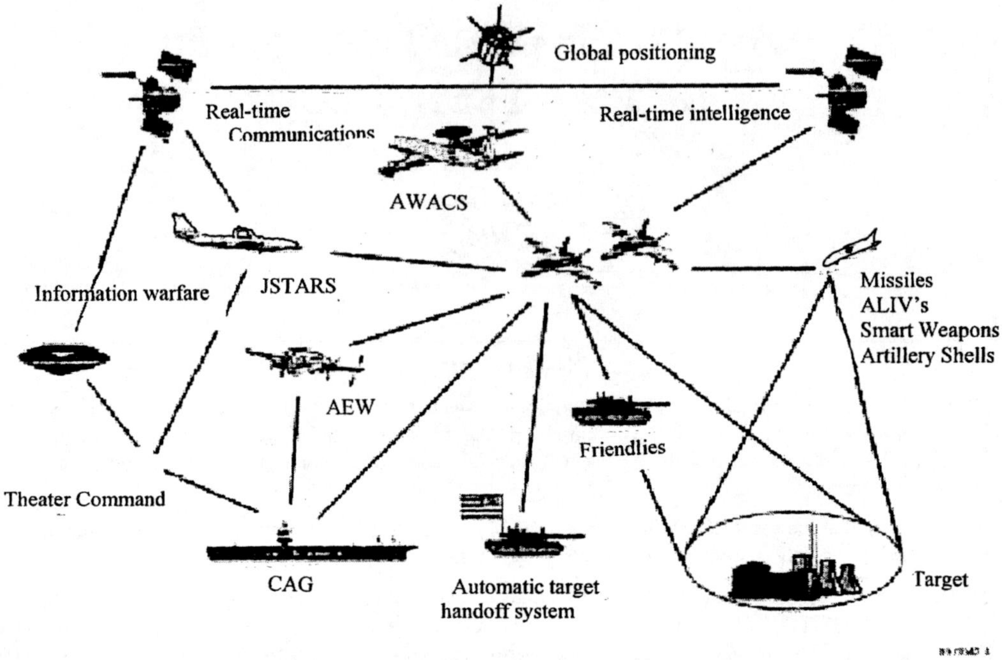

Figure 16.0-1. Concept—Sophisticated Conventional Warfare

RATIONALE

The importance of these technologies and their potential capabilities could be used (by friend and foe) to deliver conventional weapons, a single item (for nuclear), individually targeted remote vehicles (RVs), or a set of items to a level of accuracy appropriate to the destructive footprint of these munitions. These parameters could be achieved either individually or when used as an integrated or hybrid system. Accurate positioning, attitude, pointing, and control of land, sea, air, and space vehicles are essential for effective coordination of highly mobile military forces. These capabilities directly enhance the delivery accuracy and lethality of manned and unmanned guided and unguided weapons systems. In addition, other mission requirements, such as reconnaissance and detection, require accurate velocity, motion compensation, and positioning synchronization data to maintain real-time knowledge of the enemy. Accurate POSITIME sequencing and use of sensor information are key force multipliers to offset numerical superiority of an adversary and reduce casualties. Access to these technologies would amplify threats to regional stability by making available to hostile forces a much superior capability.

---

[1] *Joint Vision 2010.*

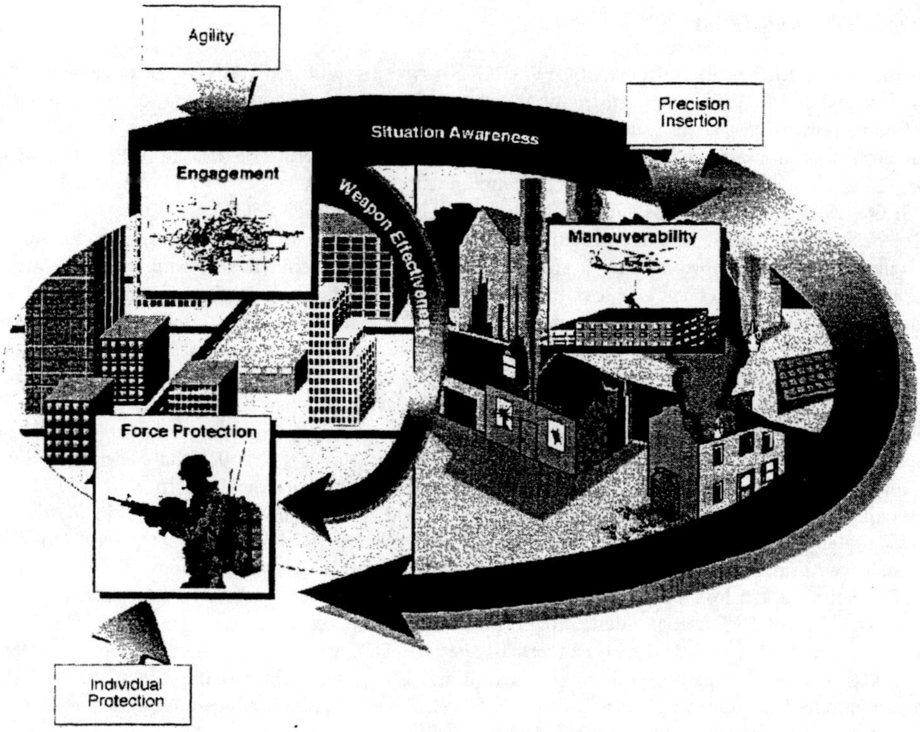

Figure 16.0-2. Concept—Military Operations in Urban Terrain

Situational awareness greatly reduces the combat stress of the warfighter. Situational awareness is a fundamental requirement for combat decision making and fratricide reduction. U.S. forces must be able to positively identify all targets in the battlespace for all combat mission areas—air to air, air to surface, surface to surface, and surface to air. Surface, in this case, includes land, sea, and subsurface. Combat identification (CID) is essential in order for commanders to effectively field fighting forces that can rapidly and positively identify enemies, friends, and neutrals in the battlespace; manage and control the battle area; optimally employ weapons and forces; and minimize the risk and occurrence of fratricide.[2]

The greatly expanded military use of the U.S. GPS has created the significant, military advantage of an accurate navigation and time reference system, but at the price that hostile forces will try to locally jam GPS signals or use GPS for their own purposes. The latter could also require the use of local jamming by friendly forces to deny use of GPS by an enemy. Either scenario will require U.S. forces and its allies to use GPS in a high jamming environment. Developing technologies for GPS antijam improvements will mitigate this issue until new developing technologies provide DoD with an affordable, nonjammable, equivalent GPS precision navigation and precise time reference for the battlespace. In urban areas, loss of GPS signals due to signal blockage and multipath problems provides a challenge to overcome.

---

2 *Joint Warfighting Science and Technology Plan: Combat Identification.*

TECHNOLOGY ASSESSMENT

Over the next 2 to 5 years, differential GPS (DGPS) systems will provide very accurate navigation within 1–2 m in localized areas. Wide-area augmentation systems (WAAS) for air traffic control purposes will provide differential corrections over a wider area than most DGPS. Local area augmentation system (LAAS) will provide differential corrections for air traffic control in the vicinity of an airport. The use of GPS, long-range aid to navigation (LORAN), and other navigation/telecommunication systems [(Russian) Global Navigation Satellite System (GLONASS), European Geostationary Navigation Overlay Service (EGNOS), European Union GNSS-2, Iridium, Teledesic, and other satellite telecommunication systems] will provide greater accuracy and minimize loss of satellite signals. Continued improvements in orbit accuracy and time could further improve GPS accuracy from 2–3 m to 1.5 m and DGPS accuracy to less than 0.6 m. However, radio navigation systems are not completely autonomous, relying on external signals that can be jammed. The military application of geodetic and geophysical databases, such as terrain, magnetic, and gravity matching techniques, offers the possibility of a more autonomous navigation capability, reducing the concern for detection and jamming of GPS and other radio aids. Such geodetic and geophysical data bases can support navigation to near-GPS accuracy when GPS service is not available (see subsection 16.3). Commercial telecommunications growth will continue, providing affordable hand-held cellular communications worldwide, including GPS/digital maps at the local area. By 2001, the Federal Communications Commission (FCC) has mandated that all cellular phones identify their locations to within 125 m for 911 emergency calls. This will significantly increase PNT usage and research investments. Hybrid inertial navigation system (INS) systems, such as GPS combined with ring laser gyroscope (RLG) or fiber-optic gyroscope (FOG) INS, will be expanded with more combinations of navigation and telecommunication functions. This will minimize the effect of GPS jamming and loss of satellite signals due to terrain or other reasons. Miniaturized GPS (GPS on a chip) will hasten hybrid INS/telecommunication applications at reduced cost (see subsection 16.1.). Maximum utilization of near and far infrared devices (i.e., thermal tapes, BUDD, and Phoenix Light) and image intensifier (I2) devices across U.S. forces, together with increased training, can provide limited improvements in CID. Further integration among the U.S. forces and its allies of new CID technologies, both cooperative and noncooperative, will present a challenge to ensure that they are reliable and affordable.

Over the next 5 to 15 years INS could be revolutionized by microelectromechanical systems (MEMS) technology, a fabrication approach that conveys the advantages of miniaturization, multiple components, and microelectronics to the design and construction of integrated electromechanical systems. Current MEMS tuning fork gyroscopes, smaller than a paper clip, have an accuracy of 100 deg/hr, with 10 deg/hr accuracy expected by the end of 1999. Given similar experiences over the past 20 years with RLGs and FOGs, it is very likely that MEMS gyroscopes with accuracy of 0.1 deg/hr are achievable within the next 5 to 15 years, with further improvements thereafter. Concurrently, RLG and FOG aircraft INS performance could continue to improve from 1.0 to 0.1 nmph circle error probable (CEP). MEMS technology could be further enhanced by the continued growth of optical computing/processing/correlating technologies (see Section 10), especially in georegistration of hybridized data from remote sensors (pseudo-imaging) using highly accurate timing data. The use of MEMS technology within RLG/FOG INS is very possible as part of the affordable INS technology evolution shown in subsection 16.1. Accurate and miniaturized "smart" gravity and magnetic detection sensors combined with GPS/INS and satellite communications will provide target recognition and real-time movement and direction of enemy forces and large mass items, such as missiles, tanks, and other large weapons. These same sensors combined with biological, chemical, and/or neutron emission sensors could help identify and track movement of biological, chemical, and nuclear emissions on the battlefield. Further cost reductions would allow localized battle areas to be "seeded" (by mini-AUVs) with these detection devices, providing critical intelligence of troop, missile, tank, and artillery movement. Accurate time, combined with multiple optical, vibration, magnetic, and gravity sensors in smart arrays, is a leveraging application of these technologies. The use of a standard time reference with ion clock technology will improve absolute accuracy to better than 10^{-15}. Further microminiaturization of low-power clock technologies will provide autonomous "flywheel" time devices in all navigation and communication equipment, providing direct-Y code acquisition and minimizing the effect of GPS jamming. These improvements will result in a precise standard military time reference for U.S. forces and its allies, providing enormous benefits in precision navigation and rapid and secure communications across the battlespace. Because of its importance, precise time and frequency (PT&F) capabilities are discussed in a separate subsection (16.5).

The use of GPS time synchronization to locate the position of 911 cellular users can be adapted by the military to improve situational awareness in urban terrain. Integrated with CID tags and localized terrain maps, mini-navigation/low probability of interception (LPI) telecommunication sensors may provide positive ID of friendly forces and thereby reduce the "fog of war" in urban terrain. Radio frequency identification (RFID) is a leveraging technology. Improvements in the land warrior system, the U.S. Army's first-generation integrated fighting system for dismounted combat soldiers, will integrate telecommunications and navigation technologies with an integrated helmet assembly subsystem including a heads-up display and image intensifier for night operations, a weapons subsystem with thermal weapon sight, close combat optics, video camera with a video capture capability, laser rangefinder/digital compass, and an infrared laser-aiming light.

Over the next 15 to 20 years nanoelectromechanical systems (NEMS) technology, a technique of building machines (such as accelerometers and gyroscopes) from individual atoms and molecules, could continue to decrease the size and cost of INS. The low manufacturing cost of this technology could further revolutionize navigation by providing an autonomous INS using multiple NEMS accelerometers (potentially eliminating the need for gyroscopes), at a cost of less than $500. For the military, this could affordably allow a nonjammable INS to be installed on every air, land, and sea vehicle; smart weapon; artillery shell; and warfighter. Integrated with assets, such as the Airborne Warning and Control System (AWACS), satellites, and AUVs, using advanced electro-optics, infrared radar, video cameras, motion-detection sensors, and sophisticated automatic target-identification algorithms and data bases, this technology will provide a very accurate view of the battlespace, particularly during military operations in urban terrain (MOUT) (see Figure 16.0-2). Expanded use of an affordable, accurate, and autonomous navigation system could better support precision insertion of combat forces and their protection, providing low fratricide and collateral damage, particularly in urban terrain. By 2025, friendly troops and equipment will enter the battlespace with their personal identifiers. The identification mechanisms could be in the form of microchips worn by or imbedded in the soldiers. The same principle could be applied to vehicles. NEMS technology promises to be among the most promising developments of the 21st century.

WORLDWIDE TECHNOLOGY ASSESSMENT

Depending on national desires and needs, nations have chosen to develop, produce, and use positioning, navigation, and PT&F technologies for military and commercial purposes. Many countries do not have an indigenous aircraft-, ship-, or spacecraft-manufacturing capability and the inherent guidance and navigation technology that are fundamental requirements. The cost for Third World countries to design, fabricate, and apply these technology areas has been a limiting proliferation factor. However, the latest transition of MEMS technology to low cost and highly accurate gyroscopes and accelerometers, coupled with advances in computer memory and speed, have significantly reduced the cost of INS. Historically, an INS was and continues to be more expensive to produce and maintain than a radio navigation system. This and other military mission considerations have led many nations to rely more on radio navigation systems for many military and commercial needs. Continuous gains in microminiaturization of sensors and computer technology will continue to drive down the cost of INS. As INS costs drop, the technology could become available to countries heretofore unable to afford this system, which has the military advantage of an autonomous, covert, and nonjammable guidance and navigation capability for aircraft, missiles, and weapons of mass destruction (WMD).

MEMS technology is already used by Japan, South Korea, and other countries for automotive airbag applications (accelerometers) and ride control and antiskid systems (gyroscopes and accelerometers). Current MEMS technology has been successful in producing accurate linear acceleration sensors, but is years away from providing an accuracy similar to a RLG INS. However, low-cost MEMS INS, compounded with the worldwide commercial use of global navigation satellite systems (GNSS) and its technology advances and the emergence of accurate terrain mapping, will allow for the proliferation and accessibility of highly accurate PNT systems, which will be a challenge to prevent. Currently, the only GNSS that exist are the U.S. GPS and the GLONASS. A regional GNSS, such as the European Union GNSS 2, is a strong possibility. The GNSS 2 may consist of 63 low-Earth-orbiting satellites, providing 5- to 10-m accuracy. The latest satellite-based cellular phone systems could also solve the geolocation problem. Increased proliferation of foreign GNSS capability could provide unbridled and accessible navigation accuracy that could enhance the delivery accuracy of Third World countries' missiles and WMD.

The radio navigation process supports a magnetic heading reference for the world. The expanding use of GPS has reversed this trend, and the future will see the conversion to true heading for geo-referencing as mapping and imaging technologies converge. Many countries are using the U.S. GPS to improve the accuracy of their weapons and the situational awareness of their operations forces. Jamming of GPS signals by the United States in the event of war would limit its usefulness to an adversary, but doing so would require the United States and its friendly forces to operate effectively in a high jamming environment. Those developing technologies such as accurate, small clocks that allow usability of GPS signals in a high jamming environment are considered critical for military use. Currently, U.S. military needs for crypto, secure telecommunications, and antijam communications are becoming increasingly dependent on the accurate time from GPS. Precise time is key to the future of navigation and communication systems. The need for an autonomous, nonjammable, common time reference source is discussed in subsection 16.5. Precise time technologies will provide accurate and stable time to be disseminated to military forces around the world. The United States has been the world leader in the development and utilization of precise time technology, including telecommunication, encryption, and data transfer.

Some countries, such as France, the UK, and Russia, have been leaders in positioning, including inertial, gravity, and magnetic sensors; radio navigation; and precise time because of their indigenous aircraft and spacecraft industry. These countries have developed their own technologies. For other countries, much of the technology capability has been obtained through U.S. licensing agreement transfers and foreign students attending U.S. universities. Now, however, inertial and GNSS technology symposia are held in Europe, Russia, and China with wide international attendance and paper presentations. Hybrid navigation system technology is also now a common topic at international navigation conferences, and the theory and practice of Kalman filters is well known. Continued cooperation with allies and international bodies is necessary to reduce the proliferation and use of accurate INS, hybrid INS, and GNSS capabilities for use by adversaries in missiles and WMD.

The development leaders of situational awareness/combat identification (SA/CID) technologies are for the most part located within the NATO countries: France, Germany, the UK, and the United States. Further details of the foreign technology assessment are provided in the respective subsections.

Country	Sec. 16.1 Inertial Navigation and Related Components	Sec. 16.2 Gravity Meters and Gravity Gradiometers	Sec. 16.3 Radio and Data-Based Referenced Navigation Systems
Australia	●●	●●●	●●
Austria	●	●●●	●
Brazil	●	●●●	●●
Canada	●●●●	●●●	●●●●
China	●●●	●●●●	●●●
Czech Republic	●	●●●	●
Finland	●	●●●	●●
France	●●●●	●●●●	●●●●
Germany	●●●●	●●●	●●●
Hungary	●	●●	●
India	●●●	●●	●●●
Israel	●●●●	●●	●●●●
Italy	●●●	●●●●	●●●
Japan	●●●	●●●●	●●●●
Netherlands	●●	●●	●●●
Poland	●	●●	●●
Romania	●	●●	●
Russia	●●●●	●●●●	●●●●
Slovak Republic	●	●●	●
South Africa	●●●	●	●●●
South Korea	●●	●	●●●
Spain	●●	●	●●
Sweden	●●	●●●	●●
Switzerland	●●	●●●	●●
Taiwan	●●	●	●●
UK	●●●●	●●●	●●●●
Ukraine	●●●●	●●	●●●●
United States	●●●●	●●●●	●●●●

Legend: Extensive R&D ●●●● Significant R&D ●●● Moderate R&D ●● Limited R&D ●

Figure 16.0-3. Positioning, Navigation, and Time WTA Summary (Continued)

Country	Sec. 16.4 Magnetometer and Magnetic Gradiometers	Sec. 16.5 Precision Time and Frequency	Sec. 16.6 Situational Awareness/ Combat Identification
Australia	●●	●●	●●●
Austria	●●	●●	●●
Brazil	●	●●	
Canada	●●●●	●●●	●●●
China	●●	●●●	●●
Czech Republic	●	●●	
Finland	●●	●●	●
France	●●●	●●●●	●●●●
Germany	●●●	●●●●	●●●●
Hungary	●●	●●	
India	●	●●	
Israel	●●	●●●	●●●
Italy	●●	●●●	●●●
Japan	●●●	●●●	●●●
Netherlands	●●	●●	
Poland	●	●●	
Romania	●	●●	
Russia	●●●●	●●●●	●●●
Slovak Republic	●	●●	
South Africa	●●	●●●	
South Korea	●●	●●●	●●
Spain	●	●●	
Sweden	●●	●●	●●
Switzerland	●●	●●●	●●
Taiwan	●	●●	
UK	●●●	●●●●	●●●●
Ukraine	●●	●●●	●●
United States	●●●●	●●●●	●●●●

Legend: Extensive R&D ●●●● Significant R&D ●●● Moderate R&D ●● Limited R&D ●

Figure 16.0-3. Positioning, Navigation, and Time WTA Summary (Cont'd)

SECTION 16.1—INERTIAL NAVIGATION SYSTEMS AND RELATED COMPONENTS

Highlights

- Inertial navigation technologies provide an autonomous, covert, and nonjammable three-dimensional position and velocity reference for land, sea, and space platforms, which will enhance the ability of the military to achieve mission goals.
- Major reduction in manufacturing complexity, size, and cost of INS will be realized by use of MEMS sensors, electronics, and radio-frequency (RF) interfaces. This will allow expanded military use of INS (for personnel, low- cost vehicles, smart artillery and ordnance, and AUVs) and expanded commercial applications, thereby providing a larger market for more nontraditional manufacturers of inertial navigation technology.
- Through better noise compensation techniques, RLG and FOG will continue to improve free inertial performance (1.0 to 0.1 nm/hr), providing better platform stability augmentation and weapon delivery.
- Future developments in nanotechnology, particularly NEMS accelerometers, may eliminate the need for gyroscopes if quantum noise measurement techniques are resolved. If achieved, this may exploit the use of NEMS within the basic building materials for tanks, vehicles, and uniforms, significantly improving situational awareness and reducing fratricide.
- Military application of INS with embedded GPS, LORAN, and data-based referenced navigation systems (DBRNS) will increase. These hybrid systems bound the time-dependent errors of the inertial gyroscopes. The resulting accuracy of position and time provides a more robust navigation system. Improvements in hybrid systems will lead to port-to-port submarine navigation without need to surface.
- Built-in redundancy through low cost, small size, lightweight, and highly reliable components will allow an affordable, throwaway logistics concept. This will enable a rapid affordable technology insertion of INS technology.

OVERVIEW

An INS is a self-contained, covert system that provides continuous estimates of some or all components of a vehicle state, such as position, velocity, acceleration, attitude, angular rate, and often guidance or steering inputs. The current major obstacle of more universal INS use is its loss of accuracy over time and high cost. Military applications include both strategic and tactical systems: missiles, AUVs, manned aircraft, satellites, aircraft carriers, submarines, surface ships, and land warfare. Targeting, surveillance, and command, control, and communications (C3) systems require high navigation accuracy. Figures 16.1-1 and 16.1-2 address the key gyroscope and accelerometer performance requirements for these military applications, as well the key commercial automotive market driver. INS technology has been enormously affected by advances in computer technology (memory and throughput), sensors, power quality, and electronics. Most current INS use optical gyroscopes: RLG or FOGs. RLG and FOG INS technology will continue to improve free inertial sensor performance from 1.0 nmph to less than 0.1 nmph, while decreasing costs.

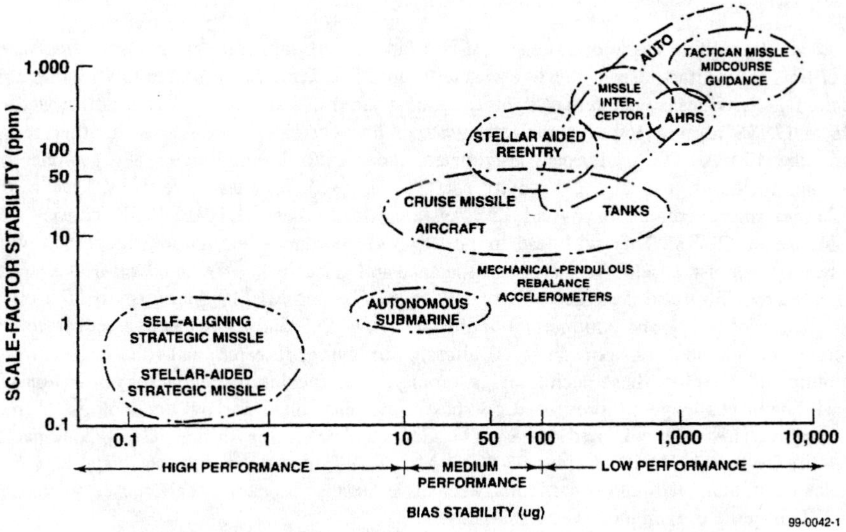

Figure 16.1-1. Accelerometer Technology Applications (shaded area is militarily critical region)

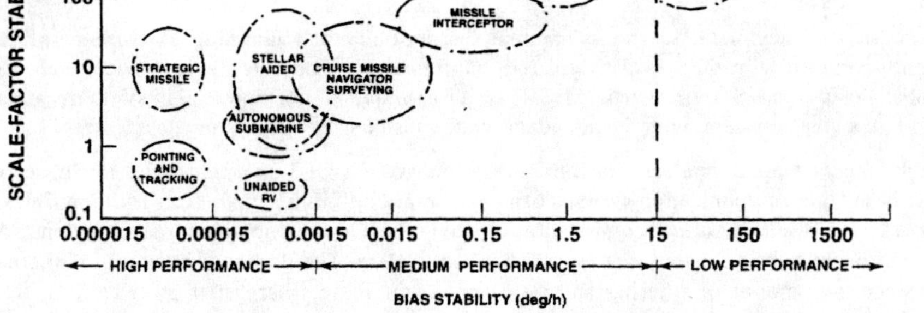

Figure 16.1-2. Gyroscope Technology Applications (shaded areas are militarily critical regions)

RATIONALE

INS technology (providing an autonomous, anti-jam, and covert capability of guidance and navigation) will continue to be critical for military use in the foreseeable future. The availability of the GNSS in a tightly coupled approach with the INS provides a system that is robust against jamming and satellite availability and more accurate than either INS or GNSS alone. However, the emergence of low cost INS, coupled with other sensors [forward-looking infrared radar (FLIR), LORAN, air data, gravimeters, radar and/or laser altimeter] and geo-referenced databases (digital terrain maps including magnetic and gravity data), will provide an alternative to GPS in a high jamming environment. Further improvements in hybrid INS systems, particularly DBRN, will increase covertness and accuracy without use of GNSS. This will lead to port-to-port submarine navigation capability without need to surface. These systems can also provide very accurate guidance and velocity data for an adversarial aircraft, missile, or WMD. The increased reliability and decreased cost of RLGs, FOGs, and MEMS gyros (gyro on a chip) will allow for an expanding list of military and commercial applications. An INS and its respective sensor components have wide applications in commercial transport and civil aircraft, surveying, research, and robotics. While this provides greater opportunities to develop these technologies rapidly, the transfer of technology for non-U.S. military advancements will be a challenge to prevent. Low cost, micromachined inertial technology is receiving wide commercial funding outside of the United States. The General Accounting Office (GAO)/National Security and International Affairs Division (NSIAD) Report 93-67 of March 1993 noted that "the technical data for nonmilitary INS is the same as for military INS and is particularly sensitive because it enables the licensee to manufacture all or part of the item." This quote is even more germane today.

Future key critical technologies are the emergence of low-cost, microminiaturized INS using MEMS (MEMS could revolutionize navigation). The commercial automotive markets are driving the MEMS technology development. Current MEMS gyroscopes are less than 5×1 cm, with an accuracy of 100 deg/hr at a cost of under $50. Industry expectations are to achieve 10 deg/hr by the end of 2000. Depending on military investments in that market, over the next 5–10 years MEMS-type gyroscopes could achieve tactical accuracy of 1.0 to 0.1 deg/hr.

Future miniaturization using NEMS sensors may be possible. The advantage would be the elimination of the gyroscope, using only accelerometers for sensing linear and rotational acceleration in a 360-deg cluster per axis. The issue is the sensitivity of the accelerometer to detect Earth's gravity because of the sensor's small mass and ability to detect quantum noise levels. Currently, the NEMS market driver is focused on medical commercial applications. This technology has been included because of its potential significant military benefits of reduced of INS cost and size. NEMS technology will lag MEMS technology by 10 years. Figure 16.1-3 shows the INS technology trends and costs projections over the next 10–20 years across multiple INS users. Figure 16.1-3 also shows that more sensor hybridization will occur over the next 5–10 years as GPS and other telecommunication functions are tightly coupled and integrated with INS. This massive production base, as well as the low cost of MEMS and NEMS sensors, could significantly reduce the cost of many military INS to less than $500. As the cost of INS decreases, their use in commercial applications will increase dramatically, especially in active robotic control. Current INS commercial applications already include stabilization of cameras (analogous to weapon sight stabilization) and automotive ride and stability control (analogous to turret stabilization). Foreign availability is increasing rapidly as the need for specialized manufacturing equipment and facilities decreases. Continued decreases in the cost and size of INS technology will affordably allow autonomous, non-jammable, and accurate INS to be installed on most DoD assets and even combat personnel, thereby providing a common coordinated reference system for the military battlespace.

The military dependency on GPS, with its inherent susceptibility in a jamming environment, has led to more use of INS tightly coupled with GPS in a hybrid INS configuration. The current tri-Service, embedded GPS/INS [embedded Global Positioning System inertial (EGI)] is an example of the movement of these separate systems (GPS and INS) into a single system, with the attendant reductions in size, weight, and cost.

Future technology advances in electronic miniaturization, such as GPS on a chip (refer to Subsection 16.3), as well as satellite-based telecommunication systems using trilateration timing signals (i.e., Iridium, Celestri, Teledesic and 911 Cellular), will result in further combination of navigation and communications functions. A "seamless" navigation system using hybrid sensors is a must for urban warfare. The degree of coupling of these external and internal sources and the amount of filtering and state vectors in these filters all play a role in determining the accuracy of the resultant hybrid system. Hybrid navigation system technology is now a common topic at international navigation conferences, and the theory and practice of Kalman filters is well known throughout the world. It is possible to procure simulators from a wide range of commercial sources, and the algorithms are published in textbooks and journals. Use of multi-MEMS INS will provide greater accuracy, survivability, and fault tolerance for an autonomous military capability.

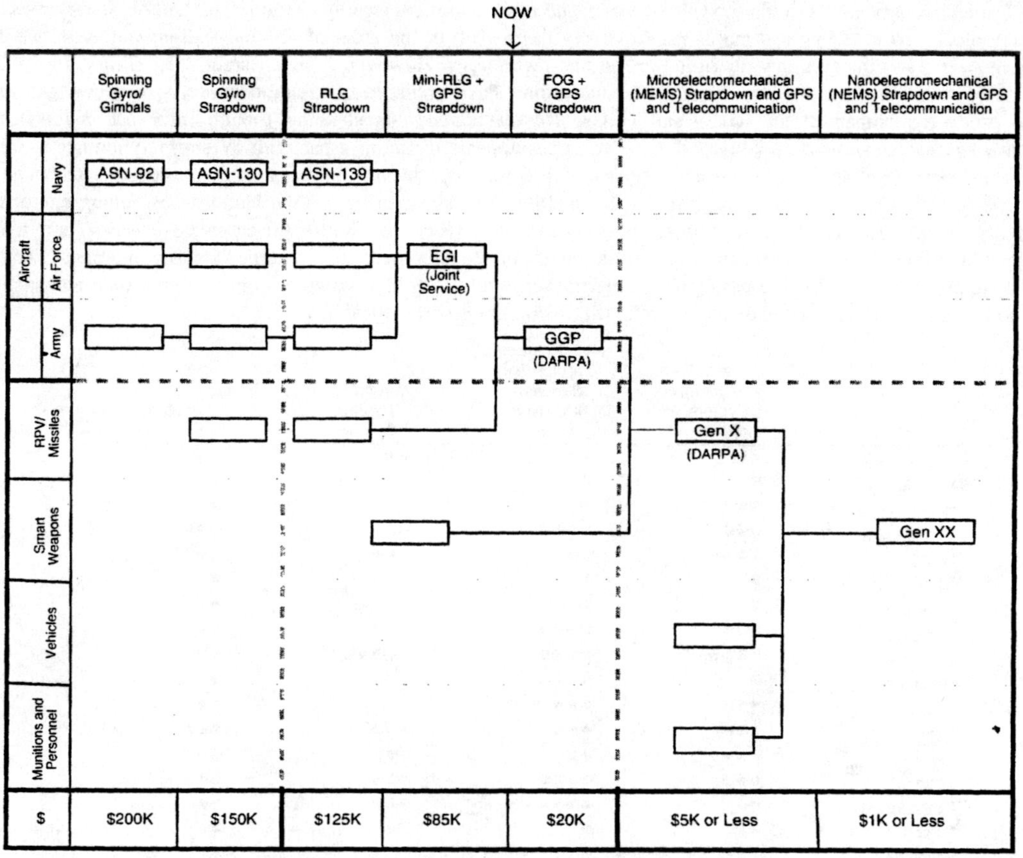

Figure 16.1-3. INS Technology Transition Trends[1]

Gyro astro-tracking devices could enhance navigation accuracy, where terrain and gravity data for geo-mapping may be difficult to obtain from certain areas and when GPS is unavailable. In addition, position errors increase for high altitude and space applications. An alternative approach to autonomous navigation utilizes an INS platform tightly coupled with a star tracker to bound the position error. Star-tracking-aided INS is particularly applicable to high-altitude and space applications. For example, a 1-nm/hr INS corrected by means of Kalman filtering with continuous star fix information can limit the position error on the order of 30 m. The tracker pointing errors, gravity compensation errors, and accelerometer instabilities are the dominant error sources. To achieve this position accuracy, tracker measurements and gravity and accelerometer compensations need to be accurate to approximately 1.0 arc sec each. These accuracies are available today at very high cost. Use of new technology trackers utilizing electro-formed optics on nickel bases with micro-precision sensor arrays, as well as improved processors, gravity

---

[1] Excludes ship/submarine INS technology evolution because of its unique performance requirements.

compensations, and future generation strapdown inertial instruments, should significantly reduce cost, size, and weight.

WORLDWIDE TECHNOLOGY ASSESSMENT

The technology gap between the United States and other nations is rapidly closing. The United States leads the world in most inertial technology and is progressively improving in the areas of accuracy, alignment, size, weight, reliability, cost, and integration with digital processing technology. However, France, Israel, UK, China, Japan, and Russia are rapidly closing the gap. Russia and the China have produced conventional inertial products and have initial production capability in RLGs and FOGs. Russia has developed some flexure rotor and magnetically suspended gyroscopes, and the quality of these gyroscopes appears to be on a par with Western equipment. Because tuned-rotor gyroscopes are inexpensive and very suitable for a space reentry guidance application, the acquisition of this type of technology has enhanced gyroscopic capability in maneuvering UAVs. Nations developing an inertial capability include Australia, Brazil, China, India, Israel, Italy, Sweden, South Africa, and Norway. Their major obstacles are access to a market of sufficient size to justify the development costs and the capital equipment costs for high-volume production. The increasing use of inertial sensors in very high volume, nontraditional markets, such as automotive and smart shells, has the potential to tilt this paradigm dramatically.

Country	Inertial Navigation System (INS)	Hybrid Inertial Navigation Systems (including GNSS)	Gyro Astro Tracking Systems	Inertial Sensors
Australia	●●	●●	●●	●●
Austria	●	●		●
Brazil	●●	●		●
Canada	●●●	●●●●	●●●	●●●
China	●●●	●●●	●●●	●●●
Czech Republic	●	●		●
Finland	●	●		●
France	●●●●	●●●●	●●●●	●●●●
Germany	●●●●	●●●●	●●●●	●●●●
Hungary	●	●		●
India	●●●	●●●	●●	●●
Israel	●●●●	●●●	●●●	●●●
Italy	●●●●	●●●	●●	●●
Japan	●●●	●●●●	●●●	●●●
Netherlands	●●	●●●	●●●	●●
Poland	●	●		●
Romania	●	●		●
Russia	●●●●	●●●●	●●●●	●●●●
Slovak Republic	●	●		●
South Africa	●●●	●●●		●●●
South Korea	●●	●●		●●
Spain	●●	●●		●●
Sweden	●●	●●		●●
Switzerland	●●●	●●	●●●	●●
Taiwan	●●	●●		●●
UK	●●●●	●●●	●●●	●●●
Ukraine	●●●●	●●●●	●●●●	●●●
United States	●●●●	●●●●	●●●●	●●●●

Legend: Extensive R&D ●●●● Significant R&D ●●● Moderate R&D ●● Limited R&D ●

Figure 16.1-4. INS and Related Components WTA Summary

SECTION 16.2—GRAVITY METERS AND GRAVITY GRADIOMETERS

Highlights

- Gravity sensor arrays will be more viable due to accurate time sequencing, computer speed, and memory advances, providing increased detection and location of submarines, mines, and mobile missiles.
- Uncompensated gravity disturbances are a large error source for INS initialization and subsequent field operation. Future gravity models will enable more accurate INS compensation.
- Use of a worldwide gravity data base based on better instrumentation and storage/access capabilities, in conjunction with on-board gravity sensors, will provide autonomous and continuous updates to INS, yielding comparable accuracy with projected INS/GPS hybrid systems.
- A developing technology to compute real-time gravity data from a moving platform may use the difference in acceleration data from an uncompensated INS and the GNSS.

OVERVIEW

This evolving and developing technology is used to measure a body's gravity field (such as Earth's), which in turn has applications for detection and localization of mass distributions, covert position determination, and inertial navigation compensation. Increasingly, gravity data will play a major and critical role in future navigation systems. Accurate geodetic and geophysical data (G&G) can improve the performance of inertial navigation systems to what may be near-GPS accuracy. G&G-enhanced INS could prove to be a significant navigation asset when GPS is not available. Specifically, future uses of G&G data will improve military weapon accuracy and increase safety of military and civilian flight by:

- Improving accuracy of navigation subsystems and stand-alone INS
- Providing accurate navigation in a hostile environment when GPS is denied
- Enabling covert terrain-following and terrain-avoidance systems
- Providing accurate attitude control of AUVs to allow for geodetic quality imagery
- Detecting underground manmade and natural structures and mass differences.

Present aircraft and ship INS use coarse models of Earth's gravity to correct for the sensed acceleration of gravity by the system sensors. Gravity anomalies that are not modeled are a major error source and limit the dynamic performance of deployed INS. Gravity meters and gravity gradiometers are used in static or mobile modes to measure gravity disturbances, deflections of the vertical, and to characterize the three-dimensional gravity vector.

G&G data is used to ground align and provide real-time, in-flight updates of local gravity to navigation systems, and these data are critical to both future precision engagement and safety of navigation requirements. Commercially, gravity meters and gradiometers are used to assist in exploration for oil, gas, or minerals by measuring the variations in the magnitude of the gravity vector or the variation in the gravity gradients. Furthermore, G&G data can be assimilated into "gravity maps" in support of data-based, POSITIME-referenced navigation systems (see Section 16.3).

Other uses include tunnel detection, buried material detection, and in arms control regimes, cargo identification, and weigh-in-motion. Covert detection and intrusion classification capabilities are militarily significant, especially for monitoring "secure" urban areas. International cooperative efforts through the International Association of Geodesy (IAG) exist for comparing absolute gravity standards.

RATIONALE

Moving-base gravity-meter accuracy is essential to generate maps for autonomous siting, INS initialization, and en-route compensation for in-flight gravity effects for WMD delivery systems, such as ballistic missiles and other long-range, unaided, inertial-guided applications including aircraft, AUVs, cruise missiles, and submarines. Applications have been developed to correct, either indirectly or directly, for the local gravity disturbances or vertical deflection on a moving base/platform. The uncompensated vertical deflection is the largest error in many INS scenarios. The indirect—and most common—compensation technique uses vertical deflection map data computed from gravity-meter surveys. The direct method uses a gravity gradiometer for real-time compensation of the vertical deflection. In the latter mode, the spatial gravity gradients are multiplied (scalar product) by the velocity vector and integrated to obtain the vertical deflection in real time.

Another application for this technology is to use the gravity meter or gravity gradiometer in a map-matching mode for accurate position determination using previously surveyed map data. Due to roll off of the high frequencies in the gravity anomaly field with altitude, these accuracies could only be obtained at low altitude, if at all. Navigation at GPS accuracy through turns and acceleration is questionable, but could potentially be recovered afterward. These map-matching techniques using sensors giving data only along the flight path are questionable for long-term navigation, as there are likely to be areas where the data does not have adequate spatial variance to achieve these accuracies. In a local area, with proper conditions for the sensor type being used, these POSITIME methods might be useful. Gravity gradiometers have a higher military value than gravity meters, since they have the ability to estimate vertical deflection in real time; however, gravity meters for airborne and marine applications are still of military importance. Gravity meters and gravity gradiometers require stabilization and the associated software to maintain a stable reference frame. The resulting hybrid system has the potential to provide the military with a non-emanating, nonjammable, totally covert system that can be used worldwide for navigation. As noted, the system will require previously surveyed gravity map data as well as a sufficiently distinct gravity signature that can be detected in the background noise. When there is not an adequate signature, the system maybe augmented with magnetic signature map matching (see Section 16.4). Another developing method of determining gravity is by computing the acceleration difference between an uncompensated INS and a GNSS. As measured by a GNSS, the computed acceleration of the platform is the pure acceleration of the vehicle, while the acceleration measured by an uncompensated INS contains the gravity vector and the platform acceleration vector. The difference is the gravity vector. Advanced filtering using optical correlators/processors and GNSS time synchronization are enabling technologies to obtain the gravity vector.

WORLDWIDE TECHNOLOGY ASSESSMENT

State of the art in mass-produced gravity meters (gravimeters) is at the 5- to 10-μgal level. A specially configured LaCoste Romberg unit has yielded gravity measurements at precision and accuracy levels of 1-μgal. A few countries have elected to be dominant in gravity meter and gravity gradiometer technologies, while others are not active in the technology because of economic considerations. Gravity meters (static mode) better than 10 μgals are manufactured and used in the United States. Canada, Germany, UK, Japan, China, Russia, and the Ukraine have developed gravity meters that have not yet achieved the 10-μgal level. In moving base instruments, several state-of-the-art gravity meters are currently being manufactured to yield mgal-level precision in the United States and Germany. All require the use of very high-grade accelerometers and are considered to be of high enough quality to be of military importance because the accelerometers are of higher grade than those found in most INS. State of the art in gravity gradiometers is on the order of Eotvos/√Hz. This sensitivity is realized in Lockheed Martin (formerly Bell) systems, which are the only mobile systems currently available. Researchers at Yale University are developing an atom interferometry-based gradiometer with a sensitivity potential of 1 Eotvos/√Hz using a 1-m baseline. A team in New Zealand is developing a cryogenic gradiometer with potential sensitivity of 0.02 Eotvos/√Hz.

Countries of concern may not use (or develop) gravity technologies because of their technical complexity, high cost, and marginal military benefits. The alternative of using data from INS and GPS (or other aids) may prove to be more cost effective as a developing technology for data base generation. Countries with proven capability to use this technology for ballistic missiles include the United States, UK, Russia, Ukraine, China, France, and Canada.

Country	Gravity Meters (Gravimeters) Absolute	Gravity Meters (Gravimeters) Relative	Gravity Gradiometers
Australia	●●	●●●	●●●●
Austria	●●	●●●●	
Brazil	●●	●●●●	
Canada	●●●	●●●●	●●
China	●●●	●●●●	●●●●
Czech Republic	●●●	●●●●	
Finland	●●●	●●●●	●●
France	●●●	●●●●	●●●●
Germany	●●●	●●●●	●●
Hungary	●●	●●●●	●●
India	●●	●●●	●
Israel	●●	●	●●
Italy	●●●●	●●●●	●●●●
Japan	●●●●	●●●●	●●
Netherlands	●●	●●	
Poland	●●	●●●	
Romania	●●	●●●	
Russia	●●●●	●●●●	●●●●
Slovak Republic	●●	●●●	
South Africa	●	●●	
South Korea	●	●	
Spain	●	●●	
Sweden	●●	●●●●	
Switzerland	●●	●●●●	
Taiwan	●	●●	
UK	●●●	●●●●	●●●●
Ukraine	●●	●●●●	●●●
United States	●●●●	●●●●	●●●●

Legend: Extensive R&D ●●●● Significant R&D ●●● Moderate R&D ●● Limited R&D ●

Figure 16.2-1. Gravity Meters and Gradiometers WTA Summary

DATA SHEET III-16.2. GRAVITY METERS (GRAVIMETERS), NONMOBILE USE

Developing Critical Technology Parameter	In next 5 to 10 years: Accuracy of < 50 µgals with a time-to-steady-state registration of less than 2 minutes under any combination of attendant corrective compensation. Continuing development of arrays.
Critical Materials	None identified.
Unique Test, Production, Inspection Equipment	SQUID sensors require superconducting temperature Dewars.
Unique Software	Algorithms and verified data for real-time gravity compensation and detection (improvement > 10 to 1) for operation using arrays.
Technical Issues	Accurate verticality and low-level vibration environment required.
Major Commercial Applications	Resource exploration. Detection of underground structures.
Affordability	Cost is proportional to usage. This is not a large-volume production technology. One of the largest manufacturers has sold only about 1,500 units in the last 40 years.

RATIONALE

Gravity meters are a POSITIME-influenced technology because of the interrelationship of gravity data with position and time and the need for verticality for sensor stabilization.

The uncompensated vertical deflection is the largest error in many INS scenarios. The indirect, and most common, compensation technique uses vertical deflection map data computed from gravity meter surveys. The direct method uses a gravity gradiometer for real-time compensation of the vertical deflection. Another application for this technology is to use the gravity meter in a map-matching mode for accurate position determination using previously surveyed map data. Due to roll off of the high frequencies in the gravity anomaly field with altitude, this accuracy could only be obtained at ground level or low altitude, if then. In a local area, with proper conditions for the sensor type being used, these methods might be useful.

The knowledge of initial conditions of gravity predetermines accuracy of self-contained, autonomous navigation systems of all types, but especially inertially equipped ballistic missiles. The knowledge of the gravity field allows accurate compensation of INS. Gravity databases are required for strategic aircraft, submarines, unmanned vehicles, and missiles. The detection of underground structures and material composition are other applications. Continuing development of gravity meter arrays using advanced correlation techniques and very accurate clocks will improve the signal-to-noise ratio (SNR) of the array system. This is considered to be a militarily critical capability.

Advanced filtering using optical correlators/processors and GNSS time synchronization are enabling technologies to obtain the gravity vector. Once the gravity map is generated, nonmobile gravity meters can be used for correlation of the grid data.

Relative to JCS Vision 2010, gravity meters have precision engagement applications. There are no special requirements for the U.S. Government to gain access to this technology. The substantial margin of capability added is critical to continued U.S. superiority.

WORLDWIDE TECHNOLOGY ASSESSMENT

Australia	●●●	Austria	●●●	Brazil	●●●	Canada	●●●
China	●●●●	Czech Republic	●●●	Finland	●●●●	France	●●●●
Germany	●●●●	Hungary	●●●	India	●●	Israel	●●
Italy	●●●●	Japan	●●●●	Netherlands	●●	Poland	●●●
Romania	●●	Russia	●●●●	Slovak Republic	●●●	South Africa	●
South Korea	●	Spain	●●	Sweden	●●●	Switzerland	●●●
Taiwan	●	UK	●●●	Ukraine	●●	United States	●●●●

Legend: Extensive R&D ●●●● Significant R&D ●●● Moderate R&D ●● Limited R&D ●

Commercial interests are advancing the development and production of this technology. State of the art in mass-produced gravity meters (gravimeters) is at the 5- to 10-μgal level. A specially configured LaCoste Romberg unit has yielded gravity measurements at precision and accuracy levels of 1 μgal. A few countries have elected to be dominant in gravity meter technologies, while others are inactive in this area because of economic considerations. Gravity meters (static mode) better than 10 μgals are manufactured and used in the United States. Canada, Germany, UK, Japan, China, Russia, and the Ukraine have developed gravity meters that have not achieved the 10-μgal level.

Countries of concern may not use (or develop) gravity meters because of their technical complexity, high cost, and marginal military benefits. The alternative of using data from INS and GPS (or other aids) may prove to be more cost effective as a developing technology for data-base generation. Countries with proven capability to use this technology for ballistic missiles include United States, UK, Russia, Ukraine, China, France, and Canada.

The following organizations have active research programs:

- ***United States***
  - Carson Instruments
  - Exploration Instruments
  - Gravity Exploration Techniques, Inc.
  - Gravity Map Services
  - GWR Instruments
  - LaCoste Romberg
  - Neese Instrument Company
  - Scinrex
  - Worden Meter Company
  - ZLS Corporation
  - μG

DATA SHEET III-16.2. GRAVITY METERS (GRAVIMETERS), MOBILE USE

Developing Critical Technology Parameter	In next 5 to 10 years: Moving platform accuracy of < 75 μgals with a time-to-steady-state registration of less than 2 minutes under any combination of attendant corrective compensation. Continuing development of arrays.
Critical Materials	None identified.
Unique Test, Production, Inspection Equipment	Test, calibration, modeling, compensation, or alignment equipment to obtain mobile accuracy. Accelerometer axis align stations.
Unique Software	Algorithms and verified data for real-time gravity compensation and detection (improvement > 10 to 1) for operation on mobile platforms or using arrays and time compensation, or both.
Technical Issues	All modern earthbound gravity meters are affected by motion from all sources (seismic, acoustic, temperature, etc.) and require isolation/compensation. The problem is compounded by mobile usage by the need for compensation.
Major Commercial Applications	Resource exploration. Underwater terrain estimation. Detection of underground structures.
Affordability	Cost is proportional to usage. This is not a large-volume production technology.

RATIONALE

Gravity meters on a moving platform are a POSITIME-influenced technology because of the interrelationship of gravity data with position and time and the need for velocity and verticality compensation for sensor stabilization.

Moving-base gravity-meter accuracy is essential to generate maps for autonomous siting, INS initialization, and en-route compensation for in-flight gravity effects for WMD delivery systems, such as ballistic missiles and other long-range, unaided, inertial-guided applications, including aircraft, AUVs, cruise missiles, and submarines. Applications have been developed to correct, either directly or indirectly, for the local gravity disturbances or vertical deflection on a moving base/platform. The uncompensated vertical deflection is the largest error in many INS scenarios. The indirect—and most common—compensation technique uses vertical deflection map data computed from gravity meter surveys. The direct method uses a gravity gradiometer for real-time compensation of the vertical deflection. In the latter mode, the spatial gravity gradients are multiplied (scalar product) by the velocity vector and integrated to obtain the vertical deflection in real time.

Another application for this technology is to use the gravity meter in a map-matching mode for accurate position determination using previously surveyed map data. Due to roll off of the high frequencies in the gravity anomaly field with altitude, this accuracy could only be obtained at low altitude, if at all. Navigation at GPS accuracy through turns and acceleration is questionable, but could potentially be recovered afterward. These map-matching techniques using sensors giving data only along the flight path are questionable for long-term navigation, as there are likely to be areas where the data does not have adequate spatial variance to achieve this accuracy. In a local area, with proper conditions for the sensor type being used, these methods might be useful.

The knowledge of initial conditions of gravity predetermines the accuracy of self-contained, autonomous navigation systems of all types, but especially those for ballistic missiles. The knowledge of the gravity field allows accurate compensation of INS. On mobile platforms, continuing development of gravity meter arrays using advanced correlation techniques and very accurate clocks will improve the SNR of the array system. Accurate time is required for compensation. This is considered to be a militarily critical capability.

Gravity meters for airborne and marine applications are of military importance. Gravity meters require stabilization and the associated software to maintain a stable reference frame. The resulting hybrid system has the potential to provide the military with a non-emanating, nonjammable, totally covert system that can be used worldwide for navigation. As noted, the system will require previously surveyed gravity map data as well as a unique gravity signature, which can be detected in the background noise. When there is not an adequate signature, the system may be augmented with magnetic signature map matching (see Section 16.4).

Another developing method of determining gravity is by computing the acceleration difference between an uncompensated INS and a GNSS. The computed acceleration of the platform, as measured by a GNSS, is the pure acceleration of the vehicle, while the acceleration measured by an uncompensated INS contains the gravity vector and the platform acceleration vector. The difference is the gravity vector. Advanced filtering using optical correlators/processors and GNSS time synchronization are enabling technologies to obtain the gravity vector.

Relative to JCS Vision 2010, gravity meters have precision engagement applications. There are no special requirements for the U.S. Government to gain access to this technology. The substantial margin of capability added is critical to continued U.S. superiority.

WORLDWIDE TECHNOLOGY ASSESSMENT

Australia	●●●	Austria	●●●	Brazil	●●	Canada	●●●●
China	●●●●	Czech Republic	●●●	Finland	●●●	France	●●●●
Germany	●●●●	Hungary	●●	India	●●●●	Israel	●●
Italy	●●●	Japan	●●●●	Netherlands	●●	Poland	●●●
Romania	●●	Russia	●●●●	Slovak Republic	●●●	South Africa	●●
South Korea	●	Spain	●●	Sweden	●●●	Switzerland	●●●
Taiwan	●●●	UK	●●●●	Ukraine	●●●●	United States	●●●●

Legend: Extensive R&D ●●●● Significant R&D ●●● Moderate R&D ●● Limited R&D ●

Commercial interests are advancing the development and production of this technology. State of the art in mass-produced gravity meters (gravimeters) is at the 5- to 10-μgal level. A specially configured LaCoste Romberg unit has yielded gravity measurements at precision and accuracy levels of 1 μgal. A few countries have elected to be dominant in gravity-meter technologies, while others are passive because of economic considerations. In moving-base instruments, several state-of-the-art gravity meters are currently being manufactured to yield mgal-level precision in the United States and Germany. All require the use of very high-grade accelerometers and are considered to be of high enough quality to be of military importance because the accelerometers are of higher grade than those found in most INS.

Countries of concern may not use (or develop) gravity meters because of their technical complexity, high cost, and marginal military benefits. The alternative of using data from INS and GPS (or other aids) may prove to be more cost effective as a developing technology for data-base generation. Countries with proven capability to use this technology for ballistic missiles include the United States, the UK, Russia, Ukraine, China, France, and Canada.

The following organizations have active research programs:

- ***United States***
  - Exploration Instruments
  - LaCoste Romberg
  - Worden Meter Company (Texas Instruments)
  - GWR Instruments
  - Neese Instrument Company
  - μG
- ***Germany***
  - Institute for Applied Geodesy
- ***Japan***
  - Kyoto Electronics Manufacturing Company
- ***Russia***
  - State Research Centre of Russia (CSRI Elektropribor)
- ***Switzerland***
  - ETH-Zurich
- ***Finland***
  - Finnish Geopetic Institute
- ***Sweden***
  - Geodetiska

Other organizations that have active research programs are Scinrex, Carson Instruments, ZLS Corporation, Gravity Exploration Techniques Incorporated, Gravity Map Services, and μG. An example of a commercial application is a recently completed aerial survey of Switzerland by ETH-Zurich, using a LaCoste Romberg system coupled to GPS.

DATA SHEET III-16.2. GRAVITY GRADIOMETERS, NONMOBILE USE

Developing Critical Technology Parameter	In next 5 to 10 years: Static platform < 0.02 Eotvos/√Hz. Continuing development of arrays.
Critical Materials	None identified.
Unique Test, Production, Inspection Equipment	Test, calibration, modeling, compensation, or alignment equipment to obtain static accuracy of sensor. Accelerometer axis align stations. SQUID sensors require superconducting temperature Dewars.
Unique Software	Algorithms and verified data for real time gravity compensation and detection (improvement > 10 to 1) for operation using arrays.
Technical Issues	All modern earthbound gravity gradiometers are affected by motion from all sources (seismic, acoustic, temperature, etc.) and require isolation/compensation. The problem is compounded by the need for measurements at different points near the test article.
Major Commercial Applications	Resource exploration. Detection of underground structures such as sink holes.
Affordability	Cost is proportional to usage. This is not a large-volume production technology.

RATIONALE

Gravity gradiometers are a POSITIME-influenced technology because of the interrelationship of gravity data with position and time and the need for verticality for sensor stabilization.

The uncompensated vertical deflection is the largest error in many INS scenarios. The indirect—and most common—compensation technique uses vertical deflection map data computed from gravity-meter surveys. The direct method uses a gravity gradiometer for real-time compensation of the vertical deflection. Data-base models developed from gravity gradiometer data are used by certain military platforms. This process was used by the Trident Missile System for stealth (nonradiating) positioning. Other applications include detection of underground structures, and nonintrusive identification of treaty-limited items. Gravity gradiometers have a higher military value than gravity meters because they have the ability to estimate vertical deflection in real time. Gravity gradiometers require stabilization and the associated software to maintain a stable reference frame.

Gravity gradiometers can detect underground structures and provide non-intrusive identification of treaty-limited items. Advanced systems will detect masses in the 300-m size at 200-m depths.

Relative to JCS Vision 2010, gravity gradiometers have precision engagement applications. There are no special requirements for the U.S. Government to gain access to this technology. This technology should be continuously monitored because of the substantial margin of capability added that is critical to continued U.S. superiority.

WORLDWIDE TECHNOLOGY ASSESSMENT

Australia	●●●	Austria	●●●	Brazil	●●	Canada	●●●●
China	●●●●	Czech Republic	●●●	Finland	●●●	France	●●●●
Germany	●●●●	Hungary	●●	India	●●●	Israel	●●
Italy	●●●	Japan	●●●●	Netherlands	●●	Poland	●●●
Romania	●●	Russia	●●●●	Slovak Republic	●●●	South Africa	●●
South Korea	●	Spain	●●	Sweden	●●●	Switzerland	●●●
Taiwan	●●●	UK	●●●●	Ukraine	●●●●	United States	●●●●

Legend: Extensive R&D ●●●● Significant R&D ●●● Moderate R&D ●● Limited R&D ●

Lockheed Martin Corporation is the major U.S. company with proven capability. Other U.S. companies and research institutions include Bell Geospace and the University of Maryland.

Commercial interests are advancing the development and production of this technology. Few countries have elected to be dominant in gravity gradiometer technologies, while most are passive because of economic considerations. Researchers at Yale University are developing an atom interferometry-based gradiometer with a sensitivity potential of 1 Eotvos/√Hz using a 1-m baseline. Using a Superconducting Gravity Gradiometer, researchers at the University of Maryland have demonstrated 0.02 Eotvos/√Hz, while Gravitech in New Zealand is also developing a cryogenic gradiometer with potential sensitivity of 0.02 Eotvos/√Hz.

Countries of concern may not use (or develop) gravity gradiometers because of their technical complexity, high cost, and marginal military benefits. The alternative of using data from INS and GPS (or other aids) may prove to be more cost effective as a developing technology for data-base generation.

DATA SHEET III-16.2. GRAVITY GRADIOMETERS, MOBILE USE

Developing Critical Technology Parameter	In next 5 to 10 years: Moving platform < 5 Eotvos/√Hz. Continuing development of arrays.
Critical Materials	None identified.
Unique Test, Production, Inspection Equipment	Test, calibration, modeling, compensation, or alignment equipment to obtain mobile accuracy of < 5 Eotvos/√Hz. Accelerometer axis align stations.
Unique Software	Algorithms and verified data for real-time gravity compensation and detection (improvement > 10 to 1) for operation on mobile platforms, using arrays and time compensation, or both.
Technical Issues	All modern earthbound gravity gradiometers are affected by motion from all sources (seismic, acoustic, temperature, etc.) and require isolation/compensation. The problem is compound by the need for measurements at different points near the test article.
Major Commercial Applications	Resource exploration. Underwater terrain estimation. Detection of underground structures such as sink holes.
Affordability	Cost is proportional to usage. This is not a large-volume production technology.

RATIONALE

Gravity gradiometers are a POSITIME-influenced technology because of the interrelationship of gravity data with position and time and the need for velocity and verticality compensation on a moving platform for sensor stabilization. Applications have been developed to correct, either directly or indirectly, for the local gravity disturbances or vertical deflection on a moving base/platform. The uncompensated vertical deflection is the largest error in many INS scenarios. The indirect—and most common—compensation technique uses vertical deflection map data computed from gravity-meter surveys. The direct method uses a gravity gradiometer for real-time compensation of the vertical deflection. In the latter mode, the spatial gravity gradients are multiplied (scalar product) by the velocity vector and integrated to obtain the vertical deflection in real time.

Another application for this technology is to use the gravity gradiometer in a map-matching mode for accurate position determination using previously surveyed map data. Due to roll off of the high frequencies in the gravity anomaly field with altitude, this accuracy could only be obtained at ground level and at low altitude, if then. Navigation at GPS accuracy through turns and acceleration is questionable, but the errors could perhaps be rectified afterward. These map-matching techniques using sensors giving data only along the flight path are questionable for long-term navigation, as there are likely to be areas where the data does not have adequate spatial variance to achieve this accuracy. In a local area, with proper conditions for the sensor type being used, these methods might be useful.

Gravity gradiometers with motion compensation provide a total covert undersea worldwide navigational capability using correlation techniques with gravity maps. The knowledge of instantaneous gravity conditions while on a moving platform is needed. Gravity gradiometers can detect underground structures and provide non-intrusive identification of treaty-limited items. Continuing development of gravity gradiometer arrays using advanced correlation techniques and very accurate clocks will improve the SNR of the array system. This is considered to be a militarily critical capability.

Gravity gradiometers have a higher military value than gravity meters because they have the ability to estimate vertical deflection in real time. Gravity gradiometers require stabilization and the associated software to maintain a stable reference frame. The resulting hybrid system has the potential to provide the military with a non-emanating, nonjammable, totally covert system that can be used worldwide for navigation. As noted, the system will require previously surveyed gravity map data, as well as a unique gravity signature that can be detected in the background noise. When there is not an adequate signature, the system maybe augmented with magnetic signature map matching (see subsection 16.4).

Another developing method of determining gravity is by computing the acceleration difference between an uncompensated INS and a GNSS. As measured by a GNSS, the computed acceleration of the platform is the pure acceleration of the vehicle, while the acceleration measured by an uncompensated INS contains the gravity vector and the platform acceleration vector. The difference is the gravity vector. Accurate time is required for compensation. A controlled vibration and acceleration environment is required to improve accuracy. Advanced filtering using optical correlators/processors and GNSS time synchronization are enabling technologies to obtain the gravity vector with better accuracy.

Relative to JCS Vision 2010, gravity gradiometers have precision engagement applications. There are no special requirements for the U.S. Government to gain access to this technology. The substantial margin of capability added is critical to continued U.S. superiority.

WORLDWIDE TECHNOLOGY ASSESSMENT

Australia	●●●	China	●●●	Finland	●●	France	●●●●
Germany	●●	Hungary	●●	Italy	●●●	Japan	●●
New Zealand	●●	Russia	●●●	UK	●●●	United States	●●●●

Legend: Extensive R&D ●●●● Significant R&D ●●● Moderate R&D ●● Limited R&D ●

Commercial interests are advancing the development and production of this technology. A few countries have elected to be dominant in gravity gradiometer technologies, while others are passive because of economic considerations. In moving-base instruments, several state-of-the-art gravity meters are currently being manufactured in the United States and Germany to yield mgal-level precision. State of the art in gravity gradiometers is on the order of 10 Eotvos/√Hz. This sensitivity is realized in the Lockheed Martin (formerly Bell) system, which is the only mobile system currently available. This system requires the use of very-high-grade accelerometers and is considered to be of military importance because the accelerometers are of higher grade than those found in most INS. Researchers at Yale University are developing an atom interferometry-based gradiometer with a sensitivity potential of 1 Eotvos/√Hz using a 1-m baseline. Gravitech in New Zealand is developing a cryogenic gradiometer with potential sensitivity of 0.02 Eotvos/√Hz. They are in the development stage, and the potential for moving base applications for either of these systems is unknown. An example of commercial application is a recently completed survey in the Gulf of Mexico by Bell Aerospace (Lockheed Martin). The survey used a three-dimensional, full tensor, gravity-gradient system derived from the Trident Missile System for positioning by use of undersea map data.

Countries of concern may not use (or develop) gravity gradiometers because of their technical complexity, high cost, and marginal military benefits. The alternative of using data from INS and GPS (or other aids) may prove to be more cost effective as a developing technology for data-base generation.

The following organizations have active research programs:

- ***United States***
  - Lockheed Martin
  - Yale University
- ***Germany***
  - Institute of Elecronic Basics of Computer Science
- ***Canada***
  - University of Vctoria
- ***New Zealand***
  - Gravitech

SECTION 16.3—RADIO AND DATA-BASED REFERENCED NAVIGATION (DBRN) SYSTEMS

Highlights

- Elimination of the U.S. GPS selective availability (S/A) by 2005 and use of differential GPS combined with LORAN and differential LORAN will provide navigation accuracy less than 1 m (6 sigma)/0.3 m (SEP) providing both friends and foes with a very precise navigation capability.
- Autonomous and common three-dimensional POSITIME grid reference will improve battlespace situational awareness by providing a precise POSITIME tag on all battlespace information collected to provide real-time knowledge of location and movement across battlespace of allied and enemy assets.
- Significant commercial and military growth and dependence on GNSS for position and time will increase as GNSS receivers decrease in cost, weight, and power.
- International GNSS (GLONAS, European Union 2, Iridium, and Teledesic) capabilities will continue to be developed as alternatives to the U.S. GPS, as a means of providing better redundancy and integrity monitoring, or both.
- Increased combination of hybrid navigation and adaptive antenna systems will significantly reduce military dependence on GPS in a jamming combat environment or during signal loss. Future increased combinations of navigation, communication, imaging, and computer functions will improve situational awareness in urban terrain.
- Data-based referenced navigation technology, leveraged by increased computer speed and memory, will have increased commercial and military usage.

OVERVIEW

Radio navigation, particularly GPS and LORAN, will continue to be used both by military and commercial users in the foreseeable future. Radio navigation continues to be the smallest and least expensive of the POSITIME systems. The ratio of commercial to military use of GPS and LORAN will probably be greater than 100:1. There is a definite trend to transition GPS from a DoD system to a commercial system. A recent Presidential Directive for DoD to remove the S/A permanently by 2006 supports this trend. With S/A off, the 10-m accuracy will be available worldwide for all commercial users, who previously were limited to 100-m accuracy. The S/A has limited this 10-m accuracy to U.S. and allied military use only.

Another issue is the emergence of DGPS, which uses a small ground station outfitted with a GPS receiver whose geographic location is precisely determined, and the difference between surveyed and GPS position transmitted to another user. This procedure can provide an accuracy of better than 5 m. In Europe, a novel technique that transmits DGPS signals on an existing LORAN C, called Eurufix, has demonstrated position-fixing accuracy of better than 3 m, even with S/A employed. Localized jamming by friendly forces will deny these accurate GPS capabilities to an enemy only if the basic GPS signal is completely jammed. Antijam GPS components and systems, such as an adaptive antenna system, combined with high-speed digital signal processing (DSP) and a closely coupled hybrid GPS/INS will optimize antenna coverage patterns to specific signal and interference environments. This will produce an antenna pattern with nulls in the direction of broadband jammers very quickly. Better time accuracy (see Section 16.5) will allow rapid GPS direct-Y code acquisition, and the use of autonomous, low-power clocks will minimize GPS jamming and loss of satellite signal. In urban areas, loss of GPS signals due to signal blockage and multipath problems is a challenge to be overcome. Like all time difference of arrival (TDOA) systems, the LORAN system accuracy can be improved by more accurate clocks. Similar to GPS receiver improvements and miniaturization, LORAN antenna and electronic miniaturization technology continues to improve LORAN receiver capability. Funding for continuation of LORAN ground stations because of the growing number of commercial applications continues to force the Federal Radio Navigation Plan to be revised, and the demise of LORAN is not as evident as it was 5 years ago.

GPS accuracy will improve from 3 m to 1 m as a result of recent efforts to link the National Imagery Mapping Agency (NIMA) and the USAF GPS tracking networks, upgraded Kalman filtering, and reduced prediction error in broadcast navigation (NAV) messages. Using fixed location sites, DGPS can further improve the GPS accuracy to less than 0.6 m. The use of DGPS, however, is currently limited to a localized area and the use of communications that may also be susceptible to jamming. Solar activity over the next 10 years is expected to cause problems in time transfer of 10–15 ns.

Technology related to radio ranging using normal tactical communication systems will be readily available for use in 5 years. The Army's enhanced position location system is representative of this technology. Communication signal-based ranging is a plausible method of positioning. Smart antenna techniques will be a catalyst, enabling commercial and military-based stations and user modes for personal communication systems (i.e., cellular telephones). This technology can provide a location accuracy approaching GPS.

Further GNSS improvements include:

- Upgrades to U.S. GPS/NAVSTAR capabilities to include removal of S/A, incorporation of unencrypted C/A codes on L2, inclusion of third civilian frequency, inclusion of new encrypted military codes on L1 and L2, increase in transmission power, and potential increase in number of space vehicles.
- Integration of European Union Galileo public/private GNSS with the U.S. GPS system.
- Improvements in space vehicle orbital definition, increased ground station update frequency, and incorporation of more accurate ionospheric correction models on GNSS accuracies.
- Use of ground-based transmitters to provide wide area DGNSS corrections.
- Use of satellites (GEO and LEO) to provide wide area DGNSS corrections, particularly to aviation.
- Impact of FAA efforts to utilize DGNSS-based en-route and precision landing guidance through the WAAS and LAAS programs and the worldwide proliferation of compatible technology.
- The use of pseudolites (airborne and ground based) for military application and precision landing support (as envisioned in LAAS).

Further improvements to GPS accuracy, as well as reducing susceptibility to jamming, will be obtained by integration with the following Digital Terrain Data Based Navigation Systems:

- Digital Terrain Elevation Data (DTED)
- Digital Feature Analysis Data (DFAD)
- The World Geodetic System (WGS 84)
- The Earth Gravitational Model (EGM 96)
- The International Terrestrial Reference Frame (ITRF).

One of the deficiencies of GPS is that map referencing is not viable unless the receiver is moving to compute directional referencing (north). A magnetic compass or an INS is required for map referencing.

The Shuttle Radar Topography Mission (SRTM)1 will collect radar data over more than 80% of the earth. This is a major step toward the multi-Service requirement for Global DTED accuracy of 30 m by 2000. Table 16.3-1 shows the projected improvements in DTED accuracy.

Table 16.3-1. Comparison of Current and Projected Digital Terrain Elevation Data (DTED) Accuracy

	Absolute Horizontal	Absolute Vertical	Time Frame
Current DTED	50 m 90% CE	30 m 90% LE	Now
Expected DTED w/SRTM	20 m 90% CE	16 m 90% LE	2000+
Future Possibilities*	5–10 m 90% CE	5–10 m 90% LE	2015+

* Requires more accurate determination of the space vehicle's attitude/altitude for images.

DFAD contains feature data equivalent in feature content to a 1:250,000 scale topographic map. Because urban warfare at the brigade and individual soldier level requires higher resolution than current DTED and DFAD can provide, the U.S. Army's Training and Doctrine Command (TRADOC) Analysis Center provides higher resolution topographic maps (30-m elevation and a 1:50,000 scale or higher).[1]

Recent improvement in the EGM 96 has decreased absolute height uncertainty from 2–6 m (1 σ) to 0.5–1 m (1 σ) worldwide. This will benefit not only GPS but also INS accuracy. Figure 16.3-1 shows hybrid INS/GPS accuracy with use of geomapping data after loss of GPS. Closely monitoring WGS and ITRF has led to improvements in the level of agreement between WGS 84 and the ITRF, with the determination that they can now be considered equivalent.[2] While the GPS reference is WGS-84, the Russian GLONASS reference is PZ-90. The transformation model is developed and is being refined by the United States and Russians.[3]

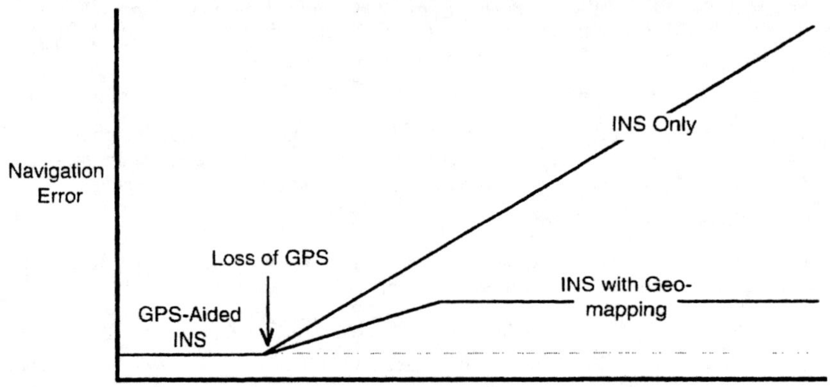

Figure 16.3-1. Hybrid INS Performance with Loss of GPS

The integration of three-dimensional digital terrain maps and other geo-mapping data [provided by NIMA and U.S. Geological Survey's National Mapping Division's Earth Resources Observation Systems (EROS)] with hybrid INS/GPS systems could subsequently provide highly accurate position, velocity, and track under dynamic and covert conditions, even after loss of GPS signals.

In addition, future improvements to three-dimensional digital map data could include global magnetic and gravity data. As a point of reference, given a gravity map having an accuracy of 1 Eotvos/√Hz) and resolution of 0.5 km, an aircraft flying at 200-m altitude at 360 km/hr constant velocity, having a 10 Eotvos/√Hz) gravimeter on board with a 0.0001 deg/hr drift rate gyro, could navigate with 5–10-m horizontal error and 5-m vertical error. All of these capabilities (by use of prestored ground and undersea terrain contour, acoustic, electromagnetic spectrum, magnetic, and gravity sensor data) will significantly increase the hybrid INS accuracy on a continuous basis to that currently provided by GPS at a rate of 1.0 Hz. PT&F form the baseline for telecommunications and navigation, and its importance to military systems is becoming more evident with the operational use of the U.S. GPS (see Section 16.5).

Other data bases, such as bathymetric maps, are used by the United States and Russia to obtain submarine position fixes without surfacing. The process of generating the map consists of registering undersea bottom contours with GPS position data on a surface ship. This then becomes the data base for future map matching aboard a submarine. There are current studies exploring the feasibility of broad ocean bathymetry that would increase the number of available sites for position fixing.

---

[1] www.trac.army.mil/trandata.htr

[2] *Refinements to the World Geodetic System*, 1984, by Stephen Malys, et al., NIMA.

[3] *GPS World*, January 1999, p. 54.

RATIONALE

Accurate positioning, control, and redundancy for platforms are essential for effective coordination of military activities. Individual system accuracy depends on mission requirements. Encrypted signals of the GPS deny non-authorized users the full capability of the systems. Null-steerable antennas are a military response to jamming. Hybrid and DBRN systems combine the best features of different navigation systems to provide autonomous, covert, nonjammable information. DBRN technology is partially derived from sensor and Geographic Information Systems (GIS) technology. DBRN technology, leveraged by computer speed and memory, resolves data ambiguities and optimizes navigational sensor and stored data. Three-dimensional position ambiguities and other properties, such as magnetic and gravity signatures, will be resolved and optimized as stored geodetic data for navigation reference using sensors such as radar altimeters, magnetometers, gravity meters, and acoustic sensors. The use of power management and phase-shift key modulation reduces the emitted signal, resulting in a decreased detectability and covert (stealth) operation. Military uses for GPS will enhance the following:

- Supply location systems
- Spacecraft navigation
- Parachute insertion
- Air vehicle attitude
- Angle of attack
- Battlefield targeting
- Helicopter hover positioning
- Gravity measuring system
- HF communications frequency management
- Encryption/decryption
- DGPS
- Ship cargo management
- Situation awareness
- Minefield positioning
- Search and rescue
- Weather balloon navigation
- Power and communication line failures
- Inertial navigator reset and mapping
- System integration of sensors
- Pseudolite positioning system
- Position reporting for high-value assets
- AUV navigation
- Imaging
- Artillery smart round
- DGPS for heading
- Nuclear reset
- Construction

FAA's WAAS corrects the standard GPS signal to provide the accuracy, integrity, and availability needed for civil aviation navigation and precision approaches (Category 1) over a very large geographical area. Some of the critical functions include corrections for navigation satellite clock, satellite orbital data, and the effects of the ionosphere on the GPS and WAAS signals and ensuring the validity of WAAS messages. WAAS uses geostationary satellites (space-based) transmitting GPS look-alike ranging signals without S/A but with integrity messages and wide-area differential corrections. The LAAS is a ground-based augmentation system providing local area DGPS corrections. DGPS is based on providing corrections of errors that are common to both ground-based and aircraft receivers in the local area. LAAS has the capability of providing integrity using pseudolites (ground-based, low-powered satellites) and DGPS for accuracy of about 1 m on final approach and taxi. LAAS complements WAAS and will operate independently. For decades, the dismounted soldier's navigation tools have consisted of maps, compass, and individual pace count. Recently, GPS technology has been added to the tool set to improve position determination. Although a significant improvement, GPS can be jammed or blocked (terrain or man-made obstructions), and GPS does not provide accurate azimuth information at slow speeds for map north referencing. Integrated navigation (INAV) technology introduces a new tool to help compensate for those times when GPS is either unavailable or unreliable—automated dead reckoning. INAV dead-reckoning capability is supplied by a dead-reckoning module (DRM), a small, low-power navigator consisting of an electronic compass and a pedometer using MEMS technology accelerometers.

WORLDWIDE TECHNOLOGY ASSESSMENT

The United States and Russia have independently developed and deployed GPS and GLONASS, respectively. These global systems and other proposed satellite systems are commonly referred to as GNSS. The United States leads the world in radio navigation and DBRN technology and is progressively improving in the areas of accuracy, size, weight, reliability, cost, and integration with digital processing technology. The gap, however, is closing quickly, as the newer technologies require, in some cases, less capital investment in the technology. The European Union GNSS 2 system could be launched starting in 2003, providing accuracy of 5–10 m. Increased computational effectiveness for a given equipment volume and weight could provide an adversary with distinct navigation payoffs: (1) enhancing navigation capabilities and (2) improving reliability and resistance to hostile actions. Signal detection and processing technology is used to acquire, synchronize, and track desired signals for measurement of navigation parameters. The substantial margin of capability added by this technology is vital to continued U.S. superiority in precision navigation and the multitude of missions dependent thereupon. Highly interference-resistant receivers for satellite navigation systems are also vital. Militarily critical signal conversion technologies contribute directly to mission effectiveness by improved anti-jam performance; increased reliability; higher precision navigation; real-time adaptive response to hostile environments; and decreased system size, weight, and cost. The know-how to achieve improvements most beneficial to military applications depends upon integrated circuit technology and the software to support it. In this area, Russia lags the United States by 3 to 5 years. The worldwide growth of high-density semiconductor design and fabrication technology has been an enabling technology. From a control standpoint, however, the design technology is widely taught in universities and available in industry worldwide. Receivers that combine GPS and GLONASS technology are available on the commercial market from U.S. and non-U.S. manufacturers. Many of these receiver designs are for timing purposes rather than navigation. All of these receivers will provide a capability that has better jamming resistance, increased integrity, and higher accuracy than GPS or GLONASS systems alone. The UK, Germany, France, Israel, and Japan are also leading nations that have developed end-use products that use signals from both GPS and GLONASS and DBRN for military applications. The acquisition of dual-use end products could allow the transfer of the necessary know-how for military applications. Russia also has developed the use of magnetic arrays to improve compensation for DBRN systems (see subsection 16.4). Bathymetry maps are used by the United States and Russia to obtain submarine position fixes without surfacing. Improvements of GPS accuracy in generating those maps will enhance submarine undersea navigation without surfacing. Several countries, including the United States, UK, and France, use terrain databases for updating cruise missile inertial navigation accuracy.

Country	Radio Navigation Including Tele-communication	Data-Based Navigation (Terrain, Acoustic, Bathymetry, Electromagnetic Spectrum, Magnetic, Gravity, Stellar Referenced)
Australia	●●	●●●
Austria	●	●
Brazil	●●	●
Canada	●●●●	●●●
China	●●●	●●
Czech Republic	●●	●
Finland	●●	●
France	●●●●	●●●●
Germany	●●	●●●●
Hungary	●	●
India	●●●	●●
Israel	●●●●	●●●
Italy	●●●	●
Japan	●●●●	●●●
Netherlands	●●●	
Poland	●●	●●
Romania	●	●
Russia	●●●●	●●●
Slovak Republic	●	●
South Africa	●●●	●●
South Korea	●●	●●
Spain	●●	●●
Sweden	●●	●●
Switzerland	●●	●●
Taiwan	●●	●●
UK	●●●●	●●●●
Ukraine	●●●●	●●
United States	●●●●	●●●●

Legend: Extensive R&D ●●●● Significant R&D ●●● Moderate R&D ●● Limited R&D ●

Figure 16.3-2. Radio and Data-Based Referenced Navigation Systems WTA Summary

SECTION 16.4—MAGNETOMETERS AND MAGNETIC GRADIOMETERS

Highlights

- Magnetometer and magnetic gradiometer technology varies with applications and cost. Anticipate more use of low-cost fiber-optic and torsion sensors for land-based usage and optically pumped technology for sea-based detection and classification.
- Magnetometer sensor arrays, a covert detection and classification technology, will be more viable because of accurate time sequencing, computer speed, and memory advances, providing increased detection and location of submarines, mines, and mobile missiles.
- Knowledge of position, GNSS time, and better computational capabilities using optical processing/ correlation will greatly enhance magnetic array detection performance.
- Newly developed potassium and helium-4 optically pumped magnetometers are demonstrating performance comparable to superconductive quantum interference device (SQUID) magnetometers at low cost. Medical research and diagnostics is major funding source for future of SQUID sensors and possibly for potassium, if sensor can be reduced in size. Magnetic gradiometers, utilizing either the SQUID or potassium technologies, nearly eliminate the natural geomagnetic background noise.
- High T_c SQUID technology has matured since its inception in 1987 to the point where nitrogen-cooled superconducting sensors are rivaling their low T_c counterpart. Use of giant magnetoresistive (GMR) sensors is projected for a number of applications for which cost, size, and power are driving factors.
- Magnetic sensor use for nondestructive testing and inspection of vehicle integrity will increase.

OVERVIEW

Magnetometers and magnetic gradiometers are of interest to the military because of their covert detection, signature classification, and position determination capability. Although magnetic sensing of direction is thought to be of secondary use by the military, there is a need to provide low-cost map referencing of magnetic north and to maintain magnetic databases using magnetic sensors. Magnetic heading can be sensed by a flux value, for instance, or computed by subtracting magnetic variation from the true heading sensed by an INS. Magnetic variation is obtained from a map data base and can be used in many formats and accuracy levels. The use of true heading vice magnetic heading by the majority of navigation applications has been vastly increased by the quantities of INS and GPS in military and commercial applications in the past 10 years. Using computational techniques, data bases with prior or real-time magnetic field data from magnetometer arrays can be used to reduce the spatial and temporal background noise in the detection of land vehicles, submarines, or mines. Increasing the signal-to-noise ratio of magnetic sensors is a major factor in improving magnetic applications by the military. Both the development technology and the production technology are military critical. Magnetic sensor types of special interest include fluxgate, SQUID, nuclear precession, optically pumped, fiber-optic, and induction coil. Figure 16.4-1 provides a comparison of the current spectral density and frequency range of these sensor types. Figure 16.4-2 provides a projected capability for these sensors within the next 15 years.

The performance possible with the more sensitive SQUID, optically pumped, and induction-coil technologies will improve, and the cost and size of these sensors will come down. At the same time, other technologies that have demonstrated limited performance in comparison to the superconducting and total-field technologies will probably become competitive. Thin-film GMR sensor technology appears at present to provide a most promising possibility (see Figure 16.4-2). Using GMR technology, nonvolatile random access memory will have lower power consumption and faster access speeds. GMR circuits, transformers, and logic gates are also viable.

For details on mine countermeasures, see Section 17.7.

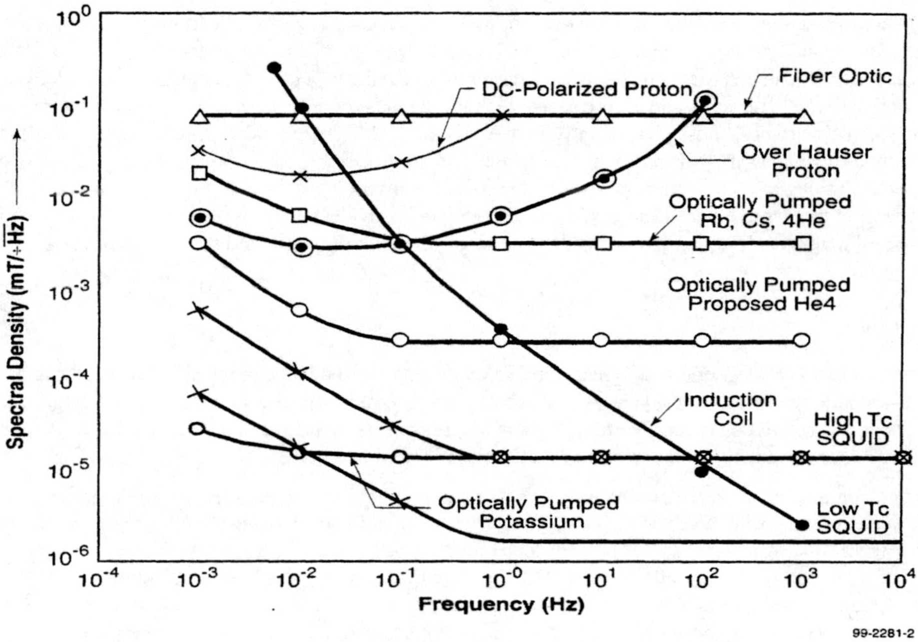

Figure 16.4-1. Comparison of the Current Spectral Density and Frequency Range of Various Sensors

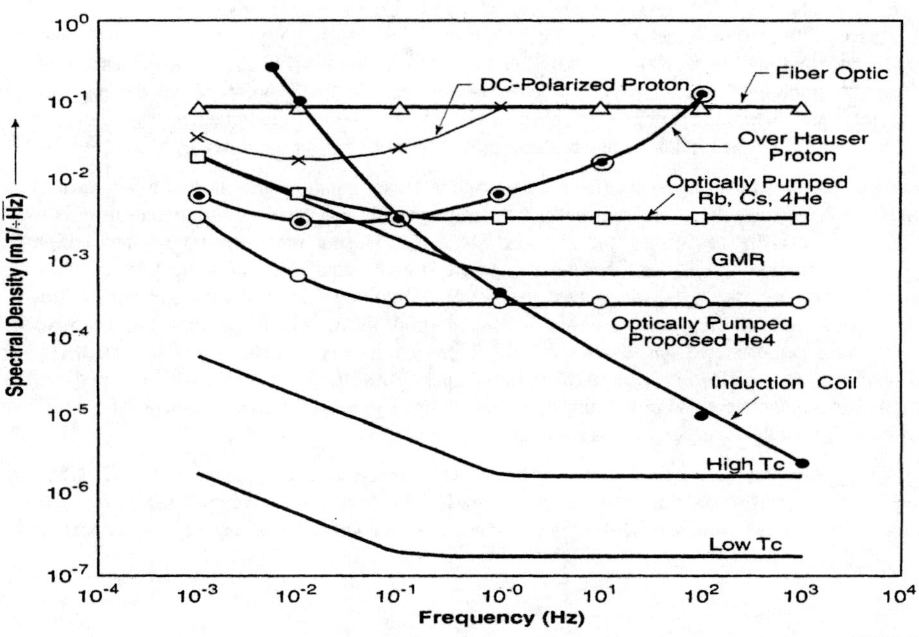

Figure 16.4-2. Comparison of Projected Spectral Density and Frequency Range of Various Sensors

On the one hand, the sensitivity of these sensors will likely increase, and applications involving multiple units of low-cost, short-range, remote sensors will evolve. Active electromagnetic detection of a target illuminated by a low-frequency magnetic or electrical-current source, currently utilized for land mine detection, will likely be utilized in a wider range of applications as underlying sensor technology and signal processing capabilities improve. Further advances in microelectronics, most significantly in the area of high dynamic range, multichannel, analog-to-digital converter technology and DSP, will enhance capabilities and significantly reduce cost and complexity of the underlying sensor technologies. These factors will make concepts involving multiple units of low-cost, short-range distributed sensors more attractive. Similarly, advancements in microelectronics will increase magnetic-sensor capabilities on-board magnetically noisy platforms through large-scale use of distributed sensors monitoring the platform noise.

RATIONALE

Magnetometers and magnetic gradiometers are key elements of magnetic anomaly detector systems for anti-submarine warfare, mine fuses, intrusion and ordnance detection, proximity detection (distance to target), underwater mine detection, and active degaussing systems. Magnetic sensors in a tactical missile can be used to detect and localize a target, such as a tank, from the background magnetic field variations.

Another application is to determine position for navigational purposes relative to a database containing previously surveyed magnetic field information. Deployment of additional magnetic observatories in open ocean areas will permit magnetic models to be developed with 1-deg RMS magnetic variation worldwide, and the possibility of reduction to 0.5 by 2015 exists. Dual-use applications include clinical diagnostics and geophysical applications. Commercial users may adapt the use of magnetometer arrays, but the main initial (developing) application is military. Ocean bottom arrays, as depicted in Figure 16.4-3, can be used to detect and classify vessels. Land-based arrays, as depicted in Figure 16.4-4, can measure time-varying natural noise for airborne compensation. Of special concern are compensation methods for platform motion and other self-induced magnetic noise. The main differences between military and commercial use are the real-time accuracy requirements for operation from a moving base and the detection and classification capability. There has not been significant progress in advancing the state of the art of magnetic sensors in the past decade, with the possible exception of the development of the potassium optically pumped sensor. The experimental results for the potassium sensor gives a magnetic sensor noise level of 10 microgamma (10^{-14} tesla rms) per $\sqrt{Hz}$. Work continues on further development of this new class of sensor. Before the advent of these potassium sensors, the instrument noise level for the best optically pumped magnetometers was about 3×10^{-12} tesla rms per $\sqrt{Hz}$, about two orders of magnitude inferior to the newly developed potassium sensors. Potassium gradiometer accuracy now equals SQUID capabilities, but without the low-temperature logistic and motion problems of SQUIDS. SQUIDS are vector devices and more susceptible to motion-induced effects. A potassium gradiometer is a scalar device, and an RMS noise of 10 microgammas/meter (or 10^{-14} tesla/meter) at 10-second periods has been achieved. Potassium gradiometer technology permits near elimination of time-varying geomagnetic noise. Laser pumping will improve accuracy by a factor of 2 to 3.

During the last decade, multiple-channel, short-baseline tensor gradiometers utilizing low-critical-temperature (low T_c) SQUIDs have been deployed from underwater tow systems and have demonstrated an effective means to detect, localize, and classify sea mines and exploded ordinance. During the same timeframe, a high T_c SQUID technology utilizing liquid nitrogen for cooling has matured. High T_c sensors are of interest for military applications because nitrogen cooling reduces logistic issues and permits smaller package sizes, compared to low T_c SQUID sensors that require liquid helium for cooling. Significant progress in this technology has been made. High T_c magnetometers have demonstrated sensitivities of 100 microgamma (10^{-13} tesla) per $\sqrt{Hz}$ at 0.1 Hz. This figure should be compared to the performance of 10 microgamma per $\sqrt{Hz}$ at 0.1 Hz that has been reported for both low T_c superconducting magnetometers and potassium optically pumped magnetometers. Further sensitivity increases are expected as high T_c manufacturing technology improves.

For tracking and homing applications in which less sensitivity is acceptable, but cost, size, weight, and logistics are important factors, a multichannel tensor gradiometer has been developed using fluxgate technology. This fluxgate gradiometer is especially useful for providing precise localization in man-portable operations.

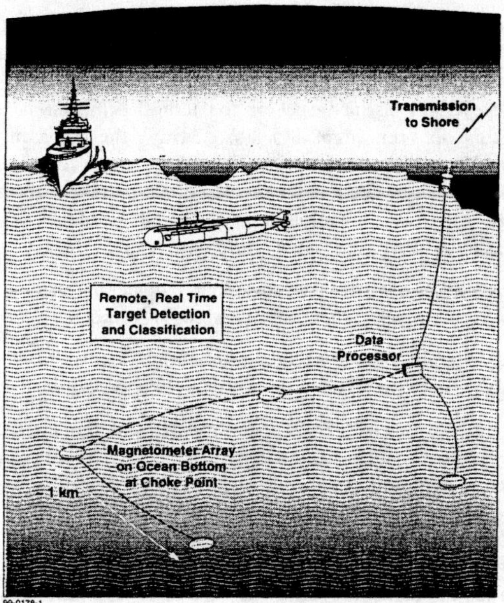

Figure 16.4-3. Ocean Bottom Arrays

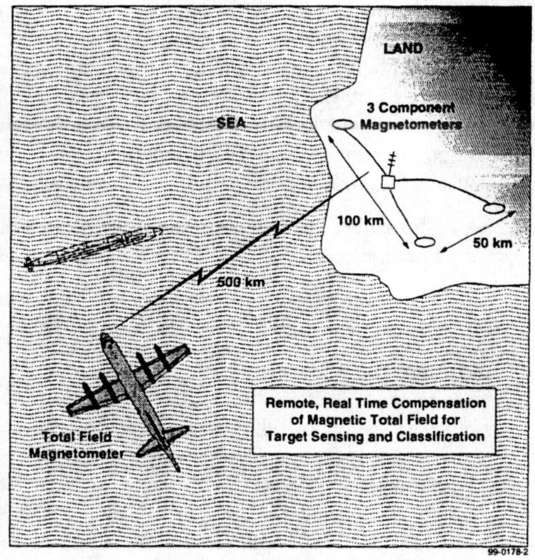

Figure 16.4-4. The Remote Real-Time Compensation Process

The major advances in magnetometry during the last decade or so have been the emergence of precise location and time information and the availability of greatly enhanced computation capabilities. As these capabilities are introduced into magnetometry systems, the effectiveness of such systems could be significantly enhanced. There are four scenarios that will illustrate this potential improvement:

1. **Fixed Array of Sensors.** When a fixed array of sensors is used on the ground or underwater, the increased knowledge of position and time are coupled to enhance the computation power because of the synergistic effect of the array. The spectral energy density functions for environmental noise can be precisely deduced. Then, in the presence of a signal of interest, the special density function of the magnetic signatures of interest can be extracted from background noise and thus provide an improved SNR than that realizable without the use of energy spectral density techniques. For instance, arrays of three component magnetometers on land or on the ocean bottom have the potential of reducing time-varying geomagnetic background "noise" by 80 percent or more. This time-varying magnetic background noise is a major impediment to detection of the magnetic anomaly signals from mines or submersibles, depending upon the operational scenario (sensor noise, geology motion, etc.). By placing three to five magnetometers at distances of 10 to 100 kilometers, gradients and amplitudes of natural pulsations may be measured. By using previously determined transfer functions (based on geologic structure and experiment), the pulsation field may be extrapolated to a distant point at sea. The predicted field for the distant point may be telemetered to an antisubmarine warfare (ASW) aircraft. These predicted fields may be subtracted from the airborne magnetometer output in near real time to significantly increase the signal-to-noise ratio and permit detection of submarine magnetic anomaly signals at greater ranges. Figure 16.4-4 is a graphic display of the remote real-time compensation process.

2. **Moving Array of Sensors.** If the magnetometry system consist of one or more moving sensors towed behind a ship or plane, the ability for precise position, orientation, altitude (or depth), etc., of the various sensors can be deduced and the background noise spectral density determined. This will enhance the signal-to-noise ratio determination of targets of interest.

3. **Target Recognition.** With the enhanced knowledge of location and orientation of the various sensor(s) when the sensor(s) and target(s) are in relative motion, the SNR will be enhanced as pointed out above. For

a typical ASW application, the sensors are remote from the target, and thus the target can be treated as a simple dipole and the "inverse problem," that is, deducing the location and moment of the target(s) of interest from the magnetic field contours, is relatively straightforward. However, in those scenarios when the target and sensors are nearby (that is, when the length of the target or the lateral dimensions of the sensor array are comparable to or less than the distance between target and sensor array), the target must be represented by dipole as well as higher order poles. In this case, the extraction of the magnetic signature from the magnetic field contours is very demanding on computation power, requiring precise information about the sensor position, orientation, etc. Thus, the future development of "smart" magnetic sensor systems requires precise position and time information as well as massive computation capabilities.

— The ability to increase knowledge of locations, time, etc., in packages that can be utilized in cost-effective magnetic sensors and sensor arrays is very crucial to these enhanced magnetometer systems.

— The development of computers with enhanced capability (both speed and memory) and with very low power consumption is crucial for the maturing of these enhanced magnetic detection systems.

4. **Submarine Electric Fields.** Dissimilar metals or differences in surface properties of the same metal give rise to "corrosion" currents flowing in salt water in the vicinity of a submarine. The currents produce a static electric dipole aligned with the longitudinal axis of the submarine. This is called the "submarine electric dipole moment" and is quite variable. The dipole moment, in turn, produces a magnetic field. This electric dipole-produced (EDP) magnetic field falls off as the inverse square of distance, unlike the well-known "hull anomaly" magnetic field, which falls off the inverse cube of distance. This means that, although the EDP magnetic anomaly has a smaller magnetic moment than the hull anomaly, at some distance the magnetic field from the electric dipole will begin to get larger than that from the hull moment. In the past, magnetometers were not sensitive enough to detect this EDP magnetic field anomaly, but the advent of the potassium magnetometer means that it may be detectable.

   The same "corrosion" currents are modulated by differences in electrical resistance between the journal bearings and the propeller shaft as rotation occurs. The modulation will be at the shaft rotation rate and its multiples. This phenomena gives rise to an oscillating electromagnetic field which propagates through the water and into the air [extremely low-frequency emission (ELFE) signal]. The high sensitivity of the potassium magnetometer promises much greater range for the detection of this signal.

WORLDWIDE TECHNOLOGY ASSESSMENT

Depending on military and commercial uses, many countries, such as France, Germany, Canada, Russia, and the Ukraine, have developed a capability in most of the differing magnetometer and magnetic gradiometer technologies. Canada leads the world in fluxgate sensors; Russia has developed a unique potassium-resonance magnetometer capability; the United States leads in SQUID development for military; and Japan, Russia, Ukraine, and Germany lead in civil clinical applications. A joint Russian, Canadian, and U.S. team has built optically pumped potassium magnetometers and gradiometers. The potassium magnetometer is based on a preliminary design originally funded by the Soviet Navy in the late 1980's. This magnetometer has achieved noise levels comparable to SQUIDs. The potassium magnetometer does not suffer from the requirement for cryogenic support systems and motion compensation techniques of SQUIDs. This potassium magnetometer should be considered a critical development because the nation deploying it for submarine detection will enjoy a noise level 3 orders of magnitude less than that of systems deployed by the U.S. Navy and others. This translates to a possible 10-fold increase in detection range. When employed as a gradiometer, it has the advantage of near cancellation of time-varying geomagnetic background noise (geomagnetic pulsations). The rms noise level of a single potassium magnetometer has been measured as 14 microgammas (14 femto tesla) at 0.1 Hz. The noise level of a 2-m gradiometer is 10 microgammas (10 femto tesla) per meter at 0.1 Hz. These measurements were made with two complete gradiometers operating side by side in the real Earth field. In an experiment conducted at Kavgolovo (a test site northeast of Saint Petersburg, Russia) it was shown that laser pumping (versus lamp pumping) could further reduce the noise by a factor of 3. Relative to ELFE, propeller-shaft rotation modulates corrosion currents in the 3- to 20-Hz range. Russia has deployed sensors on the ocean bottom and detected this phenomenon. China and Japan have limited but expanding capabilities in several magnetometer technologies. Although the United States leads in platform motion compensation, Russia appears to have the lead in the use of compensation using magnetic sensor arrays.

Country	Magnetometers	Magnetic Gradiometers
Australia	●●	●●
Austria	●●	●
Brazil	●	●
Canada	●●●	●●●
China	●●	●
Czech Republic	●	●
Finland	●●●	●●●
France	●●●	●●●
Germany	●●●	●●●
Hungary	●	●
India	●●	●●
Israel	●●●	●●●
Italy	●●	●●
Japan	●●●	●●●
Netherlands	●●	●●
Poland	●	●
Romania	●	●
Russia	●●●	●●●
Slovak Republic	●	●
South Africa	●●	●●
South Korea	●●	●●
Spain	●●	●●
Sweden	●	●
Switzerland	●●	●●
Taiwan	●	●
UK	●●●	●●●
Ukraine	●●	●●
United States	●●●●	●●●●

Legend: Extensive R&D ●●●● Significant R&D ●●● Moderate R&D ●● Limited R&D ●

Figure 16.4-5. Magnetometers and Magnetic Gradiometers WTA Summary

SECTION 16.5—PRECISE TIME AND FREQUENCY (PT&F)

Highlights

- The worldwide availability of accurate time via GNSS will increase the combination of communications, imaging, and navigation functions into multihybrid sensor systems. This will provide a common grid reference for battlespace data management. The accuracy of the ionosphere model is a limiting factor on GPS time transfer.
- Autonomous and common three-dimensional POSITIME grid reference will improve battlespace situational awareness by providing a precise POSITIME tag on all battlespace information collected. This will provide real-time knowledge of the location and movement across battlespace of allied and enemy assets.
- Accurate time is required for autonomous operation of satellite network geolocation systems and enhanced crypto/transec performance in spread-spectrum communication systems.
- The importance of PT&F has only recently been recognized in military and commercial usage because of the availability of GNSS time. Foreign sources are currently providing the funding engine for future technology improvements in PT&F, as U.S. R&D funding has declined.
- The number of U.S. Atomic Frequency Standard suppliers is declining and may be down to one within 5 years.

OVERVIEW

PT&F is the key to current and future POSITIME, navigation, communication, and imaging systems. While the need of imaging and communication systems for PT&F is becoming more critical for military use, most current emphasis for PT&F technology is to support navigation and mainly for commercial use. Both the U.S. GPS, with its inherent vulnerabilities (see Section 16.3), and Russia's GLONASS worldwide navigation systems are actually PT&F systems. Both systems employ atomic clocks in the satellites. A cesium beam frequency standard is utilized aboard each GLONASS satellite. Both systems are capable of time transfer to a precision of 10 to 30 nanoseconds, but the GLONASS system time reference is not coordinated universal time, (UTC), as maintained at the United States Naval Observatory (USNO), the designated time reference for the United States. GPS uses both cesium and rubidium frequency standards in the current satellites, known as Block II/IIA, and rubidium standards in the replacement satellites, Block IIR. PT&F and signal detection and processing technology are required to acquire, synchronize, and track the desired satellite signals for measurement of navigation parameters. Technologies contributing to superior performance include the application of analog and digital correlation filters, DSP and microelectronics. Increased computational effectiveness for a given equipment volume and weight could provide an adversary with two distinct navigation payoffs: (1) certain navigation capabilities could be enhanced by advance computer technologies, and (2) for a given available volume/weight, the navigation performance could be enhanced in terms of accuracy, reliability, and resistance to hostile actions.

The GPS capability to transmit corrections to UTC (USNO) in the navigation message and thereby transfer time is given in the GPS Interface Control Document, ICD-GPS-202, as 28 nanoseconds (1 sigma). The 1997 Master Navigation Positioning and Timing Plan, CJCSI 6130.01A, requires timing accuracy to be 100 nanoseconds or better for all DoD users worldwide. The plan further states that precise time will be valuable to users who must time-synchronize other systems such as JTIDS (see Sections 16.1 and 16.3). The disparity between the documents exists because ICD-GPS-202 was recently updated to reflect increased user requirements. These wide-sweeping capability figures can still be misleading. The capability and conditions of time transfer from GPS depend greatly upon the instrumentation used, user conditions, and period of interest. For example, the time transfer capability of a fixed site, at a well-known position, needing long-term timing, and having the ability to integrate or process data over hours or days is considerably different than for a high-performance aircraft needing time to transfer in real-time or from the individual sensors of an array to a central processor. Time-transfer operational performance of GPS and data quoted is to a fixed site. GPS capabilities for military users under different conditions should be given as a more representative capability for system planning. Transfer of this technology will assist an adversary in improving the combat capability of its platforms. The use of stable clocks in a high dynamics platform can significantly improve timing availability by providing smoothing or "flywheeling" action that preserves time synchronization made under less stressed conditions.

Systems-integration technology enables the integration of communication and multiple navigational instruments outputs through advanced digital processing to provide an extended range of operational functions and increased combat performance. These technologies support reduced size and weight of equipment, improvement in navigation accuracy, continuity of operations, reliability, survivability, and resistance to enemy countermeasures. The substantial margin of capability added is critical to continued U.S. superiority in precision radio navigation and the multitude of missions dependent thereupon.

BACKGROUND

The importance of PT&F may be clearer with a description of the overall architecture of systems involved with the generation, dissemination, and maintenance of military common time. The overall process is described below.

- **Reference Time.** The common time scale to be used by U.S. military forces and systems is that generated, coordinated internationally, and maintained by the USNO. This time scale is designated internationally as UTC (USNO), representing the actual time available as a physical signal output of the USNO master clock. USNO maintains a master clock system of various commercial atomic clocks, remote precise time reference stations, and interface to dissemination systems. The alternate master clock became operational at Schriever Air Force Base in 1996 and is collocated with the GPS master control station as DoD's primary means of global time dissemination. The alternate master clock provides system redundancy and is linked to the master clock by a two-way satellite time transfer system. A severe handicap to users of UTC time, such as in telecommunications, is the yearly insertion of a nanosecond to account for the rotational changes of the Earth due to increases in the moon's orbital path around the Earth. In the next 5 to 20 years this correction problem needs to be internationally adjudicated to resolve the communication synchronization problem.
- **Reference Time Dissemination.** The dissemination of UTC (USNO) is accomplished by a collection of methods relying upon various systems, predominately POS/NAV systems. This dissemination function is a secondary mission of these systems, and no operational systems exist that are specifically designed and used for PT&F dissemination. Dependence upon secondary mission requirements or capabilities does not support a cohesive system architecture for the many systems that rely on precise time. Since PT&F is a secondary mission, operational control, coordination, and regulation of the time disseminated is an informal agreement without the impact of operational requirements. There are limited efforts underway to provide U.S. forces an affordable, nonjammable, precise, and accurate common time reference for all military electronic systems.
- **User Interfacing.** User systems have an increasing role in the distribution and sharing of time information. The Mark XV identification friend or foe (IFF) system and its counterpart North Atlantic Treaty Organization (NATO) identification system had a major problem in the distribution and synchronization of secure communications between their user platforms, which led in major part to the cancellation of the U.S. project. This should be a clear indication that the sharing of time information by user platforms and systems will be increasingly more important as higher data rates and crypto requirements (see Section 10) become more stringent. In turn, this will create an increasingly larger problem of controlling or managing intersystem timing exchange or interoperability. Coordination and standardization of these interfaces is part of the overall coordination role needed to produce robust systems synchronized to a common time reference.

RATIONALE

The importance of PT&F to military systems as a technology area is becoming more evident with the deployment and operational use of GPS. GPS provides the means for accurate and stable time to be disseminated to forces around the world. The application of GPS for timing is increasing, particularly in telecommunications and data transfer. As a result, the technology to maintain and use precise time is becoming of increasing military importance. However, because of the small quantity of Atomic Frequency Standard clocks that are needed yearly, the number of U.S. sources may decline to one within 5 years. Also alarming is the potential demise of the precision crystal oscillator suppliers in the United States in the near term. The oscillators are a critical module of the atomic clock system. There is a need for a precise, robust protocol for setting time across networks. A robust precision network time protocol (PNTP) using broadband communication links is needed for future military systems. The following table is an example of the possible types of needs that should be recognized for PT&F.

Table 16.5-1. User Clock Precision/Accuracy Requirements and Benefits

Platform/System	Current Accuracy to UTC (USNO)	Future Accuracy to UTC (USNO)	Benefit of Improved Time Accuracy
Nominal Use			
Time Epoch (second)			
Low-Accuracy Aircraft/Land Mobile	10^{-3}	10^{-9}	Quicker network entry for comms/IFF, enhanced crypto. Improved interoperability.
Ship/Submarines	10^{-6}	10^{-9}	Quicker network entry for comms/IFF, enhanced crypto. Improved interoperability.
Communication Sites/Aircraft	10^{-6}	10^{-9}	Quicker network entry for comms/IFF, enhanced crypto. Improved interoperability.
Radar/Surveillance/Intelligence	10^{-6}	10^{-9}	Better targeting/emitter location such as TDOA. Enable spread spectrum LPI.
Frequency ($\Delta f/f$)			
Low-Accuracy Aircraft/Land Mobile	10^{-12}	10^{-15}	Quicker network entry for comms/IFF. Improved interoperability.
Intermediate Land Reference Sites	10^{-13}	10^{-15}	Comms net access. Frequency calibration source for low-power oscillators.
Long-term Autonomous Timekeeping	10^{-13}	10^{-15}	Extended autonomous periods.
Large TDMA Systems	$\sim 10^{-11}$	10^{-15}	Enhanced network synch. Higher data rates.
Precise Mode (Cesium Calibration Updating)			
Time Epoch (second)			
ECCM Comm, Radar and Surveillance Systems	10^{-9}	10^{-11}	Better targeting/emitter location such as TDOA. Enable spread-spectrum LPI.
Submarine Comm, Ship	10^{-8}	10^{-11}	Enhanced network synch. Higher data rates. Longer autonomy. Enhanced navigation.

Current military fielded systems predominantly use cesium beam standards. These well performing and reliable devices have become the workhorse of remote timekeeping systems; however, they are large units not easily incorporated into smaller mobile platforms. Many users are therefore adopting secondary clocks, such as rubidium, disciplined by GPS time for long-term performance. These GPS-dependent clocks provide good performance in both long and short time frames, and the timing signals are being incorporated into many commercial systems such as telecommunication and computer networking systems. This dependency on a GPS-disciplined clock is a military concern. The following discusses the PT&F developing technologies that are being investigated for future PT&F requirements.

1. **Time Distribution.** More accurate and stable time sources, time distribution, and time transfer methods are required by the military to provide the warfighter with a coherent tactical picture and enable rapid, accurate capability assessment. With the closer logical linkage of combat systems and C2 systems and the proliferation of GPS units within precision weapons, the control of time sources and allowable variation of time provided by those sources to systems and elements across the battle space must be better managed. For example, better monitoring techniques could be utilized to improve time stability/accuracy performance by creating algorithms that combine clock outputs (ensembling) and automatically detect phase jumps or frequency perturbations and/or compensate for them. In addition, robust selection of alternative time source(s) in the event the preferred time source becomes unavailable must be better addressed. Present implementations often use multiple redundant clocks and GPS receivers and have no means of utilizing resources from other collocated systems. There is a strong need to establish a common time reference system, available across the platforms and battlespace, that will be usable by all systems. Such a system would include flywheel clocks at individual systems to provide autonomy from battle damage or loss of external references. Methods of time distribution other than GPS, particularly at the battlespace level, need development. These can include two-way satellite time transfer, fiber-optic systems for use within a platform, and other techniques such as communications systems between platforms.

2. **Atomic/Ion Clocks (Trapped Ion Storage and Slow Atom Technology Clocks).** This developing class of atomic clocks will improve absolute accuracy to better than 10^{-15}. Initially, these clocks will be larger, heavier, and more delicate than current military and commercial devices, but they have the potential to become competitive or better in size, weight, and power with existing clocks. This is a technology area where the Europeans are making great efforts and, in some cases, lead the United States.

   a. **Cesium Fountain Clocks.** Optical techniques for the containment and cooling of atoms, which was the 1997 Nobel Prize for Physics, offers new possibilities in methods for precise clock development. Optically contained neutral atoms can be cooled to close to absolute zero and launched into an interrogation system, which can make very accurate measurements of frequency to less than 1×10^{-15} sec/day. The immediate application being investigated is a "cesium fountain" clock. A clock of this type, theoretically proposed by Zacharis in the 1950's, is being developed by the National Institute of Standards and Technology (NIST) and USNO. The French have been operating such a clock at the Paris Observatory for several years, and NIST and the French are actively developing this technology for space application.

   b. **Mercury Linear Ion Trapped Clocks.** This type of magnetically trapped and optically interrogated clock is being developed by Jet Propulsion Laboratory (JPL) and is being deployed as replacement for the hydrogen master standards in the NASA Deep Space Network. One is to be tested for possible contribution to the master clock ensemble. These units employ mercury ions suspended in a magnetically generated linear trap and optically interrogated. These systems have the potential of extremely good stability for long-term measurements. HP did preliminary commercial product development on a mercury ion clock in the 1980's, but the design did not transition to production.

   c. **Other Technology Clocks.** Research into other types of trapped ion clocks and cooled cavity resonators is proceeding. The potential for migration into usable systems is estimated to be 5–10 years. These new, high-precision standards could contribute to the definition of reference time and time-scale generation, supporting a worldwide coordinated common time and potentially an improved time scale for DoD systems. The NIST microwave ion clock is pushing 10^{-15} second, and the optical version should realize 10^{-16} second when development efforts are funded to correct the optical frequency to the microwave region where it can be used in electronic systems. Large, research-oriented clocks could play a direct role in timing centers that are involved in the best performance clocks for time-scale operation. If successful in timing center improvements, this could again press the state of the art in dissemination and distribution of timing information to users.

3. **Low-Power Clocks.** There is a very strong need for clocks that perform at or near the accuracy level of current small clocks but at significantly reduced size, weight, and power. A technology currently in development, the microcomputer-compensated oscillator (MCXO), provides the stability of an ovenized crystal oscillator at or below the input power levels of the traditional temperature-compensated oscillator (TCXO). Another promising area of development is the miniaturized gas cell clock using cesium or rubidium. This approach trades off some performance of the traditional rubidium clock to provide an atomic clock with a volume of about 25 cm^3 and a power consumption of less than 1 W. These lower power devices can offer a level of performance suitable for GPS receiver direct-Y code acquisition and communications system synchronization in a package suitable in size and power for manpack or low-power mobile platform users.

4. **Optically Pumped Clocks**. The use of laser diodes for optical pumping and cooling of atomic systems to produce clock signals is a developing technology that will be adapted to field or system usable units. Low-power, fixed-mode diode lasers offer a natural means of interrogating and controlling small, ruggedized standards for field or platform use. French and Swiss companies have already introduced prototype small optically pumped cesium and rubidium units.

WORLDWIDE TECHNOLOGY ASSESSMENT

The United States has been the world leader in the development and utilization of PT&F technology for military electronic systems, including navigation, telecommunications, and data transfer. However, this leadership role is waning as U.S. R&D funds for PT&F technology have declined. The U.S. development efforts have been traditionally funded as parts of the host programs they support. As these programs, such as GPS, reached maturity, the funding for PT&F development vanished. NIST is operating a cesium fountain clock and is also developing this clock for space applications. Foreign PT&F developments are increasing, particularly in France, Switzerland, and in the European Space Agency (ESA) and associated research centers.

Country	Time Distribution Systems	Atomic Ion Clocks	Low-Power Clocks	Optically Pumped Clocks
Australia	●●●	●●	●●●	●●
Austria	●●	●	●●	●
Brazil	●●	●	●●	●
Canada	●●●	●●●	●●●	●●●
China	●●●	●●●	●●	●●
Czech Republic	●●	●	●●	●
Finland	●●	●	●●	●
France	●●●●	●●●●	●●●●	●●●●
Germany	●●●●	●●●●	●●●●	●●●●
Hungary	●●	●	●●	●
India	●●	●●	●●	●
Israel	●●●	●	●●●	●
Italy	●●	●●●	●●●	●●
Japan	●●●●	●●	●●●●	●●
Netherlands	●●	●	●●	●●
Poland	●●	●●	●●	●●
Romania	●●	●	●●	●
Russia	●●●●	●●●●	●●●●	●●●●
Slovak Republic	●●	●	●●	●
South Africa	●●●	●●	●●●	●●
South Korea	●●●	●●	●●●	●●
Spain	●●●	●●	●●	●●
Sweden	●●●	●●	●●	●●
Switzerland	●●●	●●●	●●●	●●
Taiwan	●●	●	●●	●
UK	●●●●	●●●	●●●●	●●●●
Ukraine	●●●	●●	●●●	●●
United States	●●●●	●●●●	●●●●	●●●●

Legend: Extensive R&D ●●●● Significant R&D ●●● Moderate R&D ●● Limited R&D ●

Figure 16.5-2. Precise Time and Frequency WTA Summary

SECTION 16.6—SITUATIONAL AWARENESS/COMBAT IDENTIFICATION

Highlights

- RFID technology using secure, encrypted, millimeter waveform in the 33- to 40-GHz Ka-band will remain the primary NATO IFF capability to identify friendly forces in the battlespace for the foreseeable future. Advances in RFID technology will continue to improve performance, cost reduction, and size reduction.
- Advancements in long-range ATR databases, algorithms, and decision-aided tools, reducing the time needed to identify targets by a factor of three, will continue to complement RFID technology for positive identification of friendly, foe, and neutral targets.
- Increased usage of overhead intelligence, navigation, communication, and imaging sensor assets, using a common three-dimensional position and precise time grid reference, together with RFID/ATR systems, will provide reliable, positive, long-range identification capabilities within the lethality range of weapons.

OVERVIEW

At this time, this new section on SA/CID will only focus on the technologies that produce superior CID performance in the air-to-surface and surface-to-surface capability. Future updates of this document may address other SA/CID areas, including:

- Weather
- Air and ground traffic control
- Obstacle/ground/terrain avoidance
- Missile warning
- CID (air to air and surface to air).

SA includes all of the environmental, positional, and time conditions (past, present, and projected future) that affect the capabilities of the warfighter. Combat stress levels are inversely proportional to SA. Lack of SA will lead to adverse military decisions at all levels of command. CID plays a major role in achieving the capability to build and maintain a coherent tactical picture as discussed in the Section 16 overview and in Figure 16.6-1 below. CID is the capability to differentiate potential targets—mobile and fixed, over large areas with corresponding long distances—as friend, foe, or neutral in sufficient time, with high confidence, and at the requisite range to support engagement decisions and weapon release.[1]

The technologies include those needed for positive, timely, and reliable identification of friends, foes, and neutrals; classification of foes by class, type, and nationality; and interoperability required among the U.S. military and allied nations. The challenges are enormous, particularly in three specific areas: (1) a cooperative/noncooperative sensor systems; (2) a command, control, and communications (C3) systems—in particular, digital datalinks and radios, each of which contributes a portion to the CID solution; and (3) artificial intelligence tools that will fuse sensor and information, providing the warfighter with near-perfect, real-time discrimination between targets and non-targets on the battlefield. As such, CID is viewed as a capability, not a single system or technology. A "system-of-systems" approach is required. This subsection will consolidate the critical technologies addressed in the other MCT sections/subsections [Section 10 (Information Technology), Section 11 (Lasers and Optics), and Section 18 (Sensors)], as well as explore other technologies that will provide increasingly superior performance of SA/CID capabilities.

---

[1] *Joint Warfighting Science and Technology Plan: Combat Identification.*

This subsection will discuss three classes of technologies:

- *Sensors.* The target is characterized either noncooperatively (e.g., jet-engine modulation, high-range resolution radar, or electronic support measures) or cooperatively [e.g., MK XII (IFF) system or Battlefield Combat Identification System (BCIS)].
- *C3 (particularly digital datalinks and radios).* The target declares (either periodically or when queried) its identification and position in a reference frame that the "shooter" can correlate with its own weapon and sensor system (e.g., Link 16).
- *Artificial Intelligence Tools.* Vast amounts of data will need to be processed, correlated, stored, and displayed in real-time to be useful to a warfighter. AI tools include expert systems, intelligent agents, decision aides, modeling and simulation, and virtual reality.

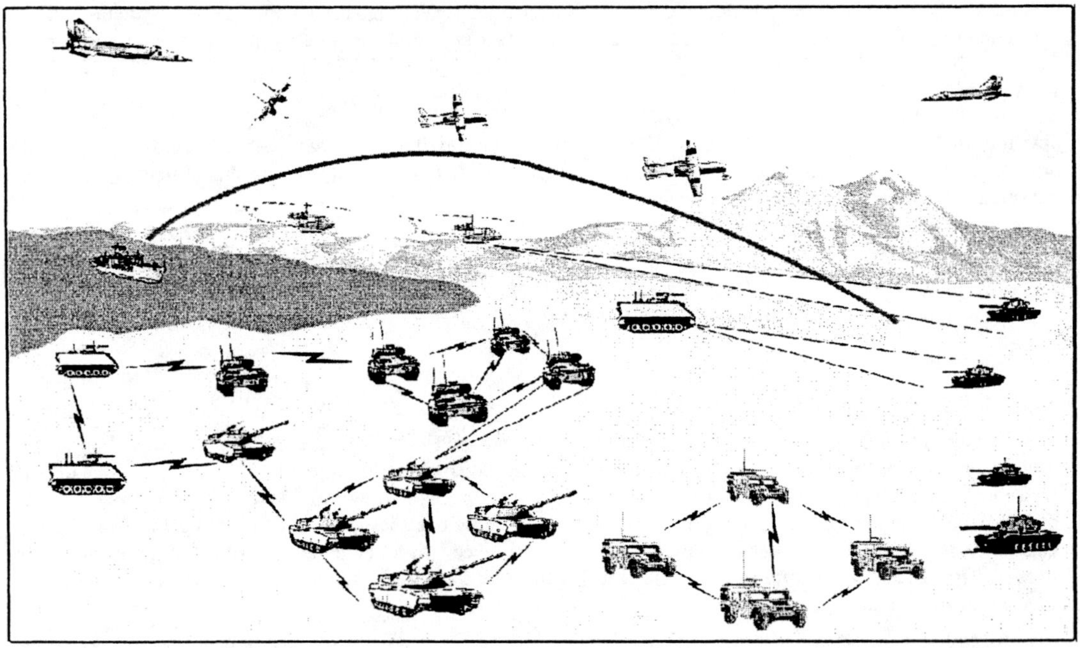

Figure 16.6-1. Concept Combat Identification

Currently, the cooperative tri-Service Mark XII RF/IFF system (circa 1970's) is the Q&A technology to identify friendly forces. Non-responses are considered unknown. Visual identification is used for neutral identification and foe identification in the air-to-surface and surface-to-surface areas, respectively. There is no long-range positive identification capability on the ground. As the lethality of weapon systems increases, and the speed and ferocity with which land battles are fought become greater, the need for systems that will aid warfighters in reducing fratricide are paramount. Making positive visual identification is difficult with allies and enemies using identical combat platforms and fighting battles under degraded natural and man-made conditions (obscurants, darkness, rain, dust, and fog). As proven in Operation Desert Storm, the confusion of a rapidly moving air-land battle using multinational forces creates a situational awareness nightmare.

Limitations in sensor resolution—coupled with variations in target aspect, state, countermeasures, and the battlespace signal propagation environment—complicate the job of target labeling. Improvements in sensors and target databases that expand the envelope of performance for these systems are necessary to increase target range, ID accuracy, and reliability.

Cooperative identification sensor systems, which only identify friends, have the advantage of being less of a technical challenge; however, they require all friendly potential targets to be equipped with the same corresponding identification equipment. This limitation will require more combined use of both cooperative and non-cooperative sensor systems.

Overhead CID technology sensor improvements that can interpret imaging and nonimaging sensor data to reliably identify the target ID in near real time are necessary. Communication improvements in secured data dissemination of SA multimedia information down to the lowest mobile echelon is required. The unprecedented amount of raw information produced by modern sensor systems and the effectiveness of C3 systems can overwhelm the capability of human operators and decision makers, requiring the need for a reliable automated decision aide tool. CID can be highly useful only when it is fully integrated with both C3 and weapon systems. CID requires effective and timely synchronization of communications systems with data from real-time surveillance, target tracking, and intelligence systems.

Affordability and exploitability are major barriers to having universal CID capability:

- **Affordability.** The cost of CID suites that are properly integrated with the weapon sight (both cooperative and noncooperative) are usually prohibitive if they are not form, fit, function and interchangeable (F3I) with an existing sensor or system. Additional functionality in the form of communications, SA, or sensing is helpful in making CID more affordable. The affordability of a system will also vary significantly depending on the environment in which it is considered. Aviation/maritime systems are generally more expensive than ground–based systems. As a result, solutions that are programmatically "affordable" for aircraft/maritime platforms are often prohibitively expensive for combat vehicles.

If the identification is determined by an off-board sensor, there is the added necessity to pass, correlate, and provide the warfighter the required information in a timely fashion. This requirement to correlate an identification label with a sensor return in the "weapon sight" is a key discriminator and a source of significant cost for the systems. Technology that eases the integration overhead of a CID-related system or reduces its component cost is required.

- **Signature Exploitability.** Noncooperative techniques of identification are most attractive to warfighters because of their ability to generate labels for foe, friend, and neutral contacts and because they can provide additional identification information on adversaries (e.g., platform type, class, nationality). For air/ maritime targets, the current capabilities of these systems are limited in range, aspect, and timeliness of reporting. The result is that the indications from this class of systems are frequently in the "unknown" or "not available" state. Improvements in sensors and target databases that expand the envelope of performance for these systems are necessary. For combat vehicles, the signal environment is such that reliable identification at maximum weapon range remains a significant technical challenge. Limitations in sensor resolution—coupled with variations in target aspect, state, countermeasures, and the battlespace signal propagation environment—complicate the job of target labeling.

Other issues are reliability and security. Unless the system is 100-percent reliable, possibilities exist for fratricide in combat. Antennas and other external devices (the BCIS uses an externally mounted transponder) may be blown off during combat, rendering the system useless. Another problem is security. If an enemy can read, jam, or duplicate the incoming or outgoing signals, the system's effectiveness becomes severely degraded. If the signals are not of an LPI nature, an enemy is likely to be able to localize emission sources and target them. It is also reasonable to expect that some of our systems will fall into enemy hands; therefore, our system must be reprogrammable. A different type of active system does not require interrogations but periodically transmits required information such as identity and status in the blind. This information "strobing" would have to be spectrally unique to prevent detection, but could simplify the overall system and allow one-half of the ID equation to remain passive. RFID tagging technology has great potential. This bar code system with a brain and voice can provide security in "secure" areas when the RFID tag is attached to the warfighter.

In the next 3–5 years, RFID technology using secure, encrypted, millimeter waveform in the 33- to 40-GHz Ka-band will remain the primary NATO IFF capability to identify friendly forces in the battlespace for the foreseeable future. Advances in RFID technology will continue to improve performance, cost reduction, and size reduction. Figure 16.6-2 shows the architecture interface concept for future U.S. Army combat identification systems to improve air-to-surface and surface-to-surface capabilities.

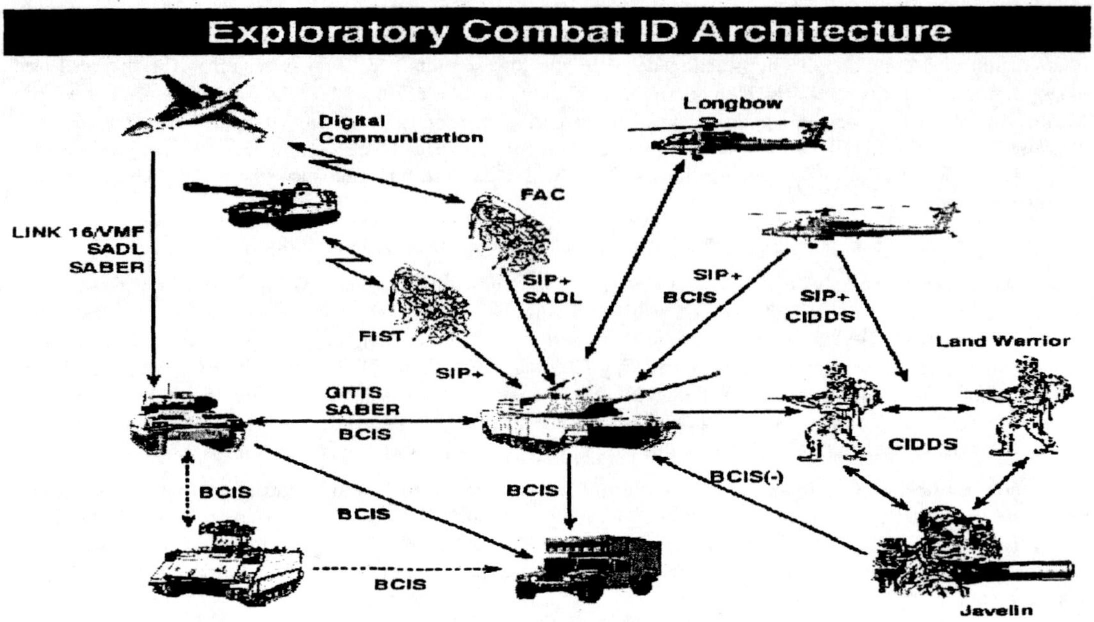

Figure 16.6-2. U.S. Army's Combat Identification Concept

- **For surface to surface.** Improvements in millimeter-wave technology, such as the U.S. BCIS, will provide identification of friends from unknowns at distances up to 14 km, day or night, with accuracy greater than 97 percent, reducing the risk of fratricide. BCIS allows the gunner or commander to make a rapid shoot/don't shoot decision at the point of engagement. Shooter platforms, e.g., tanks and fighting vehicles, are equipped with BCIS interrogator/transponder units that interrogate suspect platforms and respond to interrogations from other shooters. The interrogation process is automatically triggered by activation of the shooter's laser rangefinder, which sends an encrypted query to the targeted platform. If the target is friendly, its transponder receives the query and responds with an encrypted answer (nonshooter platforms are equipped with BCIS transponder-only units). When the interrogator receives an encrypted answer, it gives a "friend" response to the gunner/commander. If an invalid answer or no answer is received, an "unknown" response is provided to the gunner/commander, who then must continue using engagement tactics, techniques, and procedures. Responses are provided visually in the gunner's sight, as an audible tone on the intercom system, or both, eliminating the need for a gunner to remove his eyes from the target.
- **For air to surface.** The U.S. Army's single channel ground and airborne radio system (SINCGARS) improvement program [SIP(+)] is a "Don't Shoot Me System," consisting of a SINCGARS radio modified to use a GPS-coordinate-based message, enabling rotary-wing aircraft to interrogate ground-combat platforms equipped with SIP(+) radios using a "Don't Shoot Me" net feature.

In operation, the airborne SIP(+) radio transmits the interrogation message, which includes the targeted position. If an interrogation is detected by the SIP(+) ground radio and the target coordinates match, it transmits a "Don't Shoot Me" response to the airborne interrogator. The total interrogation/response cycle should take a maximum of 2.3 sec. Future improvements may reduce this to less than 1 sec.

In the next 5–10 years, advances in ATR algorithms and systems will provide a passive capability to distinguish between targets and nontargets automatically and noncooperatively. Concurrent with significant advances in information systems (see Section 10), ATR algorithms will be used more and more as a complement to RFID systems (i.e., BCIS/SINCGARS + ATR). During this period, ATR algorithm development will achieve significant growth, particularly in integration with multisensors and new ATR techniques (i.e., matched filtering, pattern recognition, model based, and clutter and edge modeling).

- **For surface to surface.** The Land Warrior System, a first-generation integrated fighting system for dismounted combat soldiers, will enhance the soldier's battlefield capabilities through the development and integration of a variety of Army components and technologies into a cohesive, cost-effective system. The Land Warrior System includes a computer/radio subsystem, a GPS receiver, a magnetic compass, a dead-reckoning navigator, VHF and UHF radios, and a video capture capability. The Integrated Helmet Assembly Subsystem includes a heads-up display and image intensifier for night operations and a weapons subsystem with thermal weapon sight, close combat optics, video camera with a video capture capability, laser rangefinder/digital compass, and an infrared laser aiming light. The system also includes protective clothing, load carrying equipment, body armor, a chemical/biological mask, and a laser detector. Improvements during this period could maximize the tactical CID performance with a view of the battlefield that exploits the time-critical combat information provided by an integrated SA and target identification (TI) system. Future potential capabilities are ATR and external near real-time (NRT) sensor cueing from overhead radar and satellites. To overcome the exploitable issues addressed and approach the 100-percent-reliable ID will require both active and passive technologies to be used. For active interrogation/responses should be multispectral; utilizing acoustics, IR, visual bands, RF, millimeter wave, and laser beams. It is also reasonable to expect that some systems will fall into enemy hands; therefore, the system must be reprogrammable.

- **For air to surface.** Improvements in communication systems (see Section 10), such as the Air National Guard's Situational Awareness Data Link Forward Air Controller (SADL FAC) system could provide information on friendly ground force positions as well as calculate the position location of laser targets. This system determines friendly ground force positions through use of the SADL (EPLRS) radio. The Mark VII Laser Rangefinder and the Precision Lightweight Global Positioning System Receiver (PLGR) generate the target location information. This information is digitally transmitted via the SADL radio to the SADL-equipped CAS aircraft. Another example is the U.S. Navy's Situational Awareness Beacon with Reply (SABER) program, which consists of a miniature UHF SATCOM transceiver and an integrated GPS receiver to provide a combined SA/direct-target ID solution for ground platform, fixed-, and rotary-wing aircraft performing air-to-ground missions. A beacon is installed on friendly ground platforms to enable them to periodically transmit an identification code, geolocation (position) information, and other host platform data either directly to line-of-sight receivers using ultrahigh frequency (UHF) radios or indirectly via an UHF satellite link using a low probability of intercept/low probability of detection (LPI/LPD) waveform called "Collection of Broadcasts from Remote Assets" (COBRA).

In the next 10–20 years, further improvements in information systems will provide a new class of high-data-rate, networked radio communications as part of command and control systems. These radios will have the ability to deliver data within specified time lines using prioritization and quality of service routing techniques. The combination of higher data rates, more reliable networking, and quality of service routing will enable these networks to provide significant, timely information on friendly situations, to include friendly locations and identifications. This information will contribute directly to Combat Identification. A particularly critical enabling set of technologies for these new high-data-rate radio communications systems will be low probability of intercept (LPI)/low probability of detection (LPD) waveforms. Wavelet overlay technology is particularly critical. Use of multisensors integrated with ATR algorithms and data bases and nanotechnology (providing more affordable position and precise time tagging) will also significantly increase SA/CID reliability and reduce target ID time to near real time. For example, the U.S. Air Force Spacecast 2020 study suggested an exotic space-based target recognition scheme in which a laser beam from a satellite would scan the battlefield, and the reflected energy would be analyzed by sensors on the spacecraft. By comparing the spectra, identification would be possible. Friendly tanks and aircraft could be chemically coated to produce a characteristic spectrum when excited by the space-based laser, proving a degree of IFF capability. Using techniques likened to "licking" and "tasting" to identify objects on the ground, the licking would be done by a laser beam fired from a satellite, which would be equipped with sensors that would "taste" the spectrum of the radiation reflected back from the target. By comparing this with a database of know tastes it would be possible to identify an object. Friendly tanks and aircraft could be coated with a chemical that produces a characteristic spectrum when excited by energy of a certain frequency or other characteristic.

- **For surface to surface.** The range and reliability at which identification can be reliably accomplished can be improved with increased use of overhead asset data and image technology and more use of smart RFID tagging. All weapons could include semi-passive interrogation systems that read the identity from a tag or label of some type, such as an RFID, on the vehicle or person. Integrated with a passive system, ATR pattern recognition could discriminate enemies from friendly and combatant from noncombatant. Using ATR and pattern recognition logic could assist in threat determination of noncombatants, based on discriminators such as vehicle type, color, and motion, or observations that personnel are carrying weapons, moving in a tactical manner, etc.

- **For air to surface.** A fully interoperable NATO C3 system (such as Link 16/22) could provide (1) friend identification automatically (for all friendly combat participants on the network); (2) a medium for passing hostile/neutral identification generated from other sensors/sources; and (3) a medium for passing friend identification (for those platforms not on the network) generated from other sensors/sources. Addressing the interoperability issue, the United States is migrating towards a J-series family of datalinks to include Link 16 for air operations, Link 22 for maritime operations, and variable message format (VMF) for ground operations. For air, maritime, and ground weapons, this could play a significant role in sorting friend from foe or neutral in the battlespace.

By 2025, friendly troops and equipment will enter the battlespace with their personal identifiers. The identification mechanisms could be in the form of microchips worn by or imbedded in the soldiers. The same principle could be applied to vehicles. An active or passive system could identify friendlies by reading a label attached to an object. As the battlenet sensors detect each target in the battlespace, they apply a physical label to the target. For example, a particle beam imprints coded information on the exterior of specially painted vehicles or irradiates the clothing of exposed personnel. Labels placed on targets could be magnetic, optical, or electronic, and can be sized down to the molecular level. The label contains data that includes the type of target, date time group, and military unit controlling the vehicle or person. Sensors would be required to interrogate an unknown transponder, analyze the response, and determine if the response came from a friendly system or a designated hostile system. If the interrogator receives a response that does not correspond to known friendly systems or fails to receive a response at all, the interrogator activates a separate series of identification methods involving discriminators such as material composition, acoustic, electromagnetic, thermal, or vibration signatures. For situations involving a mixture of hostile forces and noncombatants in an environment where no external evidence distinguishes the two (a riot or urban disturbance, for example), the system may need only distinguish between friendly "tagged" personnel and others.

RATIONALE

U.S. forces must be able to positively identify all targets in the battlespace for all combat mission areas—air to air, air to surface, surface to surface, and surface to air. Surface includes land, sea, and subsurface—otherwise known as ground and maritime. The CID need is essential for commanders to effectively field, at any time, fighting forces that can rapidly and positively identify enemies, friends, and neutrals in the battlespace; manage and control the battle area; optimally employ weapons and forces; and minimize the risk/occurrence of fratricide.[2]

Lack of positive target identification and the inability to maintain SA in combat environments are the major contributors to fratricide. Vehicle commanders, gunners, and attack pilots cannot distinguish friendly and enemy thermal and optical signatures at the ranges at which they can be acquired. Our weapons can kill beyond the ranges where we have clear ID. Our tactics lead us to exploit our range advantage over the enemy. During limited visibility or in restricted terrain, units in proximity can mistake each other for the enemy because of short engagement windows and decision time. We do not have a means to determine friend or foe, other than visual recognition of our forces and the enemy's.

When the enemy and our allies are equipped similarly, and when the enemy uses U.S. equipment, the problem is compounded. Simple, effective fire and maneuver control measures and plans, good SA, and disciplined engagements are absolutely necessary. The long-range vision of CID is to enable U.S. forces to positively identify all targets in the battlespace for all combat mission areas. Our fighting forces must be able to rapidly and positively identify enemies, friends, and neutrals; manage and control the battlespace; optimally employ weapons and forces to increase economy of force; lower combat attrition and increase enemy losses; and minimize the risk and occurrence of fratricide. To achieve this will require significant improvements in these critical technologies: sensors, C3, and AI tools.

WORLDWIDE TECHNOLOGY ASSESSMENT

The development leaders of SA/CID technologies are for the most part located within the NATO countries. To coordinate national combat identification work, France, Germany, the UK, and the United States established a four-nation Combat Identification Working Group (CIWG) in October 1992. A final report in January 1998 recommended Ka-band (33–40 GHz) Q&A as the most cost and operationally effective technology for interoperable target identification for the long term (2001+). This was approved, and the four-nation CIWG is working to develop a standardization agreement (STANAG) for NATO target identification.

---

[2] *Joint Warfighting Science and Technology Plan: Combat Identification.*

Under a program being jointly conducted by the U.S. Air Force's Wright Laboratory, the Massachusetts Institute of Technology/Lincoln Laboratory, Sandia National Laboratories, the U.S. Naval Air Warfare Center, the U.S. Army Research Laboratory, and 15 other university and industry organizations in the United States, signature-prediction codes are being developed. These codes will cope with frequencies ranging from B to K band and with targets consisting of electrically conducting material, homogeneous and isotropic frequency-dependent materials, radar-absorbing nets, and nonmetallic structures made from glass fibre and other semitransparent materials. The task of compiling the more than 10 million signatures and images required for noncooperative target recognition development and operational fielding will take several years, even with fast computers running around the clock.

Country	Sensors	C3	Artificial Intelligence
Australia	●●	●●●	●●
Austria	●●	●●●	●●
Belgium	●		
Canada	●●	●●●	●●
China	●●	●●	●
Finland	●		●
France	●●●●	●●●●	●●●●
Germany	●●●●	●●●●	●●●●
Israel	●●●	●●●	●●●
Italy	●●	●●●	●●●
Japan	●●	●●●	●●
Russia	●●●	●●●	●●
South Korea		●●	
Sweden	●		●●●
Switzerland			●●
UK	●●●●	●●●●	●●●
Ukraine	●●	●●	
United States	●●●●	●●●●	●●●●

Legend: Extensive R&D ●●●● Significant R&D ●●● Moderate R&D ●● Limited R&D ●

Figure 16.6-3. Situational Awareness/Combat Identification WTA Summary

BIOMEDICAL CRITICAL TECHNOLOGY

F. Spalt
(Editor)

Highlights

- Endemic organisms, the major cause of infectious disease in deployed forces, can be dealt with using new technologies.
- Human genome sequencing will reveal disease and stress susceptibility factors.
- Pathogen genomics permits rapid detection/identification of agents.
- Immune system enhancers and novel antibacterial and antiviral compounds promise protection from or defeat of biological hazards.
- Interfacing biomedical technology with communication and information systems is enabling for global surveillance of endemic disease.
- Telemedicine provides opportunity to stabilize and enhance survivability of traumatically injured persons.
- Growth factors enhance wound repair and neural tissue regeneration.

OVERVIEW

The biomedical issues confronting the Defense community in the 21st century differ markedly from those of the past two decades because of rapid advances in biotechnology and medicine, emerging and re-emerging infectious diseases, and increasing environmental health risks. At the same time, rapid progress in the technologies of computers, information storage and retrieval, and miniaturization affect access to medical histories of defense personnel; this information can be used to prevent disease and enhance operational capability. The reduction in numbers of combat personnel, the threat of urban warfare in police actions, and the changing nature of the threat from biological and chemical warfare by rogue nations all contribute to the major role of medical care in the military community of the future. These changes require the re-examination of existing systems for health-care maintenance, individual and group protection, and management of injured persons in a manner that conforms to contemporary command, control, communication, computer, intelligence (C4I) capability. The Tactical Medical Coordination System (TacMedCS) provides a basis for the anticipated applications for the technologies likely to develop in the next two decades. The resulting structure will provide real-time information to unit commanding officers (CO) and enhance the development of dynamic battlespace orientation, mission planning, asset allocation, and intelligence.

Among the health threats to military personnel are infectious disease; chemical and radiological exposure; and environmental stresses, including sleep deprivation, exposure to temperature extremes, and loud noises. The most common sources of infectious diseases are endemic or epidemic diseases in the deployed area. Because of recent incidents involving biological agent capability by nation states (e.g., Iraq) and non-national terrorist groups (e.g., Aum Shinrikyo in Japan), the threat of contracting an infectious disease through biological agent exposure has increased markedly. Several task groups in the United States have examined the current response capability of U.S. military, civil defense, local police, and emergency medical personnel to such an incident and have identified several needs. Among these are the production of novel antitoxins, vaccines, immunization adjuvants, and antibiotics to perceived threat agents. The threat agents include those identified by the Australia Group (AG) and others identified by the U.S. defense community. Advanced technologies related to training in the management of infectious disease, trauma, stress, and treatment are additional requirements.

RATIONALE

This biomedical science section addresses those technologies in the biomedical arena that will affect the operational readiness and sustained force projection capability of the U.S. Army, Navy, and Air Force. Section 4.1 considers the etiological factors of infectious disease and susceptibility of persons to disease and stress. Biodiversity plays roles in the evolution of disease and in developing seed stocks that prevent total destruction of a food material. Biodiversity also provides protection to the human population from potentially catastrophic events. Nonlethal weapon effects are considered in this section. The effects of new technologies on treating, managing, and protecting individuals and groups from infectious disease are considered in Section 4.2. Innovative approaches to training first responders to biological or chemical attack are also included. The protection of the forces from conventional weaponry, and the treatment of traumatic injuries, including damage to the nervous system, are presented in Section 4.3. Telemedicine approaches, treatment of wounded forces at sites distant from medical facilities, and affordability issues are also discussed. Technologies enabling for medical command strategies are presented in Section 4.4. The time line between development of the technology and the capability of inserting that technology into a system (i.e., application readiness) and the issues of "affordability" are considered in each section. The report provides a WTA that compares the capability of developed and developing nations to use the leading-edge biomedical technologies for support of their military operations. All the technology items identified here are driven by broad requirements and applications and all have substantial non-military commercial applications and support infrastructures. The industrial sectors supporting the biomedical science thrusts include health/medicine, pharmaceuticals, food processing, cosmetics, computers, communications, and electronics.

Identification of Developing Biomedical Technologies Related to Military Applications

The "Developing Biomedical Technologies" have been categorized into three functional sets: infectious diseases and protection, trauma, and stress. The direct relevance of the developing critical biomedical technologies to military applications and needs is then indicated.

Developing Biomedical Technologies as related to infectious disease, trauma, and stress:

Infectious disease related

- Rapid genomic analyses of biological agents and of human susceptibility to disease, toxicants, and stress;
- Rapid diagnosis of infectious disease by genomic/antigenic analyses and sensors;
- Database for infectious diseases worldwide;
- Microchip encoding of medical history;
- Containment and transport of infected persons;
- Remote diagnosis and surgery; and
- Novel antivirals, antibiotics, vaccines, and drug development.

Trauma related

- Blood coagulation factors;
- Artificial organs, including blood, and growth factors for neural repair; and
- Novel prosthetic materials.

Stress related

- Performance enhancing or sustaining chemicals (vigilance/attentiveness);
- Analytical measures and metrics for human performance;
- Iconographic and audio-visual displays and ergonomic tools;
- Nonlethal weapons including physiological/psychological assault; and
- Ingested/injected materials to identify location of persons.

Relevance of these technologies to military applications and needs:

Topics	Section
Affordability and medical care	4.3
Analytical tools to assess impact of information or disinformation (books, media, community civil defense training) on responses to biological/chemical (B/C) incidents or accidents. New article in *The New England Journal of Medicine* by Timothy Johnson ("Shattuck Lecture—Medicine and the Media," 339:87–92, 1998) addresses most issues.	4.2
Bioeffects of nonlethal weapons	4.1
Bioequivalence methods to compare effect of toxicants on performance (e.g., the degree of function compromised by a person intoxicated with an industrial solvent or fuel in terms of compromise by alcohol intoxication or by hypoxia). Related to aging, stress, gender, etc.	4.3
Biomarkers of toxicant exposure including sensors	4.3
Facilities and technologies to test efficacy of B/C agent protection (Phase II and Phase III, including physical protection, biological and chemical protection); includes side effects of protection on the performance of personnel	4.2
Genomics of personnel and of agent	4.1
Organ replacement by use of organ culture and polymeric materials	4.3
Personal medical history microchip (personnel medical history of persons pre- and post-deployment) [Institute of Medicine Report on Gulf War]	4.1, 4.2
Preparation of soldiers for deployment	4.2
Reinsertion of the soldier following trauma or illness, including reducing permanent disability	4.2, 4.3
Remote detection of vital signs and locator of living forms	4.3
Repair of the central nervous system	4.3
Sensors (see biomarkers)	4.1, 4.2
Small deployed units	4.1-4.3
Stanch abdominal bleeding and treat extensive burns	4.3
Surveillance, protection, and response to B/C agent, including differentiating between intentional release of B/C agents and disease in newly deployed areas	4.1, 4.2
Telemedicine for the care of wounded personnel in remote or combat locations	4.3
Training of health-care personnel for incidents/accidents involving B/C agents, nuclear material, and directed-energy sources	4.2
Vaccines, antivirals, antibiotics, and vaccine platforms (e.g., viral modification to deliver epitopes and/or DNA components of interest)	4.2

Country	Etiological Factors	Defeat or Management of Biological/ Chemical Attack	Management of Trauma, Stress, and Treatment	Tactical Medical Command/ Control
Australia	●●●	●●●	●●●	●●●
Belgium	●●	●●●	●●●	●●●
Canada	●●●	●●●	●●●	●●●
China	●●	●●●	●●	●●
Cuba	●	●●	●	●●
Czech Republic	●●	●●●	●●●	●●
Egypt	●	●●	●	●
France	●●●●	●●●●	●●●	●●●
Germany	●●●●	●●●●	●●●●	●●●
Hungary	●●	●●	●●	●●
India	●●	●●	●●	●●
Iran	●	●●	●	●
Iraq	●	●●	●	●
Israel	●●●	●●●	●●●	●●●
Italy	●●●	●●●	●●●	●●●
Japan	●●●●	●●●●	●●●●	●●●●
Malaysia	●●	●●	●●	●●
Netherlands	●●●●	●●●●	●●●●	●●●●
North Korea	●	●	●	●
Norway	●●●	●●●	●●●	●●●
Poland	●●	●●	●●	●●
Russia	●●●	●●●	●●	●●●
Singapore	●●	●	●●	●●
South Korea	●●	●●●	●●	●●
Sweden	●●●●	●●●●	●●●●	●●●●
Switzerland	●●●●	●●●●	●●●●	●●●●
Syria	●	●	●	●
Taiwan	●●	●●●	●●●	●●●
Ukraine	●●	●●	●●	●●
UK	●●●●	●●●●	●●●●	●●●●
United States	●●●●	●●●●	●●●●	●●●●

Legend:
Capability in technology elements ●●●● most ●●● many ●● some ● at least one

Figure 4.0-1. Biomedical Technology Systems WTA Summary

WORLDWIDE TECHNOLOGY ASSESSMENT

The technology assessment shown in Figure 4.0-1 describes the capability of the identified nation to implement new or novel solutions to needs faced by the military sector in that nation. The military applications of biomedical sciences are related in large part to the civilian applications; therefore, the technologies are dual use. The industrialized nations are the world leaders in medical technology developments, and their ability to contribute innovative solutions to biological and chemical threat agents is therefore higher than that of less developed countries. Several databases related to biomedical sciences are available via the Internet. These databases include the Human Genome database (i.e., nucleic acid sequences of the human population), protein sequences, and anatomical reconstruction of the human from magnetic resonance and whole-body sections (e.g., Visible Man and Voxel Man). As a result, scientists and technical persons in all nations can access these databases. Nation states can purchase equipment to meet needs or can identify new strategies by accessing the worldwide web (WWW). These states, however, may not have the capability to innovate new solutions to the problems because of lack of a research base or an infrastructure with limited financial, technical, or human resources. On the other hand, those nations with leading technical assets (human, financial, material, and infrastructure components) are capable of adding to the knowledge base at a rapid rate. They have a marked advantage in developing vaccines, new drugs, and novel body coverings that can mitigate or eliminate the effects of biological threat agents. The capability is therefore a function of leading-edge technology, rather than of access to available data. The WTA tables reflect the technological capabilities of nations to advance understanding of and response to threat agents, rather than the capability to access the information on the Internet.

SECTION 4.1—ETIOLOGICAL FACTORS

Highlights

- Genomics of pathogens are being determined.
- Susceptibility factors of human genome are being characterized.
- Worldwide database of endemic diseases is being compiled.
- The interface between human susceptibility, virulence factors, and worldwide endemic disease will now make possible prediction of risk, required treatments, and a human factors element of readiness of deployed forces.

OVERVIEW

The biomedical components of etiological factors in disease include genomics of biological agents, human genomics, global surveillance of infectious disease, and nonlethal weapons. The management of physical injury is considered in Section 4.3.

The development of clinical infectious disease in the human subject is the result of the interaction between the human host and the pathogen. The genome of the pathogenic organism and the susceptibility of the subject to the organism are factors in disease manifestation. The pathogenicity of viruses and of bacteria is a function of the genomic composition of the microorganism. There are common genomic sequences shared by biological agents which increase the pathogenicity of these organisms. These sequences, identified as pathogenicity islands (PAIs), have been characterized for several agents on the Australia Group list. Knowledge of the sequences can be used both to construct deoxyribonucleic-acid (DNA)- or ribonucleic-acid (RNA)-based sensors or to develop antiviral or antibiotic pharmaceuticals that will delay or reduce the pathogenic effects of the organisms. Individual susceptibility to a pathogen is a function of the nutrition, neuropsychopharmacological characteristics, fitness, and the genome of the person. One example of the genomic component of resistance was the finding that those persons whose predecessors came from areas of the globe that experienced bubonic plague were more resistant to human immunodeficiency virus (HIV) infections than persons from other areas. One explanation for this observation was that there was a common genetic factor that protected individuals against both diseases and that the plague permitted survival of such protected populations.

Progress made in biotechnology and the understanding of the genetic code of humans will enable the susceptibility of persons to disease and stress to be determined by the year 2003. Since the genetic material of the human is determined at conception, it is possible to identify and characterize genes responsible for the inherited predisposition to decreased human performance and clinical disease. By the year 2003 it will be possible to have gene-based screens and diagnostic tests for alterations in genes; to develop the informatics needed to collect, store, analyze, and integrate the resulting molecular patterns; and to obtain epidemiologic and clinical data of persons at birth. With this information it will be possible to identify every major human gene that predisposes people to disease; to use the knowledge to help persons at risk; and to deal with the psychosocial, ethical, and legal issues associated with inherited disease. A portion of the material in Section 4.3 (Affordability—National Clinical Information System) is relevant to this section.

RATIONALE

The deployment of forces is frequently associated with increased incidence of illness in persons who have not previously resided in the region. Viral or bacterial disease is present in a high proportion of persons who report for sick call. As the number of forces to be deployed diminishes and the requirement for sustained performance increases, sick call can be anticipated to reduce force projection capability. The global surveillance of infectious diseases provides a database that can be used for immunization or other protection of individuals before deployment.

The identification of important enzymes/gene products in the human, relevant to protection from B or C induced injury, can be utilized in defense strategies. The rapid genotyping of individuals is possible through the use of kits or chips that allow for genotyping for biomarkers of disease or chemical susceptibility. This has to be done only once in an individual's lifetime unless new, relevant genes are identified. The genotyping of individuals, along with a catalog of the manifestations of the various GP alone, and in combination, could lead to more appropriate (and potentially reduced risk) job assignments. It will also allow intervention/protection strategies to be tailored to the individual, rather than the traditional one-size-fits-all approach.

Nonlethal weapons offer the opportunity to manage crisis situations in a manner that can inflict minimal traumatic injury and enhance survivability of U.S. forces. As U.S. forces are called upon to perform an increasing number of "police and peacekeeping actions" and as the proportion of these actions occur in urban settings, the requirement for nonlethal means of control increases. This technology has direct relevance to the world scene as it appears in the early period of the 21st century.

A. SUSCEPTIBILITY TO INFECTIOUS AGENTS

1. Genomics of the Pathogenic Organisms

The genome of many pathogenic organisms on the AG list has been characterized. Within the genome there are sequences of nucleic acids that enhance the pathogenicity of the organisms. These sequences are virulence genes and PAIs. The PAIs and virulence genes may allow the bacteria to adhere more firmly to mammalian cells or to otherwise enhance the survival and multiplication potential of the bacteria in the host (human or animal). The sequence of the PAIs in pathogenic *E. coli*, for example, is similar to sequences of cell proteins (adhesins) on the external surface of cells and to those used in cell-cell binding. The knowledge of the PAIs, virulence genes, and sequences unique to each of the pathogenic organisms allow rapid detection of these agents in samples of tissue (including blood) from infected patients.

The ability to amplify and sequence nucleic acid fragments from infected tissues has given rise to a new approach in the understanding of emerging pathogens. Until the past decade it has been necessary to culture tissue samples to recover and identify the infectious agent. This requirement meant that appropriate culture media and conditions for culture had to be known. With the polymerase chain reaction (PCR) technology and with a gene data bank of infectious organisms, it is now possible to obtain very small samples of tissue and biological agent. The nucleic acid component of the agent can be amplified and matched to other known agents. This technology has been used to identify a bacterium involved etiologically in Whipple's disease. In a related technology, "representational difference analyses" (RDA), samples of healthy and diseased tissues from a single subject are taken and the genomic material from the healthy tissue subtracted from that of the diseased tissue. This method has been used to identify Kaposi Sarcoma HV (KSHV), a virus responsible for Kaposi sarcoma in patients infected with HIV. The known genes can also be used in the construction of multi-array sensors used to detect pathogenic agents in the military or medical environment.

The knowledge of the genomics of bacteria, viruses, and fungi permits rapid identification of infectious materials, allows the development of new immunogens without injecting material that has a potential to cause infection, and allows the development of new antibiotic and antiviral compounds. It permits the identification, by PCR and RDA, of pathogenic organisms not previously recognized and is expected to provide evidence of new microbial genera. See *Science*, 282: 219221, 1998.

2. Genomics of Personnel as Related to Disease or Toxicant Susceptibility

Large, population-based studies are being conducted to evaluate disease risks associated with the combined effects of genetic status and environmental exposures, including those related to lifestyle and diet. Using minute quantities of DNA in cells obtained from a simple mouth rinse, blood, or tissue, it is possible to detect gene mutations, the functions or effects of which may point the way to environmental, nutritional, hormonal, and other factors that contribute to diseases in humans. As more information about human genes becomes available, there will be opportunities to test the importance of newly discovered genes not only for their relation to susceptibility, but also for clues to environmental agents that affect health and performance. The tools to examine these complex interactions between genetic susceptibility and environmental exposures are being developed for studies that could greatly advance our understanding of how inheritance, lifestyle, and environment combine to cause disease. This knowledge will lead to new strategies for disease prevention.

The development, function, and behavior of humans have been shown to involve genetic and environmental elements. In recent research efforts, biomedical science has been giving attention to inherent differences among individuals and groups of persons. These differences, at the genetic level, are collectively called genetic polymorphisms (GPs), and can affect the way that individuals and groups respond to various physical, chemical, and biological stresses. GPs generally manifest themselves as differences among individuals with regard to their ability to metabolize drugs and other substances and to their susceptibility to disease and other stressors. As a general rule, each human population is estimated to possess 3 to 10 genes that affect its susceptibility to disease or toxicants.

Some GPs can be identified only by phenotyping the individuals, i.e., by examining blood or other tissues for the presence or absence of specific enzymes or receptors. Other GPs may be identified by looking at specific DNA sequences in the cells or at RNA products.

Polymorphisms can manifest themselves as:

- The presence or absence of a specific enzyme that metabolizes xenobiotics;
- Differing relative concentrations of a specific enzyme, with associated differences in levels or rates of metabolism;
- Modified enzyme/gene product with a different level, or spectrum of activity than the wild type;
- The presence or absence of a specific gene product that controls a vital body function; and
- Differences in the ability of a gene to be induced to produce a product.

The polymorphisms can manifest themselves as differences in enzyme activity, differences in cell receptors, and differences in response to stress or injury. Different polymorphisms confer increased or decreased sensitivity to various external insults. These polymorphisms do not confer absolute resistance or sensitivity to an insult, but affect the probability that the individual (or group) with that polymorphism would be more or less sensitive to the disease/damage. For example, the presence or absence of various cytochrome P450 and GST isoforms will increase or decrease the susceptibility of that individual to different types of smoking or chemical-related cancers, whereas other isoforms would increase the susceptibility for contracting the disease. Genetic polymorphisms and environmental factors interact in the development of resistance or susceptibility to disease.

The human genome project will inform the medical community about the specific genes related to the development of disease, the susceptibility to shock following traumatic injury, and the susceptibility of individuals to intense stress associated with exposure to temperature extremes, intense noise, sleep deprivation, and infection.

In the next decade it is likely that polymorphisms that affect performance, sensitivity to biological and chemical agents and specific drugs, and sensitivity to environmental conditions and situations will be identified. The individual genotype will affect sensitivity to biological and chemical agents, susceptibility to infection by specific organisms, activation or detoxification of chemical agents, and modulation of chemical or biological agent exposure. Examples include P450 and glutathione S-transferase (GST) polymorphisms that are associated with resistance or susceptibility to smoking-induced cancers; cell surface major histo compatibility (HLA) molecules that confer resistance or sensitivity to infection by specific microorganisms; and acetylase variants that rapidly or slowly metabolize (to activate or inactivate) certain toxins. GPs also affect the response of individuals to drugs that are important in the prophylaxis of disease and treatment of injury or fatigue. For example, polymorphisms for rapid or slow metabolism of drugs can lead to a propensity for adverse reactions, decreased drug efficacy, or to the need to use lower doses to achieve the desired effect. A correlation between genetic background and human performance will provide a capability to determine where, on the spectrum of susceptibility or resistance, military personnel may be most suited for defense-related tasks.

3. *Global Surveillance of Infectious Diseases*

There are three components to the surveillance database: endemic and epidemic disease; emerging and re-emerging diseases, particularly of viral etiology; and antimicrobial drug resistance.

There are several agencies that gather data on worldwide endemic disease. These include the Center for Disease Control (CDC), World Health Organization (WHO), and Pan-American Health Organization (PAHO). The military services, through organizations such as Naval Medical Research Units (NAMRU) and Walter Reed Army facility, have interests in collating data on endemic disease. A comprehensive database of such information can function as a clearinghouse for defense and public health requirements, as well as an early warning system for outbreak of infectious diseases, whether caused by organisms endemic to an area or disseminated by terrorist or nation/state entities. The information would serve to alert pharmaceutical firms and the defense community to increase production requirements of vaccines and antibiotics.

The database includes

- Information gathered from agencies currently involved in monitoring disease;
- Genomic sequences of viral, bacterial, and fungal pathogens;
- Genomic sequences of pathogenicity islands of AG agents and other relevant organisms; and
- Antigenic epitopes characteristic of AG agents and other relevant organisms.

Multi-array sensors consisting of probes for the relevant genomic sequences or epitopes could then be constructed and used in the surveillance system.

B. NONLETHAL WEAPONS EFFECTS

Military peacekeeping, humanitarian efforts, and missions other than war have become increasingly common. In such operations, many dangers exist to the troops, yet the use of lethal force is often not justified or acceptable. DoD recently issued a policy statement for nonlethal weapons (NLW), in which such weapons are defined as "weapon systems explicitly designed to incapacitate personnel or materiel while minimizing fatalities, permanent injury, and undesired damage to property and the environment" (DoD Directive 3000.3).

Bioeffects and Nonlethal Weapons

Bioeffects include any effect an internal or external stimulus has on part or all of a biological organism. Examples of bioeffects include DNA mutation, loss of equilibrium, stimulation or inhibition of sensory input, emotional response, nausea, fear, increase in heart rate, behavior avoidance, cellular damage, altered metabolism, confusion, loss of consciousness, convulsions, and death. They can be as innocuous as a recognition of a pleasant odor or as harmful as stopping of the heart. Bioeffects specialists include medical doctors, physiologists, psychologists, behavioral scientists, veterinarians, anatomists, neuroscientists, biologists, epidemiologists, theoreticians and others, all concerned with the effects of any stimulus (biological, chemical, or physical) on part of all of a biological organism.

In addition to achieving a nonlethal goal, i.e., incapacitation, operational commanders are also interested in the parameters of the incapacitation. Some relevant parameters include dose for main desired effect; can the effect be tuned; time until initial effect; duration of effect; synergy with other factors; reversibility versus irreversibility of effects; side effects to targets; undesired collateral effects; environmental effects; and susceptibility to countermeasures.

Policy acceptability is a topic in which bioeffects have two major roles. The first role of bioeffects concerns the policy that antipersonnel NLWs should "minimize permanent injury." The second role of bioeffects in setting NLW policy concerns the long-term medical consequences of exposure to the NLWs for everyone exposed, including the operator, the target, and bystanders. If occupational exposure standards exist for the particular agent being used, as they do for many types of noise, radiation, and chemicals, then these standards should be followed when possible. If the exposures are sufficiently novel that no health standards exist—for example, certain types of directed energy—then standards need to be developed.

The Variability of Human Responses and the Probabilistic Nature of Bioeffects

Because of biological variability there will always be uncertainty in predicting the biological responses to nonlethal weapons. This is true for target, operator, and bystander effects. Even among a consistent population of humans, such as a group of young adult males, there will be a variability in responses to the same stimulus. When the variance of the population increases, for example, by adding persons of differing sizes, ages, weights, frailty, health, and sex, so will the variability of the population response to most NLWs. Therefore, biological responses to nonlethal weapons are probabilistic.

Although bioeffects are certainly variable, it should be noted that they are not the only source of variability in the use of a NLWs. The energy emitted from the weapon itself (e.g., a directed-energy weapon) is variable to some extent. The transmission of the energy from the weapon to the target is affected by variations in beamforming, aiming, or focusing of the energy, as well as by intervening conditions such as rain, wind, temperature, terrain, and structures. Coupling of the energy to the target can be passively affected by the target's size, orientation, and clothing, as well as by active countermeasures purposively employed by the target. All of these sources of variability precede the biological variability and probabilistic nature of response.

Acoustic Energy as a Nonlethal Weapon

In his work *Life of Marcus Crassus,* Plutarch described the use of bells and drums as a psychological NLW. More recently, rock music was used to annoy Manuel Noriega in Panama. Current ideas for NLWs use high-intensity infrasound. A special pressure chamber, the infrasound test device (ITS), has been constructed to examine the effects of infrasound at different frequencies on both anesthetized and awake subjects. The utility of acoustic energy as an NLW is likely to be defined in 5 years.

WORLDWIDE TECHNOLOGY ASSESSMENT

The identification of genomic maps is the result of multinational efforts, with the United States, UK, France, Germany, Switzerland, and Japan acting as the primary participants. The large majority of the genome database is available on the WWW. The database includes bacterial and viral organisms as well as livestock, agricultural plants, and humans. The databases are maintained at sites in the United States, France, Japan, Germany, Switzerland, and UK. Some genomic sequences have been patented for protection of intellectual property that may have applications for pharmaceutical production. The 100,000 genes of the human genome are expected to be completely sequenced by the year 2002. The information obtained from combining susceptibility factors with virulence factors in biological agents makes possible the construction of a formidable biological weapon. The relative capabilities shown in the WTA table reflect the technical capability of the identified nation to contribute to the human genome and pathogen genome database. They do not reflect access to the web. Technical capability to add to the genome database provides a temporal advantage to identify and deploy countermeasures. Large national or federal laboratories with leading capabilities to address the issues of the genome include the National Institutes of Health (NIH) and the National Institute of Standards and Technology (NIST) in the United States, the National Cancer Center and the Chemical Research Center in Japan, and the National Health and Medical Research Institute (INSERM) in France. Large private sector groups in the developed world also are actively characterizing the human genome and the genome of pathogens.

There is not a single information database that describes endemic disease worldwide at the present time. Information is compiled from CDC, WHO, PAHO, and NAMRU data sets. Efforts to create a single database are in progress.

See Technology Data Sheet for WTA chart.

SECTION 4.2—DEFEAT OR MANAGEMENT OF BIOLOGICAL AND CHEMICAL ATTACK

Highlights

- Enhanced protection achieved by super-antigens, synthetic immune or antibody systems, vaccines, antibiotics, and antivirals.
- Preventing secondary organ target infection following viral exposure is now a realization.
- Multi-array sensors detect B/C agents by genomic and immunogenic probes.
- A process has been initiated to training first responders to B/C incidents and attacks.

OVERVIEW

The etiological basis of infectious disease in otherwise physically fit persons is exposure to sufficient amounts of pathogen or toxin that will overcome natural or induced protection against such agents. Traumatic injury or environmental toxicants (chemical or physical) may increase susceptibility to pathogens. The protection of persons from illness and the management of illness involve the determination of the causative factors, then the administration of appropriate medications and nutrition, and isolation of those individuals likely to transmit illness and compromise group function. For infectious disease the identification of the infectious organism and appropriate antivirals and antibiotics for the organism is important. The majority of infectious diseases affecting military personnel is associated with organisms in the environment of the deployed soldier. From a historical perspective, disease resulting from the intentional release of biological agents is the exception. Emerging and re-emerging viruses and bacteria, increased drug resistance in the general population associated with the increased use of antibiotics, and increased deployment of forces to third world countries where nutritional standards and sanitary conditions are wanting increase the risk of exposure to infectious disease.

RATIONALE

Rapid and accurate detection and identification of biological and chemical agents are the key elements to survival from exposure to these agents. Two needs related to an appropriate response to a biological incident are (1) determination of where and when an incident occurred and (2) identification of appropriate technologies to protect exposed persons. Sensor systems provide the capability to detect and identify biological and chemical threat agents. Biomarker systems provide the capability to detect an individual's exposure or susceptibility to biological or chemical threats. Biomarkers also provide a capability to assess changes in an individual's fitness for duty or vigilance in a timely manner. This section considers pretreatment of personnel prior to exposure to infectious agents or toxins, the detection of biological and chemical agents in the environment, and the treatment and management of persons following exposure to infectious or chemical agents.

There will be a window of approximately 12–72 hours between the time persons are exposed to biological agents and the first appearance of clinical signs and symptoms. This window provides time to detect, identify, and initiate countermeasures that will interdict the appearance of clinical signs. Toxicant exposure may have acute, mid-term, and long-term effects on an individual's fitness for duty and vigilance. The performance component of persons in high-performance aircraft and closed-environment combat units (tank, submarine) will be reduced acutely following exposure to critical levels of toxicants. Early identification of toxicant exposure with the use of biomarkers can assist in maintaining combat readiness. The advanced distributed systems (ADS), particularly those used in platforms with minimal numbers of combat personnel, depend upon retention of high-performance capabilities by all personnel. On a mid- to long-term basis the exposure to toxicants may compromise cognitive function and thereby endanger groups of persons. Biomarkers will assist in detection of acute, mid- and long-term sequellae of toxicant exposure. The susceptibility biomarkers may also serve in selection of persons having increased resistance to chemical agents and toxicants. The human genome project, described in Section 4.1, is expected to identify persons with low and/or high susceptibility.

The ability of first responders to recognize victims of a biological attack is limited by good medical practices associated with differential diagnoses. In normal medical situations the medical care provider evaluates the most likely cause of symptoms and complaints (differential diagnosis). The differential diagnostic process does not place intentional release of agent as a likely factor in current systems because biological warfare has not been and is not currently a major cause of illness in the public health domain or in the military. In addition, the presenting signs at early periods following biological agent exposure appear similar to those associated with influenza and other less serious threats. The training of first responders is therefore a critical element in developing an appropriate management regimen.

A. ENHANCED PROTECTION AND TREATMENT OF PERSONNEL PRIOR TO AND IMMEDIATELY FOLLOWING EXPOSURE

These technologies include materials that may be provided to personnel before deployment (e.g., vaccines), physical equipment used at the time that a biological or chemical agent is encountered [face masks, mission-oriented protective posture (MOPP) gear, aeroprecipitation of the agents], treatment of exposed persons entering a contaminated area (antibiotics, antivirals). The physical protection of persons from biological agents requires the use only of respiratory and eye protection filter systems, while that for chemical agents requires MOPP gear. The physical protection from biological agents is embodied by the filters that remove particles having an average diameter between 0.1 and 15 microns from the air flow system to the person. This has been addressed in Section 3 of Part II of the Militarily Critical Technologies List (MCTL). One potential new physical method of protection against biological and chemical agents is aeroprecipitation.

Biologically Based Protection

New technologies are in development that will afford biological protection against biological agents. These include novel unconventional vaccines, antibiotics, antitoxins, superantigens, immune modulating viruses, synthetic immune systems, synthetic antibody-like drugs, chemical agent neutralizing drugs, and antiviral drugs that interdict viral disease after a person is exposed to viral agents.

Vaccines

Vaccines that protect against a subset of the biological agents are currently available, but vary in efficacy and time required to achieve protection after exposure. Methods to increase the rate at which immunological protection can be achieved are described below.

Antibiotics

The development of sulfa drugs in the 1930's and of penicillin and others in the 1940–1970's was a major gain in the protection of persons (military and civilian) following exposure to infectious bacteria. The use of antibiotics remains the primary mode of treatment of infectious disease of bacterial etiology, even as new problems emerge as a result of the broad distribution of these drugs. Drug-resistant organisms have become a significant concern in the treatment of bacterial infection in immuno-compromised persons [e.g., with nutrition difficulties, acquired immune deficiency syndrome (AIDS), and transplant recipients]. Associated with the rise of drug resistance is the increase of nosocomial infections in patients entering hospitals. The rise in drug resistance is observed as new pharmaceuticals are developed. The use of combinatorial chemistry, multivalent drug regimes, and rational drug design based on knowledge of the immunogenic and genomic properties of infectious agents are three approaches used to interdict disease from bacterial agents.

Antitoxins

Active or passive immunization or the administration of toxin binding agents are two methods of protection in the time period immediately following toxin exposure.

The above strategies may not apply in certain circumstances because (1) antibiotic-resistant organisms are relatively easy to produce; (2) not all pathogens produce toxins; (3) while vaccines are available for many pathogens, they have limited effect against active disease; (4) conventional vaccines can take several weeks to produce using the hybridoma-based technologies; and (5) vaccine development for a novel, unknown pathogen requires isolation and culture or generation of a large library of genomic DNA of the putative pathogen.

New approaches for the development of antiviral, antibacterial, and antifungal materials include the following:

- Development of super-antigens to achieve rapid immune responses;
- Synthesis of antibody-like drugs constructed through combinatorial approaches;
- Development of immune modulating constructs that increase production of immuno-modulators in affected persons;
- Development of compounds that block the entry of viruses into secondary target sites, the replication of viruses in the secondary sites, and the packaging of the viral materials into mature viruses.

These approaches are flexible in that they could be developed very rapidly with minimal laboratory requirements and in self-contained modular units.

Super-antigens (Part III, Section 3, Biological Technology). Enhanced prophylaxis can be achieved by use of active vaccination against biological agents or infectious agents endemic in deployed areas. Vaccination with super-antigens or advanced adjuvants (i.e., improved antigen presentation) can up-regulate the immune system before deployment and provide protection in a 4-day period rather than the usual 10 days after vaccination. The enabling technologies include genomic sequencing of all known threat agents and infectious disease organisms; multi component, multivalent vaccination systems to upregulate the immune system; and development of immune response modifiers, including interferons and interleukins, having the potential to enhance immune response under crisis situations. The sequencing of PAIs can reduce the total number of vaccines needed.

Immune modulating viruses. Retrovirus coding for interleukin 12 (IL-12) or interferon gamma (IFN-g) could be administered directly to the lungs as an aerosol. This would enhance the immune response in the location of agent delivered by inhalation, i.e., the lungs. This method is currently under development and proof of principle may be achieved within a year. This strategy may be most useful during early infection.

Synthetic immune systems. These systems involve the delivery of retroviruses containing a pseudo-gene coding for a oligonucleotide derived from a SELEX-based method (see below). Gene therapies for cancer and various disease states characterized by a mutation-associated dysfunction of a key enzyme or receptor (for example, cystic fibrosis) have been used experimentally for the last 5 years. Genes coding for immunomodulators could also be used. The main reason for this approach is that large-scale oligonucleotide synthesis is not readily achieved, and in any case repeated administration would have to be performed to ensure efficacy. Since the body would express the retroviral gene construct continuously, a single dose would provide protection for an extended period of time. An alternative approach could use a phage-based selection system (Gao et al., *Proc. Natl. Acad. Sci.*, 94, 11777–82, 1997). The approach in principal is similar to SELEX; however, selection and enrichment occur directly through phage expression and selection without PCR enrichment steps. Both technologies are extremely robust, do not require extensive participation by the host immune system, and can be developed very rapidly with minimal laboratory equipment.

Synthetic antibody-like drugs. One promising strategy utilizes a PCR-based technology called SELEX ("Systematic Evolution of Ligands by Exponential Enrichment," Gold et al., *Proc. Natl. Acad. Sci.*, 94, 59–64, 1997). Very specific DNA fragments can be made against a pathogen in a very short period of time. The pathogen does not have to be cultured or strictly isolated from surrounding infected tissue. The fragments bind to the pathogen very tightly and can be made to include other elements that either alert the immune system to the pathogen or inactivate the pathogen directly. This technology is very robust but is also quite new.

Chemical agent neutralization. The SELEX- and phage-based synthetic immune systems can be used to prepare a combatant for exposure to a wide array of intoxicants. A major hurdle to be overcome would be the construction of appropriate transcription control elements based on the presence of an array of intoxicants because it would not be desirable to have the response systems turned on unless the intoxicants were present in sufficient concentration.

Interdiction of Viral Entry, Replication, and Maturation in Secondary Target Cells

For the virus to attack secondary target sites successfully it must enter the cell, the nucleic acid may travel to the nucleus for replication, and then the newly synthesized viral nucleic acid must be coated by the viral protein capsid. Cysteine protease inhibitors can block entry of the virus to cells, inhibitors of cytoplasmic nuclear transport can inhibit the second step, and inhibitors of molecular chaperones can inhibit the nucleic acid-viral protein coat interaction. Finally, mature viral particles may only be released following initiation of apoptotic event. Compounds that inhibit apoptosis may protect the patient from rapid development of clinical signs.

Recent studies indicate that the development of compounds that inhibit specific hemagglutinins and neuraminidases may protect individuals from virulent influenza viruses, a major cause of illness in the military and civilian populations. There are virulent and avirulent influenza viruses. Hydrolytic enzymes play a major role in the infectivity of influenza viruses. The virulent type replicates in all organs, while the avirulent type only replicate in gut and respiratory tract. Hydrolytic cleavage of hemagglutinin is required for virulence. Hemagglutinin (HA) of virulent organisms cleaved in all organs, but HA of avirulent cleaved only in gut and respiratory tract.

New sets of antibiotics that inactivate PAIs may have utility in individual and group protection. These antibiotics can be antisense materials or chemicals similar to traditional antibiotics. It is important to develop antibiotics that have not been used in the general population to preclude the further development of drug resistance in the population.

Assessment of Drug Toxicity on Microchips

All the above technologies involve the development of new drugs to be administered to affected persons. Two major expenses in drug development are (1) the determination of the efficacy of the drug in protecting persons from biological or chemical agents, and (2) determining whether the novel drugs have adverse side effects. A new technique is emerging that will allow initial screening of a drug's actions on a microarray of DNA or peptide fragments. In this method, living cells are exposed to the drug of interest and the altered expression of genes is analyzed on the microarray. Because many of the drugs that are developed have undesirable side effects and because these side effects involve increased expression of specific proteins, the treated cell can be monitored for overexpression of messenger ribonucleic acid (mRNA) for the undesirable proteins. This would provide a critical step in reducing costs of drug development. The method has been described in Friend et al., in *Nature Medicine*, 4: 1293, 1998.

Aeroprecipitation

Aeroprecipitation technologies can be used to defend against the dispersal of biological agents as aerosols. The net electrostatic charge on the particles can be used to cause aggregation of the infectious agents, thereby reducing the number of particles that may affect the target population. The other physical properties of dispersed aerosols facilitate developing countermeasures that result in aggregation or precipitation of the biological agents shortly after dispersal.

B. DETECTION AND IDENTIFICATION OF BIOLOGICAL OR CHEMICAL AGENTS

Some chemical and biological agents may not result in an overt clinical presentation until days or weeks after an attack. It is possible to envision a scenario where a number of intoxicants and biological agents may be present because of endemic disease rather than because of an intentional release of weaponized agent. In either case a subset of the exposed population may suffer due to genetic predisposition.

Biopolymer-based sensors are available to detect and identify infectious agents, chemical agents, and biomarkers. Until the mid 1980's the majority of sensors depended on macroscale systems that utilized changes in optical density or changes in enzyme reactivity when a biological agent or chemical agent entered the field of the sensor. The optical density and enzyme reactivity changes were relatively nonspecific for agents, and the sensors therefore had utility as detectors but not identifiers of specific agents. With the advent of monoclonal antibody (MAb) and PCR technology it became possible to recognize specific infectious biological agents by the proteins on their surface (MAb) or by their genomic material (PCR). It became possible to identify toxins or chemical agents by the use of MAbs or receptors that specifically interacted with the agent. A new technology has arisen that permits miniaturization of the sensor through the production of multi-array sensors. The multi-array sensor recognizes either surface antigens or genomic sequences on a single detector surface. Upon interaction between the immobilized antibody or complementary genomic sequence on the sensor surface with the biological or chemical agents an optical/electrical signal is generated. The multi-array sensor may contain 1,000 detector elements on a single 4-cm^2 surface. This technology permits rapid detection of the AG biological agents when appropriate false negative/positive controls are embedded on the sensor surface. The genomic sensor could detect bioengineered organisms that have pathogenicity factors incorporated into otherwise innocuous organisms. The miniature size of these systems, low energy requirements, and specificity as agent or biomarker identifiers now permit development of a smart suit with multiple sensors; a data fusion system that can organize the input from the many sensors into coherent information remains to be developed.

Another new technology permits the rapid synthesis of microscale amounts of chemicals. This technology is a fusion of microfluidics and chip technology (see Robert Service, *Science*, 282: 399–401, 1998). The apparatus has dimensions of 5 × 8 cm and can synthesize or sequence small amounts of DNA for inclusion as genomic probes on a multi-array sensor. Companies producing these microlaboratories include Affymetrix, Perkin-Elmer, Caliper, Aclara Biosciences, and Orchid Biocomputer. This new technology will have a major impact on affordability because almost all of the processes are automated and therefore require many fewer technicians. See also Biological Technology, Section 3, Part III of the list of MCT.

A second general technology for assessing whether a population has been exposed to a biological or chemical agent is a probe that would identify an endogenous biomarker (see Section 4.3) at concentrations below that at which any pharmacological effect becomes evident. The detection of low molecular weight chemical molecules will probably be more tractable than the identification of high molecular weight compounds or microorganisms.

Probes/detectors for known chemical agents:

- May be implanted to measure blood levels or the appearance of the chemical or a secondary product; or
- Can be attached to clothing or respirators to measure ambient levels.

Probes/detectors for known biological agents comprise several types:

- High-affinity bio- or immuno-chemical probes for specific antigens or cell products can be used to monitor body fluids and lavages for the presence of specific biomarkers of infection.
- High-efficiency probes can be used to monitor ambient air for specific microorganisms.
- Sensors that identify specific toxins/soluble infectivity factors/cell-surface invasive elements can provide advance warning and enable appropriate protective responses.
- Sensors that identify the body's early response markers to pathogens or toxins can be used to provide warning and initiate responses.

One system for early detection of an extreme physiological stress or infection is an artificial antibody/antitioxin-producing cell: an implanted reservoir containing specific antibodies, antitoxins, or antibiotics that are connected to implanted sensors. This sensor will identify specific molecules entering the circulation (or lungs or other tissues) and trigger the reservoir to release its contents and also transmit an outside signal to inform others of the exposure. This could lead to remote triggering of reservoirs in individuals who have not yet encountered the insult, but are potentially at risk.

Sensors for Biological and Chemical Agents

Sensor systems are discussed below and in Biological Technology, Section 3, Part III of the list of MCT.

a. *Synthetic reporter systems.* Recent approaches in combinatorial chemistry permit the placement of thousands of unique protein or DNA fragments onto a microchip (multi-array sensor). Such a device could be used to identify intoxicants in urine, saliva, lung lavage, etc., or to identify organic markers of oxidative stress (for example: nitrotyrosine) or the presence of transcription factors related to stress. Nucleic acid sequences about 20 nucleotides long and complementary to the nucleic acid of pathogenic agents may be immobilized on these sensors and serve as probes for biological agents. These systems are likely to be extremely robust and technologically feasible. Since thousands of targets can be evaluated simultaneously by placing a drop of liquid on the chip, the chances of not detecting an AG agent or a biomarker for stress are reduced relative to conventional analytical or molecular biological approaches. These systems may also be semi-quantitative, enabling biokinetic modeling and thereby allowing determination of when an exposure occurred. This information may be useful in determining the extent and course of the exposure within a population. Such information could be valuable in choice of treatment modalities and in predicting how severely units will be affected. The development of robust systems using combinatorial chemistry approaches is feasible at this time.

b. *Cell-based reporter systems.* Biological or chemical agent exposure could be identified by observing the changes in gene expression occurring in cells. Genes involved in cellular death (apoptosis), DNA damage (p53, p21, RB), and response to oxidant stress (NF-kappa-B, c-jun, OX-1, HSP-70) are up-regulated early during stress. The transcription of these genes can be identified with a fluorescent or electrochemical reporter system. A packet of cells could be carried on individual soldiers and the cells contact with a chemical or biological agent read with either a hand-held spectrophotometer or a ultraviolet light source.

c. *People-based reporter systems.* A similar approach as that suggested in (a) above could also be used to monitor people. In this scenario the increased synthesis of biomarkers of infectious disease, chemical, or stress exposure will be observed in peripheral blood cells. The cost of observing gene regulation is currently high in terms of number of personnel, time, and machine-maintenance cost.

C. TRAINING OF HEALTH PERSONNEL FOR INCIDENTS OR ACCIDENTS INVOLVING BIOLOGICAL AGENTS

During the past decade the acquisition and management of chemical and biological warfare weapons have become a concern to the international community. The economic situation in the former Soviet Union (FSU) raises concerns about the spread of B/C weapon technology to developing nations. The low cost of obtaining and weaponizing chemical/biological agents raises concerns about the potential use of such agents by lesser developed countries (LDCs), by surrogate groups operating at the behest of nation states, and by terrorist groups.

Several panels appointed by President Clinton examined the ability of the United States and other industrialized nations to respond to the threat of use of weapons of mass destruction (WMD). The reports of these panels resulted in a set of Presidential directives that included

a. increased international cooperation in the area of WMD;

b. new approaches to prevent weapon acquisition;

c. new approaches to crisis management;

d. development of a critical infrastructure to coordinate responses; and

e. increased transportation security.

Items c, d, and e involve increasing public awareness (including medical professionals; hospital administrators; and law enforcement, fire department and emergency medical response teams) to the potential threat of a developing incident/accident and appropriate responses to this threat. A new infrastructure system for Emergency Preparedness is now accessible on the WWW at http://222.apgea.army.mil/ops/ dp/missions.html

Emergency Preparedness in CONUS–Integrated Civilian and Defense Activities

Processes will be identified for informing residents when a release of biological agents has occurred or is imminent. Appropriate responses to the threat and ways to limit the exposure of adjacent communities to the biological agent will be developed. The critical elements of the process include the training of first responders and the development of mechanisms to control the dissemination and spread of the agent with minimal social disruption and panic following release of information to the public. The agencies involved in this task include the National Guard, Federal Emergency Management Agency (FEMA), and the Departments of Justice, Health and Human Services, and Defense (DOJ, DHHS, and DoD).

At the present time (1998), the Chemical and Biological (CB) Hotline serves as an emergency resource for first responders to request technical assistance. The Nunn-Lugar II act specifically calls for "establishment of a designated telephonic link to a designated source of relevant data and expert advice for the use of state or local officials responding to emergencies involving a weapon of mass destruction or related materials." The Hotline's intended users include trained emergency response personnel: emergency operators and "first responders" (the firefighters, police, and emergency medical technicians who arrive at the scene of a CB terrorist incident). Other users may include the state emergency operations centers and hospitals that may treat victims of CB agent exposure.

- The CB Hotline is staffed by trained operators servicing the phones 7 days a week, 24 hours a day. Operators use extensive databases and reference materials in addition to immediate access to the nation's top subject-matter experts in the field of CB agents. Areas of specialty include CB agent identification, medical treatments, and information on military and civilian defense equipment.
- The CB Hotline is a joint effort of the Coast Guard, Federal Bureau of Investigation (FBI), FEMA, Environmental Protection Agency (EPA), DHHS, and DoD. The National Response Center (NRC) is the entry point for the CB Hotline. The NRC receives basic incident information and links the caller to the DoD's and FBI's chemical, biological, and terrorism experts. These and other federal agencies can be tapped within a few minutes to provide technical assistance during a potential CB incident. If the situation warrants, a federal response action may be initiated.
- Use the local established policies and procedures for requesting federal assistance before contacting the CB Hotline. State and local officials can access the CB Hotline in emergency circumstances by calling 1-800-424-8802.
- Chemical and Biological Defense Command (CBDCOM) Public Affairs Office * Phone (410) 671-4345 * Facsimile (410) 671-5297.

Emergency Preparedness–Military Response Teams

Following the experience of the Gulf War, the Pentagon designated a platoon-size Army force, the technical escort unit (TEU) to have specific training and equipment to permit detection and identification of biological agents in deployed and combat areas. This TEU of 182 persons from the 310th Chemical Company and the Marine Corps 375-member Chemical Biological Incident Response Force (CBRIF) respond to biological agent threats. A biological integrated detection system (BIDS) has been developed and fielded for the 310th Company. The M93A1 Fox nuclear, chemical, and biological reconnaissance system vehicle is a component of the detection system. The Army has a "Domestic Preparedness Program" intended to strengthen the ability of federal, state, and local emergency personnel to respond to BW threats. A new National Guard initiative, National Guard Rapid Assessment and Initial Detection (RAID) organization is in the formative stages.

WORLDWIDE TECHNOLOGY ASSESSMENT

Traditional sensors for biological and chemical agents are available to many nations. The biological agents of greatest concern to the defense community are naturally occurring in many regions of the world. In addition, many benign organisms have genomic sequences that are similar to those of the potential agents. The design of useful sensors must therefore differentiate between threat and nonthreat organisms and must identify a threshold level of concern. The advanced technologies described in this section as having the capability to identify threshold levels and distinguish between threat agent and benign organisms currently have limited distribution. The primary nations developing these technologies include the United States, France, UK, Germany, Japan, Netherlands, Switzerland, and Israel. Many of these nations have large pharmaceutical capability. Multinational efforts relating to the development of sensors for biological threat agents have emerged: the ITF 24. The United States, Canada, Sweden, Germany, France, and UK are among the nations performing leading research in the area. Smaller nations have acquired niche capabilities in the sensor and protection arena; these include Israel, Poland, and Czech Republic.

The immune, antibiotic, and antiviral protection technologies are most developed in the nations with leading-edge pharmaceutical industries: the United States, Switzerland, Germany, France, UK, Japan, Italy, Sweden, and Denmark.

SECTION 4.3—MANAGEMENT OF TRAUMA, STRESS, AND TREATMENT

Highlights

- Biomarkers will permit early detection of high-level toxicant and stress exposure.
- Biological growth factors enhance rate of wound healing and neural regeneration.
- Telemedicine promises the rapid removal and treatment of persons with traumatic injuries to enhance survival and reinsertion into units.

OVERVIEW

As a result of the advances associated with the human genome project (see above) a needle-stick in the finger of a person provides a drop of blood to test for (1) signs of emerging disease, (2) unusual sensitivity to chemicals in the environment, or (3) stress. Sensitive tests look for the presence of telltale proteins that may be secreted from cells. This may be evident hours to days before clinical disease or functional decompensation becomes evident. A computer is used to connect to a database of DNA samples that can be compared to the patient's DNA. If the patient's DNA matches DNA sequences in the database that are consistent with early-stage disease or chemical sensitivity, treatment protocols can be initiated immediately. Such strategies can also be applied to alerting a physician that a flight line operational person may be exposed to toxic levels of fuel or other chemicals. Biomarkers of cell responses to drug therapy are also now being used to rapidly screen new drugs in development for early signs of toxicity. Biomarkers of intense emotional stress have been identified. Evidence that a new drug is causing the synthesis of proteins in cells that signal cell death or dysfunction (biomarkers) can short-circuit expensive tests required to demonstrate toxicity. Because the new technologies of combinatorial chemistry and drug design produce thousands of compounds per week, it is necessary to devise tests that provide rapid indication of utility and of toxicity. Biomarkers may provide information related to both of these needs. Telemedicine technologies and novel methods to treat traumatic injury are included in this section.

RATIONALE

Infectious disease and traumatic injuries are the primary causes of disability in persons of military age. The detection of predisposition to agents, toxicants, and stress will be important to evaluations of combat readiness, fitness for duty, and training. This section considers the biomarkers which may be of assistance in determining profiles of human susceptibility to disease and toxicant exposure. Mechanisms that facilitate the removal of persons contaminated with biological or chemical agents from the contaminated area are considered in the Bioscience section of Part 3 of the MCTL. The present section discusses the interface between medicine and telecommunications (telemedicine) in the management of persons at a site distant from primary medical care facilities. This approach is of increasing importance in situations where U.S. forces are deployed in small peacekeeping units at a site distant from primary care. The removal of physicians from front-line combat areas has also increased the need for telemedicine approaches. The final portion of this section discusses control of internal bleeding, cell and organ transplants, and the increasing role of growth factors in wound healing and nervous system repair.

A. BIOMARKERS OF TOXICANT EXPOSURE AND OF PHYSIOLOGICAL STRESS

A biomarker is an indicator of exposure of persons to particular agents, toxicants, or physiological stress. There are three types of biomarkers:

- *An exposure biomarker*—the presence of an exogenous compound or its metabolite signaling that the person has been exposed to the toxicant;
- An effect biomarker—the detection of an endogenous component or of a change in functional capacity of a person or other indication that an individual has been exposed to a toxicant or a biological or chemical agent or to physiological stress; and
- A susceptibility biomarker—a genetic indicator that an individual is particularly susceptible to specific toxicants or stresses.

The "exposure biomarkers" are typically metabolites of the toxicant and their presence in body fluids (serum, urine, saliva, and spinal fluid) usually follows an acute exposure. They may be detected minutes to hours after exposure in most cases.

The "effect biomarkers" are usually biomolecules such as proteins or hormones produced in the body of the exposed person. Examples of such "effect biomarkers" are "acute phase proteins," "heat shock proteins" (HSP), and epinephrine. The change in these compounds may be detected in minutes to days after exposure to the toxicant or agent.

The "susceptibility biomarkers" are usually components of the genome of the individual. Examples include the enzymes such as glutathione S-transferase (GST) and cytochrome P450 (P450) that function to detoxify compounds in the environment. Persons deficient in one of the GST isoenzymes have a fourfold higher incidence of lung cancer, if they smoke, than persons who have that subunit.

There are endogenous chemicals that are produced by the body under conditions of stress such as intense noise, temperature extremes, industrial chemical exposure, fear, and fatigue. These compounds include the high molecular weight heat shock proteins, interferons and other immunomodulating factors, acute phase proteins, and low molecular weight hormones such as adrenalin and corticosteroids. Sensors that detect changing levels of stress biomarkers can be used in distributed information systems to assign tasks and alert command about rapid changes in readiness and vigilance.

Exposing personnel to elevated concentrations of organic chemicals or industrial solvents for extended periods of time may compromise the health of a subset of the population. Acute effects may degrade the operational capability of units in which these persons function. Historical examples include the "Agent Orange" and "Gulf War Syndrome" incidents with associated morale issues and liability claims. A potential future example is exposure to jet fuel. Jet Propulsion 8 (JP-8) is the standard military fuel for U.S. and NATO forces. JP-8 as a parent fuel will be used by our military forces well into the 21st century. In FY 97, the Air Force alone used 2.2 billion gallons of JP-8, 77 percent of which went for direct aviation use. Similar in chemical composition to the fuel used in U.S. civil aviation (Jet A-1), JP-8 was introduced in Europe in the late 1970's, and its reduced volatility contributed to fewer aircraft explosions and fires. This property, in addition to lower benzene content than its predecessor, JP-4, led to a general opinion that the fuel, in vapor form, was safe from an occupational exposure standpoint. The report of the Committee on Toxicology, National Research Council, entitled "Permissible Exposure Levels for Selected Military Fuel Vapors," recommended that exposure levels to JP-8 not exceed 1,000 mg/m^3 as a short-term exposure limit and 350 mg/m^3 as an 8-hour time-weighted average). Irritant dermatological, neurobehavioral signs (headache, disorientation), and complaints of belching or tasting fuel have been reported by persons who have been exposed to aerosols of JP-8 for long periods in some occupational settings. The optimal strategy, from an occupational health standpoint, is to identify biomarkers that signal when a person is on the verge of developing clinical signs following exposure to a toxicant. Such a signal will permit the temporary removal of the affected individuals from exposure.

Isotope-dilution gas chromatography (mass-spectrometry method), has been used analyze levels of exposure of Air Force veterans who worked in Operation Ranch Hand to Agent Orange (by its dioxin contaminant). The method now provides the opportunity to analyze occupational exposures to industrial chemicals in the workplace.

Sensors and Probes to Detect Changes in Biomarkers (see also the sections on Disease Detection, Diagnosis, and Treatment)

Sensors may be developed to identify these markers so that the psychological fitness of individuals can be monitored. The response of the body to a physiological stress, foreign chemical, or infectious agent involves the increased or decreased synthesis of specific molecules by cells/tissues in the affected person. The substance produced by the subject could be an antibody (which could take from several days to months to appear); a cellular protein or mRNA, the concentration of which is increased or decreased in response to the insult; or a hormone that is produced as an effect of, or in response to, the insult.

The measurement of biomarkers of exposure allows for prophylactic measures to be taken. The ideal system would be one that identifies the insult and initiates prophylactic or preventive measures before, or simultaneous with, any toxic response by the body.

B. TELEMEDICINE

Telemedicine is the use of electronic information and communications technologies to provide and support health care when distance separates the participants.

The technologies included in this area are multimedia medical databases, virtual reality presentations (visual, auditory, haptic), telepresence surgery, life support for trauma and transport (LSTAT), microsensors and transmitters, and sensate liners in which both coverings contain sensors.

These technologies offer to the military the potential of

- Sustaining life of wounded persons during transport to hospital facilities and subsequent reinsertion into force;
- Detection of exposure of persons to chemical/biological agents and subsequent protection of cohorts from contamination by exposed individuals;
- Permitting remotely located, experienced surgeons to assist medical evacuation teams in performing surgery and reducing abdominal bleeding and shock;
- Determining susceptibility of deployed forces to infections and toxicants in an area where deployment will occur; and
- Determining the etiology of changes in health status of persons during deployment.

The civilian biomedical community will utilize these technologies while providing service to remote areas and when transporting persons experiencing traumatic injury (e.g., automobile and sports accidents). The industrial sector will utilize these technologies to monitor employees working in hazardous areas.

Telemedicine Applications

The Information Age has direct impact on the medical industry, replete with robotics, telemedicine, telepresence surgery, remote manipulation, and dexterity-enhanced surgical techniques—even three-dimensional visualization technologies. As a result, a paradigm shift has occurred in medicine, empowering surgeons and patients. Creating the operating room of the future that is capable of integrating and facilitating this infrastructure of applications is a task on the drawing boards of technology developers, research institutions, and innovators worldwide. One goal is to create an operating theater capable of freeing the surgeon from peripheral distractions during surgery, delivering information to the surgeon's fingertips, and supplying speech-recognition technologies and robotics applications to assist in the process. The primary drivers in this technology will be the medical, civil defense, and national defense communities.

Telemedicine Systems

Telemedicine technology requires a sophisticated telecommunications system that includes video, audio, and haptic interfaces from medical evacuation platforms and forward field medical units to areas that may be several thousand miles away. In this process, satellite systems are involved as are continuous communications. The fielded units are anticipated to contain smart systems that can facilitate the remote information processing between experts at a distance and medics on site.

Telemedicine trials in psychiatry, home care, dermatology, radiology, cardiology, and renal dialysis monitoring have been tested successfully in the civilian health-care delivery system. Radiology has been employed extensively in this mode, utilizing transfer of images between distant sites. Neurological diagnoses and pharmacological control of seizures have been accomplished through transmission of electroencephalographic (EEG) patterns by telecommunications. Dermatological images of nonpigmented lesions have been transmitted to distant sites for diagnostic purposes. This approach has relevance to the military community in the treatment of skin burns and potential diagnosis of chemical burns and blistering.

The applications of telemedicine technologies to radiology have increased as a result of marked progress in computer-based imaging. During the past 20 years, the anatomical and functional imaging of human organs has taken a quantum leap forward. The methodology of X rays has now been complemented by computerized tomographic (CT) systems including magnetic resonance imaging (MRI), positron emission tomography (PET), or single photon emission computerized tomography (SPECT). Organs deep within the body can now be biopsied by long, thin needles guided safely to their targets by CT or ultrasound scanning; in many cases, this capability has eliminated the need for general anesthesia and an open surgical procedure. Adaptations of MRI permit the placement of catheters within 1 mm of any region in the brain. This can allow specific delivery of materials to organs if the vascular bed of the organ is permeable to the material. In the next decade it is likely that metabolic imaging techniques will provide information about the disruption of cellular signaling pathways or specific patterns of gene expression.

Using data from the National Library of Medicine's "Visible Male" from the Visible Human Project, researchers have developed and tested imaging software that enables technicians to combine X-ray, CT scan, and MRI data to create a three-dimensional rendering of any organ in a patient, as well as the surrounding organs. By manipulating the three-dimensional image, the surgeon can see the relationship of one organ to other anatomic structures prior to surgery. This information enables the surgeon to determine the best way to conduct the operation.

The application of computer technologies to enhance and manipulate images has been illustrated most dramatically in the pictures beamed back to Earth from orbiting satellites or from interplanetary probes. Extensions of this technology have profound implications for analyzing and refining patterns detected in medical images. Neural networks are a component of some artificial intelligence technologies that can be "trained" to recognize patterns. The demonstrated potential of these new technologies to enhance diagnosis, prognosis, and treatment has compelling implications for medical evacuation missions or for reinsertion of the soldier into the unit.

C. NOVEL APPROACHES TO TREAT TRAUMATIC INJURY, STANCH ABDOMINAL BLEEDING, TREAT EXTENSIVE BURNS, AND PROVIDE ORGAN REPLACEMENT

Projectile injuries to the abdomen frequently result in extensive loss of blood through hemorrhage. Consequences of hemorrhage are a marked reduction in oxygen-carrying capacity of the blood; rapid changes in blood fluid volume; and associated changes in brain, kidney, and liver function. Reperfusion of the patient with cells and/or blood substitutes is often accompanied by damage to the central nervous system and other organ systems. The generation of free radicals and associated oxidative stress are among the factors that are etiologically involved in tissue damage following traumatic injury and reperfusion.

The following are among the developing technologies that can be used to treat traumatic injury, severe hemorrhage, and their sequellae:

- Rapid detection of changes in vital signs of personnel;
- Mass casualty respirator capability;
- Tissue viability enhancers including pharmacological intervention to reduce oxidative stress, development of synthetic reperfusion fluids that have low hepato-toxicity, pharmacologic strategies to lower tissue oxygen requirements of tissues in shock;
- Stem cell and organ culture systems to increase the rate of organ regeneration and repair;
- Development of cell growth factors to enhance regeneration of damaged tissue; and
- Development of nerve growth factors to enhance rate of neural regeneration.

Specific Elements of Traumatic Care

a. Vital Signs

Rapid detection of vital signs can be greatly enhanced by future technologies to the point that "tricorder" technology should actually exist. In its current form, the "tricorder" is a radio-frequency radiation device that will detect pulse/respiration. It has a distant life-signs scanning capability originally built for detecting vital signs through chemical warfare protective clothing, and can detect life signs through brush, structures, or collapsed buildings. Although this could be critical in finding the wounded, particularly in terrorist blast scenarios or natural disasters, it also has offensive parameters, i.e., detecting enemy personnel, and could easily be used as a "life form" detector in a tactical situation.

b. Mass Casualty Respirator

Mass casualty respirator capability to sustain survivability has recently been explored. This technology consists of high-frequency ventilation capability from a central source, but with multiple stations such that overall ventilation will be sufficient for 80–95 percent of casualties.

c. Tissue Viability

Tissue viability is dependent on cellular respiration, which in turn depends on supplying oxygen and removing wastes, and is a function of blood flow and perfusion. The availability of oxygen for tissue use is also governed by the perfusion fluid and cellular metabolic level. Blood substitutes, stroma free hemoglobin, and blood banking capability offer near-term vehicles of delivering oxygen to tissues and thus promote tissue salvage. Tissue salvage includes manipulating oxygen dissociation to allow oxygen release to hypoxic tissues. A more radical means of tissue and body salvage is an overall reduction in metabolic activity, thus decreasing oxygen demand and preserving available oxygen for tissue use.

Another mechanism likely to be available in the battlefield by 2003 is the use of hyperbaric oxygen (HBO). Oxygen breathed at increased pressure will follow gas solubility laws (3 vol% per atmosphere in plasma); hence, at standard treatment pressures, oxygen no longer depends on red blood cell concentration or hemoglobin. Tissue oxygen levels in compromised areas can approach or exceed levels found in normal tissue, provided there is minimal perfusion. It could also be used in conjunction with blood substitutes or volume fluids. In battlefield and terrorist scenarios, HBO may have multiple actions. Preliminary studies demonstrate that HBO has a protective effect in reperfusion injuries. It is indicated as adjunct treatment for crush injuries, burns, cyanide poisoning, and severe carbon monoxide poisoning. HBO has been recommended as an adjunct treatment when available for injuries due to phosgene and blister agents. Multiple studies show that HBO increases oxygen levels in compromised tissue enough to allow polymorphonucleocytes to phagocytize bacteria. This may be important in certain biological agent treatments and in preventing infection in burn cases.

The use of antioxidants to counteract radical formation during reperfusion injury is a near-term methodology. Research into decreasing lymphocyte adhesion to vascular endothelium and consequent inflammatory reaction is key to reperfusion injury treatment. Pharmocologic means of modifying biochemical responses in reperfusion injuries are also being studied.

d. Growth Factors

The new horizon in treating wounds is related to the understanding and development of growth hormones, cytokines, and integrins that trigger wound healing and angiogenesis. There appears to be an independent critical mechanism that involves the oxygen molecule that as of yet is not understood. The process of wound healing from a biochemical standpoint should advance quickly over the next 10 years, yielding a palate of pharmaceutical enhancements to heal traumatic injuries sooner.

In stabilizing patients, surgery and antibiotics will continue to play a major role. New technologies concerning telemedicine consultation and advice (discussed in other areas of this report) offer to the field a means of handling more complicated cases. However, in a mass triage situation, time may be limited for this capability. In the surgical arena, scanning or similar technology should also develop over this time period to allow tissue viability determination that is independent of blood flow. Such a technology would allow the surgeon to prevent overly aggressive debridement.

e., f. Cell and Organ Culture and Organ Transplant

New cellular technologies are emerging that may allow for growing stem cells, growing organs, or stimulating organ regeneration. The successful reports in November 1998 of stem cell culture from fetal tissues promises availability of cells to replace all tissues, including neural tissues. A second technology report appeared in November 1998, indicating that nuclei from mature human cells could be caused to dedifferentiate by inserting the nuclei into cytosol of cells from cattle. Such dedifferentiated cells can also behave as stem cells. Pluripotent cells have the potential to differentiate into any type of cell or organ system. This is particularly promising in spinal cord contusion or partial transection where immature neural cells act as a bridge between host nerve tissue. It also promises benefit to aging patients with neuro-degenerative diseases such as Alzheimer's disease, Parkinson's disease. and dystrophies. Another concept is an organ bank with cryogenically preserved replacement organs for transplantation. Over the next 20 years, advances in standard organ transplantation and organ storage will also develop, as well as a means of reducing organ rejection. Artificial skins, grown from cell tissue cultures, and synthetic membranes are already entering the marketplace. Continued development is certain.

The first successful cloning of an adult mammal was achieved in 1996 in Scotland using sheep. The feat was replicated in 1998 using adult mice (*New York Times*, July 22, 1998). The implication of these findings is that the nuclei of cells that have differentiated can be induced to serve as primary genetic material for future generations of animals. The nucleic acid sequence in the future generations will be identical to that of the nucleic acid in the cell nucleus used at the start, hence the future generations are clones of the primary donor. The cloning of cattle can result in a highly predictable milk yield and quality from subsequent generations and has significant economic impact. The capability to clone humans has social and ethical consequences that remain to be addressed, and the technique is therefore controversial. However, this technical demonstration also has implications for the large production of organ tissues from individuals. Although the technology is not available at this time to specifically program the generation of a liver, kidney, spleen, etc., from an adult nucleus, such directed programming is likely to appear by 2010. In this manner the donor of the nucleus may be a service person who suffered extensive damage to an organ; the regrown organ may be implanted with little to no risk of tissue rejection or of infections with latent or slow viruses.

Sustain Function and Repair of the Nervous System

Functional integrity of the nervous system is a sine qua non for military personnel. The nervous system is critical for sensory perception and motor function, information analysis, and decision making. This area encompasses neural and psychiatric aspects of brain, spinal cord, and peripheral nerve function. Traumatic injury to the spinal cord is a primary concern, both because at the present time there is little clinical treatment to be offered to the patient with spinal cord transection, and as a consequence the massive loss of function will remain for the duration of the patient's life. The financial cost to relatives and to society in caring for the patient is extensive. The incidence of spinal cord injuries in the civilian sector is approximately 40 persons per million population, with the rate of spinal injuries for males 2.5 times greater than that seen in females. The costs of caring for persons in the first year following injury is approximately $200,000, and each subsequent year has a cost of approximately $30,000. Veterans constitute 22 percent of the spinal cord injury population. Although hemisection or transection of the spinal cord is a civilian and military problem, the military cost relates to the inability to reinsert the affected soldier into a combat setting. As the fighting force is reduced in size and the ability to deliver firepower increases, spinal cord injuries loom as major problems in the next decade. Other neural and psychiatric sequella related to combat injuries include infectious diseases of the nervous system, effects of toxicants on the nervous system, development of seizures following traumatic injury to the head, increased neuronal cell death due to hypoxia from blood loss, induction of necrotic or apoptotic events in cells leading to neural degeneration, and stress-induced changes in the central nervous system with associated loss of synaptic connections.

Spinal Cord Repair

The past 15 years have seen the emergence of new technologies that may facilitate repair of hemisected spinal cord. Three major contributions have been (1) the recognition that placement of bridging matrices across the transected section of cord will serve as a bridge across which the nerve axons may pass; (2) placement of biological growth factors in the vicinity of the regenerating nerves appears to enhance the growth of the cut axons across the transection; and (3) biochemicals or biomimetics may be able to limit the "scar" forming growth of the glial cells, thereby removing a barrier to axonal growth across the transection.

The bridging materials that have shown efficacy in experimental animals include fibrous biopolymers, biomimetics, and synthetic polymers. The ends of the polymers are placed in each edge of the cut surface of the cord. Neuronal axons will grow along the surface of the polymer and become inserted into the spinal cord on the other side of the cut cord. Most frequently the axons penetrate only 1–3 mm, thereby limiting the regeneration of the cord.

Several proteins have shown utility in enhancing survival of neurons in spinal cord following injury and in facilitating axonal outgrowth. Included in this category are brain-derived neurotropic factors. Other proteins have demonstrated utility in reducing edematous swelling of neural tissue following traumatic injury. Since edema is accompanied by increased cell damage and death, these compounds are considered neuroprotective. Insulin-like growth factor I is an example of a compound with protective action. Insulin-like growth factor also reduced the marked increase of neuronal nitric acid synthase, an etiological inducer of nerve damage.

In neural regions other than the spinal cord it has been shown that in humans, hypoxic damage is associated with the phosphorylation of a "mitogen-activated kinase activating death domain protein" and migration of the protein to the nucleolus of neurons. The phosphorylation is catalyzed by stress-activated kinase Jun kinase 3. The correlation of hypoxic damage of the nervous system with an enzymic event suggests that low molecular weight inhibitors may be developed that can interdict the hypoxic damage. Finally, during early development it has been shown that proteinaceous materials (GAP 43) are required for growth of retinal ganglion cell axons in the visual system. It is likely that such signaling molecules are required during neural regeneration in certain brain regions. With the advent of combinatorial chemistry it is anticipated that new classes of compounds will have applications in neuroprotection and regrowth of nerves across the lesion.

Civilian Sector Relevance

Penetrating abdominal wounds often accompany automobile and motorcycle accidents and gunshots. The same technologies used to treat military personnel are applicable to the victims of vehicular and gunshot incidents.

D. SYSTEMS TO DETECT VITAL SIGNS AND REMOTE DETECTION OF LIFE FORMS

Microneural bionics, microchips designed to translate sensory input and mimic nerve efferent signals, are used to treat patients with lost hearing or vision. Current use in otolaryngology allows hearing in children who are born with hearing deficits. Ophthalmologic uses have been demonstrated in recently blind individuals, allowing light perception. It is not inconceivable that similar technology may not only restore some sensory deficits from injured warfighters, but could be developed to allow one to recognize nonhuman sensory input (infrasound, ultraviolet, and infrared spectrum) or enhance normal sensory perception.

E. EVIDENCE-BASED MEDICINE AS RELATED TO AFFORDABILITY

Medicine as practiced in 1998 is based on the interpretation by a clinical care giver (e.g., physician) of clinical signs and symptoms through a process of differential diagnosis. The database of information available to the physician is derived from information provided by a patient, experimentation in clinical trials, personal experience, scholarly reading of the literature, continuing medical education, and informational briefings. In these situations, the data available is relatively large and the physician relies in large part on that which can be processed in a period of minutes. The processing of this information is a function of recall ability, access to current information, state of readiness, and vigilance. The degree to which the physician can monitor patient compliance with prescribed medication and therapy varies with the clinical setting, diagnosis, and cost of follow-up.

The availability of new computer-based systems and interfaces with the physician and other care givers can facilitate establishment of a database warehouse and provide a coherent format for taking patients' medical histories. It also offers the ability to track distribution of medication; "ticklers" regarding upcoming visits; and responses to medication, including allergic reactions. The new technologies offer the opportunity to obtain evidence-based information and provide evidence-based management if the number of subjects is sufficiently large. They also provide the basis for developing a data chip that will contain the medical history and deployment history of each individual in DoD.

Kaiser Permanente (KP), a large health-maintenance organization, has developed "National Clinical Information System (NCIS)," an effort to achieve information acquisition, retrieval, presentation, and management.

NCIS comprises Oceania (structured notes); Pegasystems (information flow system); Oacis (Pharmacy knowledge base); IBM (integrated user interface); Ontyx (Lexical technology and dictionary); and SAIC (clinical data repository in Oracle). One of the outcomes expected with NCIS is the determination of most useful treatment modalities and management of patients. KP is the largest private, not-for-profit, health-care organization in the world, with 9 million patients, 10,000 MDs, 350 medical offices, 30 hospitals, $15 billion in annual revenues, and $10 billion in assets. In many respects KP resembles U.S. military health-care system (patients, physicians, cost issues). To improve health-care delivery, a system must incorporate new technology and knowledge. Associated with the new information and technology is the need to change the culture and behavior of physicians and health-care providers as well as the behavior of patients. The approach requires a medical component to examine technology effectiveness. A management focus examines treatment tradeoffs and must invest in most effective outcomes.

Care management (evidence-based management), focuses on:

- Component-based care delivery;
- Eliminating inappropriate physician variation (diagnostic testing, pharmaceuticals); and
- Developing and adopting breakthrough methods (e.g., outcome driven, technology enabled).

The new model for health-care delivery should identify the role of the physician, establish a system responsible for patient adherence, and determine outcomes accountability. All these points based on large N of patients, large number of physicians, geographical distribution, and interaction with Blue Cross/Blue Shield cooperative arrangement. These large N values allow one to look at statistical outcomes of each treatment paradigm.

Issues Related to Affordability

One critical point is the rapid change in technologies that will become available to physicians and health-care professionals in the next two decades. The utilization of these new technologies can improve health-care delivery. In those cases where new diagnostic modalities and pharmaceuticals improve treatment outcome, new standards for appropriate medical care in the 21st century will emerge. These technologies may also provide an ever-increasing capability to produce new compounds that affect human behavior and performance. The issue of affordability rises to the surface because the new pharmaceuticals and diagnostic tools are likely to be expensive. Past experiences suggest that what is not affordable at time *a* may be at time *b*. The factor of affordability in the military community relates to the actual cost of the new item, the benefit achieved by the patient in terms of time away from the task, the compromise in function if treatment is not initiated, and the number of persons affected. These considerations lead to three action items:

- Provide incentives for new technologies so they can be adapted;
- Measure affordability as related to outcome; and
- Real cost effectiveness will increase in biotechnology and informatics because of rapid growth of fields and affordability not necessarily correlated with best business practice.

Many of these items have parallels to the TacMedCS discussed in Section 4, Overview, above.

WORLDWIDE TECHNOLOGY ASSESSMENT

The identification of molecules that may serve as biomarkers of toxicant exposure is in a rapid growth phase. A limited number of such molecules, synthesized in the body of affected persons, have been characterized. The genes coding for the biomarkers discovered to date have been sequenced. As indicated above in the genomics section, such sequences are available on the WWW.

The technology of telemedicine is contingent on electronic, communication, satellite, and computational technologies. The nations that have these technologies are primarily first world states.

The emerging technologies identified above for management and treatment of traumatic injuries are most developed in nations with very advanced medical treatment capability. Because they are costly, the national culture and resources determine the extent to which these technologies will be implemented. The rate of acquisition of these technologies is likely to expand rapidly during the next decade in the United States, UK, France, Germany, Japan, Sweden, and Switzerland.

See Technology Data Sheets for WTA charts.

SECTION 4.4—TACTICAL MEDICAL COMMAND AND CONTROL

Highlights

- TacMedCS manages care from field through treatment.
- Medical care, medical informatics, and medical communication are integrated.

OVERVIEW

Medical command and control is the integration of disparate technologies including electronics, communications, information systems, and trauma care. It is evolving into a system providing both specific medical information as well as procedures that have utility across all services. The consequences of the advances in this technology have profound dual-use applications. These include emergency services, occupational medicine, rural health-care delivery, and disaster response. These broad applications will make the technologies more affordable. These changes will continue to evolve during the next decade in response to:

- Maneuvers initiated from the sea and otherwise over the horizon;
- Fast strike and move out;
- Discontinuous fronts and nonlinear battle areas;
- Autonomous small unit operations; and
- Urban warfare.

Each of these mandates a medical status and care system which can operate with integral purpose while drawing resources and capabilities from a variety of sources not collocated.

Associated with these shifts are changes in the medical response strategies to locate and treat wounded, provide accurate diagnosis/triage, and evacuate to an optimum treatment facility. The new technologies available to the corpsman will enable more accurate information about the wounded soldier, a more complete selection of options to treat the patient, rapid communication with forward medical units to prepare them for receiving the patient, evacuation options, and additional procedures to stabilize the patient. This paradigm requires three elements of information management: accurate information, real-time information, and data integration.

The technologies available to the corpsman and other health-care providers in the combat arena are described in Sections 4.1, 4.2, and 4.3. The consequences of these new technologies will be to (1) maintain the health of all individuals in the theater, (2) protect deployed individuals from chemical and biological attack, (3) afford treatment that will minimize future disabilities, and (4) return the individual to his or her parent unit in a rapid manner. The information technologies supporting the medical technologies and assets are discussed below.

Tactical Medical Coordination

A Tactical Medical Coordination System (TacMedCS) has been developed. TacMedCS is a multicomponent system for efficient management of patients and their associated clinical/identification information from the field through evacuation to definitive treatment.

TacMedCS has three components:

- Patient ID/clinical data chip (component 1);
- Hand-carried data input/transmission device (component 2); and
- Operations area command and control suite (component 3).

Component 1 (EPROM chip) contains identification and pertinent clinical history; data transmitted when interrogated by a Component 2 device; and stores limited treatment information. Component 1 automatically measures and records vital signs, past/present medical history including biomarkers and other clinically important information.

Component 2 (handheld reader/transmitter/recorder) interrogates the EPROM chip using an RF signal; transmits GPS-derived location and clinical data to the command and control suite; stores data internally for later download; writes data back to the patient EPROM chip; and has the potential to utilize different formats for storing information (barcode). A future component 2 is likely to contain capability for transmitting real-time images, auditory displays, and matrixed information to component 3.

Component 3 (clinical command and control center) provides command and control real-time data coordination and management, along with patient tracking and status monitoring (includes visual display). Component 3 is staffed with high-level medical experts and is remote from the combat situation. Its function is to advise the corpsman regarding the most appropriate short-term treatment, recommend procedures for transfer of the patient, prepare the receiving facility for the arrival of the patient, and initiate care.

The TacMedCS is patterned after the Tactical C4I and overnight delivery services currently available nationwide. The system utilizes adaptable, off-the-shelf technology in a unique combination. The new system provides real-time data integration and coordination, patient identification, and immediate information concerning clinical status, evacuation, and treatment choices. It is adaptable and conforms to contemporary C4I requirements.

The TacMedCS has advantages over current approaches. It provides a battlefield medical command and control system and an operational medical informatics system for the integration of diagnostic, treatment, and evacuation information; makes real-time information available to unit COs for developing dynamic battlespace orientation and asset allocation and intelligence; makes real-time information available to MEDEVAC assets for mission planning; and makes real-time information available to definitive care assets for utilization planning, patient tracking, and triage. This approach has applications to a wide range of civil sector needs, and therefore the cost and generation of new ideas will be shared with the broader medical community. This will increase affordability to the defense community.

WORLDWIDE TECHNOLOGY ASSESSMENT

Leading-edge tactical medical command and control is a newly developing area of expertise. It relies on innovations in information gathering and processing, structured medical hierarchy, and integrated communication systems. The early stage of development has provided many opportunities for innovation, but few if any current models are that of TacMedCS.

See Technology Data Sheet for WTA chart.

DATA SHEET 4.4. MEDICAL INFORMATICS

Developing Critical Technology Parameter	A tactical medical coordination system comprising a patient ID/clinical data chip, a hand-carried data input/transmission device, and an operations area command and control suite.
Critical Materials	EPROM chip with ID and pertinent medical history; a component that interrogates the EPROM chip with an RF signal, transmits GPS location and clinical data to command suite, and stores data as barcode; a component to coordinate/manage data in real time, track the patient, and monitor status.
Unique Test, Production, Inspection Equipment	Validate EPROM chip and data storage system for reliability.
Unique Software	N/A.
Technical Issues	Integration of the three components. An operational system exists in the NASA program.
Major Commercial Applications	NASA control of astronaut data systems. Rural health-care delivery. Home care delivery to persons with chronic illness requiring daily or more frequent monitoring of status (i.e., renal dialysis).
Affordability	The system is relatively low cost in the current configuration.

RATIONALE

Early detection of biological effects on individuals exposed to low-level toxicants can permit initiation of protective actions. This will increase vigilance, readiness, and fitness for duty of affected persons. Biomarkers of exposure and of susceptibility will permit early warning and management of persons. Joint Vision 2010 concepts of full dimensional protection and full spectrum dominance would benefit from improved tactical medical command and control. It would enhance alertness and fitness for duty of highly trained personnel. The identification of biomarkers of low-level toxicant exposure is an emerging and rapidly developing field. The Army has established an effort in the Military Operational Medicine Program at Fort Detrick, Maryland. There is rapid access of the federal sector to the available data.

WORLDWIDE TECHNOLOGY ASSESSMENT

Australia	●●●	Belgium	●●●	Canada	●●●	China	●●
Cuba	●●	Czech Republic	●●	Egypt	●	France	●●●
Germany	●	Hungary	●●	India	●●	Iran	●
Iraq	●	Israel	●●●	Italy	●●●	Japan	●●●●
Malaysia	●●	Netherlands	●	North Korea	●	Norway	●●●
Poland	●●	Russia	●●	Singapore	●	South Korea	●●
Sweden	●●●●	Switzerland	●●●●	Syria	●	Taiwan	●●
Ukraine	●●	UK	●●●●	United States	●●●●		

Legend: Extensive R&D ●●●● Significant R&D ●●● Moderate R&D ●● Limited R&D ●

BIOLOGICAL CRITICAL TECHNOLOGY

F. Spalt
(Editor)

SECTION 3. BIOLOGICAL TECHNOLOGY

Highlights

- Biotechnology has sustained a growth rate of doubling the base of knowledge every 18 months to 2 years. This trend appears to be accelerating in the human genome, multi-array sensor, and biomaterials areas.
- The economic engine for biotechnology is the non-military commercial sector, including medicine/health, agriculture/food processing, cosmetics, and transportation.
- Military exploitation and harnessing of biotechnology is in the embryonic stage. The impact on the military is anticipated to be pervasive and fundamental in the areas of human performance, sensors, biomaterials/nanofabrication, and individual and group performance.
- Significant advances in biotechnology capabilities should be anticipated from developed and less-developed nations.

OVERVIEW

The rapid growth of biotechnology since 1970 has provided many items and systems that are useful to the civilian and the military sectors. These products and processes relate to improvements in human health by production of new pharmaceuticals, including antibiotics, vaccines, and performance sustainers. Sensors capable of detecting infectious agents, toxins, and chemical agents have been developed and miniaturized using insights provided by biotechnology. During the past decade, innovations in monitoring human performance, vigilance, and fitness for duty have emerged from the use of imaging systems that permit viewing brain activity on-line and correlating the electrophysiological activity of the brain with human performance. Advances in genetic engineering, molecular biology, and polymer chemistry have created new opportunities in the fabrication of electronic circuits, molecular motors, and structures having high strength and low weight. The shelf life and palatability of foods has been markedly improved by discoveries in biotechnology.

The rapid growth in all these areas has been accompanied by federal and private sector investment in the United States and other developed nations. Some of this advanced technology has been acquired by developing nations, including those hostile to U.S. interests, potentially enabling adversaries to diminish the superior military capability of U.S. forces. This section describes the developing biotechnologies anticipated to be realized in a 20-year time frame. The roles of biological systems in the production of electrical energy are included in the Energy Technology section of Part III of the Militarily Critical technologies List (MCTL).

The rate of economic growth of the biotechnology industry is the most rapid of all industrial sectors, and this reflects the perceived importance of this industry on product development, processes, and systems. Between 1996 and 1997, the market capitalization of this industry increased 60 percent—from $52 to $83 billion. The revenues increased 15 percent, and the sales increased 16 percent. The largest applications of medically related biotechnology were in the United States.

Each of the technology items identified in this section is being driven by broad requirements and applications. All are expected to develop a substantial set of non-military commercial applications and attendant manufacturing and support infrastructures. The industrial sectors supporting the biotechnology thrusts include health/medicine, pharmaceuticals, agriculture/food processing, and transportation. The military utility of specific biotechnologies for applications in national security systems has not yet been fully explored and developed.

RATIONALE

In this section, the developing technologies have been categorized into four groups:

1. Human Performance Enhancement
2. Biological Sensors
3. Biomaterials and Nanofabrication
4. Individual and Group Protection.

These categories reflect the proposed and actual military application suggested in the Joint Vision 2010 documentation and the supporting Service planning guidance. Table 3.0-1 provides a matrix of the 22 developing technologies and identifies where they are covered within the four technology areas.

Table 3.0-2 identifies the military applications of these developing technologies. Appendix A supports these summary tables by showing the individual Service needs related to each of the identified biotechnologies.

Because the technology developments are funded primarily for dual-use pharmaceuticals, materials, and equipment, nations that currently possess leading-edge capabilities in biotechnology and the biosciences are anticipated to be the most economically developed nations with particular strength in pharmaceuticals, food processing, medical technology, and cosmetics. High-value-added products requiring these technologies are in the pharmaceutical, food, and cosmetic arenas. The rapid progress currently seen in the Human Genome Project is likely to give rise to an explosion of new products affecting human health, performance, and disease resistance. New diagnostic procedures for disease susceptibility and anticipated life-span enhancement will emerge. Since the sequencing of the genome of bacteria, viruses, and fungi is progressing at a high rate and results are correlated with genomic data from the Human Genome Project, the genetics of disease susceptibility in various human populations will emerge. A current example is the finding in 1998 that resistance to Human Immunodeficiency Virus (HIV) in persons of European ancestry is associated with resistance to plague developed in these populations during the epidemics of the 14th to 17th centuries. The genetic factors responsible for virulence of certain bacteria, viruses, and fungi are currently being identified, and the pathogenicity islands (PAIs) are being mapped.

Table 3.0-1. List of Developing Technologies and Concepts by Technology Group

	Technology Groups			
Developing Technology	Human Performance Enhancement	Biological Sensors	Biomaterials and Nanofabrication	Individual and Group Protection
A. Recognition/perception	X			
B. Memory	X			
C. Visual/auditory/olfactory	X			
D. Cognition	X			
E. Electrophysiological monitor and brain activity	X			
F. Brain imaging	X			
G. Human performance maximization	X			
H. Nanofabrication		X	X	
I. Sensors/molecular recognition		X		
J. Biomedical imaging and automation				X
K. Biomaterials		X	X	
L. Encapsulation		X		X
M. Human Genome Project	X			X
N. Pathogen Genome Project		X		X
O. Atomic Force Microscopy (AFM) and Near Field Scanning Optical Microscopy (NSOM)		X	X	
P. Increased disease resistance				X
Q. Thin films [charged-coupled device (CCD)]		X	X	
R. Microelectromechanical Systems (MEMS)		X	X	X
S. Blood substitutes and biocompatible clotting matrices				X
T. Locator of persons				X
U. Water purification	X			X
V. Biomarkers for toxicant/stress exposure	X			X

Table 3.0-2. Developing Technologies and Concepts Related to Biological Systems

A. RECOGNITION/PERCEPTION Grandmother cell Gamma phase 40 Hz Hebbian circuit	**F. BRAIN IMAGING** Magnetic Resonance Imaging (MRI) Functional MRI (fMRI) Positron Emission Tomography (PET) X-ray digital
B. MEMORY In vivo gene regulation (CREB)	**G. HUMAN PERFORMANCE MAXIMIZATION** Biological Response Modifier (BRM) as performance sustainer Drugs, diet, and equipment that sustain performance Analgesia without depression Pain control chemicals Circadian rhythm synchronizers Neuroprosthetics and chemicals (growth factors that enhance nerve repair) Exoskeletai supports Immune response enhancers
C. VISUAL/AUDITORY/OLFACTORY Visual perception HMD Visual display NVD Iconographic display 3-D display (audio, visual, tactile)—simulation and virtual reality Odor detectors Miniature auditory implant	**H. NANOFABRICATION** Protein/Lipid/Nucleic acid self assembly Switching devices (MEMS) AFM and NSOM devices
D. COGNITION Artificial intelligence Information warfare Axonal guidance-model for data fusion DNA computing	**I. SENSORS/MOLECULAR RECOGNITION** Dynamic interactive arrays based on antibodies, nucleic acid sequences, or receptors Swarm systems Human vital systems card (physiological/metabolic indicator; triage; alertness; and ID)
E. ELECTROPHYSIOLOGICAL MONITOR AND BRAIN ACTIVITY EEG Visual evoked potential Brain stem auditory evoked potential Eye movement	**J. BIOMEDICAL IMAGING AND AUTOMATION** Telemedicine—pod for rapid transport/containment of wounded Information systems—worldwide current status of infectious diseases Medical history of personnel on card Train and assist Corpsmen by telecommunication Distributed decision-making Integrate medical into C2 architecture

WORLDWIDE TECHNOLOGY ASSESSMENT (See Figure 3.0-1)

The Human Genome Project is an international effort with complete exchange of information and databases for humanitarian purposes. The development of multi-national industries in the area of biotechnology has increased markedly as a result of tax and tariff structures that favor such alliances. The international regulatory standards applied to pharmaceuticals, food, and biomedical materials further encourage international cooperation and technology transfer. The biological and biomedical sciences sections of graduate and postgraduate education in the United States have the highest numbers of foreign students. This reality alone enhances transfer of technology to all nations, particularly less-developed countries. The low economic cost required to establish biotechnology facilities should result in rapid transfer of production capabilities to countries that do not lead in technologies. Shipping costs and capital investment are minimal. Precursor material and end products are low in weight. Production is not capital intensive compared with silicon-based electronics and analogous processes. Figure 3.0-2 provides an estimate of the capability of the identified nations in the development of products using biotechnology as a means of production. Although the less-developed countries will be able to acquire these technologies, the time interval between achieving production in the most advanced nation and least developed nation is anticipated to be on the order of 2 to 5 years.

Because of the breadth of effort and level of funding, the United States leads the rest of the world in aggregate discoveries and progress in biotechnology. However, other industrialized nations carefully track biotechnology progress through multiple open and corporate sources. The rapid dissemination of information through the world wide web and the rapid access of summaries of newly published articles through MEDLINE, Chemical Abstracts, and Science Citation Indexes increase the transfer of information across national borders at rapid rates. The delay between publication and appearance in electronic distribution is less than a month. Thus, the transition of a technology into products can be accomplished with equal speed in any relatively industrialized nation. Countries that have demonstrated such capability include China (for internal consumption), France, Germany Japan, Israel, Sweden, and the United Kingdom. Other nations, such as Switzerland, have moved extensive development and production facilities out of the country for economic reasons. Russia has a residual aggregate knowledge base from the days of the Union of Soviet Socialist Republics (USSR), and this is moderately enhanced by international partnering with industries in the United States and other nations. However, in Russia, the transition of the knowledge base into products continues to decline.

Table 3.0-2. Developing Technologies and Concepts Related to Biological Systems (Continued)

K. BIOMATERIALS bacterial Rhodopsin phthalocyanine Bioceramics Body armor and personal protective systems Neuronal assemblies on chip Molecular switches (see Technology Item H) Blood substitutes (see blood substitutes (see Technology Item S) Molecular motors (MEMS—see Technology Item R)	**Q. THIN FILMS (CCD)** Artificial retina Information storage
L. ENCAPSULATION Vaccines Performance enhancers Food materials and nutrients	**R. MICROELECTROMECHANICAL SYSTEMS (MEMS)** Switching devices Sensors
M. HUMAN GENOME PROJECT Determine susceptibility of human populations to disease Determine genetic susceptibility of individuals in military to toxicants, resistance to infectious diseases and sleep deprivation, and stress. All these are components of allelic profiling. Drug design	**S. BLOOD SUBSTITUTES AND BIOCOMPATIBLE CLOTTING MATRICES (see Technology K)** Topical Fibrin
N. PATHOGEN GENOME PROJECT Pathogenicity islands (PAIs)	**T. LOCATOR OF PERSONS** Transmitter Excreted biocompatible material in urine
O. ATOMIC FORCE MICROSCOPY (AFM) AND NEAR FIELD SCANNING OPRICAL MICROSCOPY (NSOM) (see Technology Item I) Nanoscale manufacture quality control	**U. WATER PURIFICATION** Fluid capture and retention
P. INCREASED DISEASE RESISTANCE Super vaccine Recombinant vaccines Inhibit virus entry, intracellular transport, and maturation necessary for viral infectivity Multi-component, multi-valent vaccination Novel antibiotic with restricted distribution	**V. BIOMARKERS FOR TOXICANT/STRESS EXPOSURE** Glutathione S transferases P450 Acute phase proteins

Note for Table 3.0-2: *The bolded categories identify the developing technologies. The concepts under these categories provide examples of specific elements/applications of the technology.*

Country	Human Performance Enhancement	Biological Sensors	Biomaterials and Nanofabrication	Individual and Group Protection
Australia	●●●	●●●	●●	●●
Canada	●●●	●●●	●●●	●●●
China	●●	●●	●●	●●
Cuba	●	●●	●	●●
Czech Republic	●●	●●●	●●	●●
Egypt	●●	●	●	●
France	●●●	●●●●	●●●●	●●●●
Germany	●●●	●●●●	●●●●	●●●●
Hungary	●●	●●	●●	●●
India	●●	●●	●●	●●
Iran	●	●	●	●
Iraq	●	●	●	●
Israel	●●●	●●●	●●●	●●●●
Italy	●●●	●●●	●●	●●
Japan	●●●●	●●●●	●●●●	●●●
Malaysia	●●	●	●●	●
Netherlands	●●●●	●●●●	●●●●	●●●●
North Korea	●	●	●	●
Norway	●●●	●●	●●	●●
Poland	●●	●●	●●	●●
Russia	●●●	●●●●	●●●	●●●
Singapore	●●	●●	●●	●
South Korea	●●	●●●	●●	●●
Sweden	●●●	●●●●	●●●●	●●●●
Switzerland	●●●●	●●●●	●●●●	●●●●
Syria	●	●	●	●
Taiwan	●●●	●●●	●●●	●●
Ukraine	●●	●	●	●●
United Kingdom	●●●●	●●●●	●●●●	●●●●
United States	●●●●	●●●●	●●●●	●●●●

Legend: Capability in Technology Elements: ●●●● Most ●●● Many ●● Some ● At Least One

Figure 3.0-1. Biological Technologies Area Worldwide Capability Assessment

SECTION 3.1 HUMAN PERFORMANCE ENHANCEMENT

Highlights

- Selection of personnel for task suitability will become increasingly accurate and sensitive based on identification of specific biomarkers for toxicant susceptibility, stress, and sustained performance.
- Performance monitoring and warning will enhance the ability of the military cohort to sustain and maximize mission goals of individuals and teams.
- Pharmacological intervention will enable military units to extend and sustain effective operational capabilities.

OVERVIEW

In the future, one of the most significant applications of emerging biotechnologies will be to enhance operational warfighter performance by improving perception, information processing, decision-making, and task execution capabilities. For the Militarily Critical Technologies List (MCTL), biotechnology can be used to increase these facets of human performance in three areas:

- **Physiological monitoring.** Psychological monitoring technologies, using miniaturized electrodes and data fusion systems that require low energy, will detect and assess changes in warfighter cardiovascular and neurological processes in real-time to drive system adaptation.
- **Information display and control.** Information display and control technologies will present and organize information or control systems to optimize warfighter awareness, decision-making, workload, and reaction times.
- **Pharmacological intervention.** Pharmacological intervention technologies, such as drugs or environmental changes, will improve the physiological processes underlying warfighter performance.

These areas will provide for the prescreening of personnel to determine genetic and non-genetic prognosticators that may indicate a susceptibility to stress and toxicants and for the analyses of paradigms for complex behaviors involving swarms and hives. If successfully pursued, these areas will be critical in maintaining and increasing our military superiority.

RATIONALE

The soldier is the basic unit of the combat force. The intrinsic nature of the person and the training, equipment, and support components provided are decisive elements in determining success in military operations. In this decade, it will be possible to understand the genetic basis of human function. As a result, it may become possible to determine how the individual soldier is likely to respond to chemical and biological threats and other fear-inducing situations. The particular susceptibility of individuals to specific biological agents may be able to be determined. It is also likely that new vaccines and therapies will be developed to help cope with threats in the environment. Knowledge of the individual susceptibility to disease and the mechanisms by which such susceptibility can be managed are critical to the maintenance of force structure and mission success. This awareness is even more essential as the number of people engaged in military actions is reduced and the active duty forces are requested to participate in an increasing number of police actions.

New technologies have also increased our understanding of the way in which humans respond to information. Iconographic displays associated with auditory and haptic information markedly reduce the time required to detect a change in the environment that may be hazardous (e.g., the spatial location of an incoming air-to-air missile or the direction of fire). Distributed information systems have enabled modeling of combat or rescue scenarios before the actual mission and have improved the identification of potential threats in a given setting.

Interfacing the human genome database with computer-assisted perception will prepare the soldier and the command for threats in a particular setting and, thereby, improve mission performance.

TECHNOLOGY DESCRIPTIONS

Physiological Monitoring (Technology Items A,B,C,D,E,F,G,M,V in Table 3.0-2)

Technologies that can non-obtrusively measure core physiological processes in a military environment are evolving. Applied research is successfully producing models that use those measures to determine operator cognitive workload, vigilance, and fatigue reliably in order to drive task allocation between the operator and the system or change information presentation to capitalize on available resources. The physiological parameters that can be correlated with functional performance changes include auditory and visually evoked potentials, broadband brain activity (e.g., using complex electroencephalographic output with an extended numbers of channels), eye movements, heart interbeat intervals, respiration rates, and blood oxygen levels. Imaging technologies, especially functional Magnetic Resonance Imaging (fMRI), Positron Emission Tomography (PET) using ^{15}O or ^{18}F labeled derivatives, and magnetoencephalography, can reveal brain areas that are active during specific tasks.

These technologies promise real-time analyses of performance for automation decisions or rapid screening tools for measuring job fitness. Interpretation of EEG-monitored brain wave patterns to indicate intended operator actions will occur online with near instantaneous reaction times and will allow the elimination of motor response errors (termed "brain activated control"). Technologies requiring physiological inputs, such as brain activated control, will capitalize on the development of remote (i.e., non-contact) sensors and analytical tools like fMRI. In addition, physiological changes could be inferred from changes in task accomplishment or loading by using models of human performance. Monitoring operator state for impending unconsciousness will serve to ensure safety by permitting the automatic transfer of operator control to the system computer.

The exposure of warfighters to common physiological stressors, such as toxic chemicals, sleep deprivation, jet lag, and anxiety, can result in both acute performance decrements (like reduced vigilance or fatigue) and chronic health problems (generally resulting from an immuno-suppression that increases susceptibility to diseases of the nervous system, liver, and lungs). Today, the military can identify an affected individual only after his performance has been critically impacted. By that time, simple interventions are usually too late to correct mission performance or prevent disease. Military readiness could be improved dramatically by tracking physiological measures that can provide an early warning of impending performance problems. The biochemical markers of acute exposure to stress include the enzymes glutathione S-transferases and cytochrome P450. Biomarkers that reveal longer term exposure to conditions that compromise function or lead to unacceptable risk for error in highly skilled individuals are also of critical importance. Recent advances in the behavioral sciences, neurophysiology, biochemistry, medical imaging, and ergonomics provide the opportunity for developing early physiological monitoring techniques to detect situations that will degrade human performance. The development of unobtrusive monitors for these early warning markers will increase the likelihood of effective performance and reduce the risk of failure. A substantial improvement in the effectiveness of weapon systems that depend upon human interactions for success can then be realized. These capabilities will have classic and multiple dual-use applications and linkages across large industries.

Information Display and Control (Technology Items C,D,G in Table 3.0-2)

In the realm of biotechnology, the presentation of information or the employment of control modalities can be designed to capitalize on specific interactions with human perceptual, cognitive, or motor systems. It can be expected that the human-machine interface will be incrementally transformed into an interactive fusion that is driven by changes in neurophysiological states.

The modeling of human information processing will be precise enough to design truly intuitive interfaces and to delineate the differences between computer and human information processing in order to optimize human-machine interactions and the allocation of mission functions. This improved understanding of mental processes will lead to enhanced situational awareness and decision-making by fundamentally changing the structuring of information interactions. Real-time monitoring of operator cognitive and perceptual resource consumption (i.e., workload), coupled with performance measures, will tailor information formatting to sensory modality and the assignment of tasking between the human and machine. This will provide a set of new boundaries in human information processing tailored to the individual performer.

Changes in information display and system control technologies will revolutionize the human-machine interface. Binocular, full color, wide field of view helmet mounted displays (HMDs) will integrate sensor and data fusion symbology into the visual scene and employ eye-tracking technology to allow extremely fast and accurate target designation, even at night. Thin films of biomaterial or biomimetics on the HMD may provide the visually detected data and symbology (see Biomaterials). Three-dimensional (3-D) audio and haptic interfaces will significantly decrease reaction times by providing redundant target localization cues. Robust speech recognition, developed at General Motors and Delco, will permit hands-off operation, even under the most stressful conditions. Significant decreases in material weights will decrease the fatigue and muscle strain associated with current systems. Together, these technologies will create a true virtual environment capability to allow windowless vehicles (e.g., Citadel Ships) or optimized remote control.

The development of virtual audio, visual, and haptic realities and the emerging networking of multiple and diverse systems allow many individuals to interact at remote stations. These technologies provide new opportunities for training skilled personnel, planning and testing responses to crisis situations, developing new operational strategies, conducting mission rehearsals, and evaluating the proposed solutions in low risk environments. This multi-system approach, using virtual realities and remote interactive systems, is the basis of advanced distributed simulation (ADS). A support system infrastructure for developing standards to be used in ADS has been created and is called distributed interactive simulation (DIS). DIS exercises support a cluster of virtual entities (human-in-the-loop simulators), live entities (operational platforms and test/evaluation systems), and constructed entities (automated simulations).

Pharmacological Intervention (Technology Items A,B,C,G,M,U in Table 3.0-2)

In the future, pharmaceuticals or manipulations of the environment will be used routinely to increase the cognitive and physical prowess and endurance of U.S. warfighters. This section provides a sampling of the expectations for military medical interventions to maximize performance.

The nature of the biological mechanism that allows a person to discriminate a subject cognitively from its background has been termed the "binding problem." Currently, two fundamental theories address the binding problem:

- There exist nerve cells that store all the collective neural inputs for a solitary object through unique neurons (e.g., a cell that is used to recognize "grandmother"). Accordingly, different "grandmother cells" would be required for recognition of all items in long-term memory.
- There exist gamma phased 40-Hz circuits that are programmed to become activated when a target to be recognized appears. Sets of individual 40-Hz circuits may collectively comprise Hebbian circuits. If the memory is comprised of 40-Hz circuits, the introduction of selective 40-Hz signals from external sources may stimulate or suppress the memory system.

Evidence exists to support each of these physiological processes in different scenarios. The next decade is likely to see novel approaches to the design of pharmaceuticals that can enhance recognition speed and accuracy. Compounds, such as AMPAkines, that increase both working and long-term memory functions will significantly improve warfighter situation awareness, decision-making, and task execution under stress.

The response of the warfighter to the battlefield environment is determined by training, physiological factors (e.g., strength and age), and the genetic factors of the combatant. The advent of the 21st century will provide increased information on the genetic contribution to the effectiveness of the warfighter. The sequencing of the entire human genome is expected to be complete by 2005. The genome will be characterized by rapid gene sequencers. One system in development prepares 400,000 bases per hour and analyzes about 8,000 bases per 16 hours. This system can determine the sequence of about 2 million bases per several months. A second strategy uses variants of mass spectrometry to sequence DNA. Matrix-assisted laser desorption ionization (MALDI) mass spectrometry has proved to be an important tool in synthetic- and bio-polymer characterization. The molecular sizes that can be sequenced by this technique are limited in 1997, but the method currently has utility for determining mutation sites in DNA fragments. The information to be obtained from characterization of the human genome is likely to provide an understanding of performance, vigilance, and susceptibility to disease.

Classical gene therapy approaches allow new genes or functions to be introduced into cells or individuals to correct genetic disorders. While optimal vehicles for gene therapy still need to be developed, this limitation is likely to be solved in the foreseeable future, opening the way to many applications. In addition to correcting genetic defects, the same technology could be used to enhance performance or allow new capabilities. An extension of this notion is the use of similar technological approaches to control specific genes—turning them on or off at particular times. As one example, many of the genes responsible for programmed cell death (apoptosis) have been identified. Selectively altering their expression could greatly extend the life of desired cells or help to prevent their premature death, and, in combination with use or expression of growth factors, make possible selective growth control and controlled tissue regeneration.

Combinatorial genomics has also recently been developed. Similar in concept to combinatorial chemistry, this technology shuffles portions of genes to give a vast number of new combinations, which are then screened for a desired function. Enhanced function, or even new functions, can be generated by repeating the cycle many times to "evolve" optimized recombinants in vitro.

Research characterizing the expression of specific genes has determined that several genes, most notably the cyclic AMP-response element binding protein (CREB) gene, have products that enhance memory function in humans. The binding protein may be regulated by phosphorylation. As the human genome is further characterized, it is likely that other gene products will be associated with memory. Knowledge of the gene sequence can facilitate the development of drugs that interact with critical regions of the gene to increase production of proteins that enhance memory. Conversely, anti-sense molecules that effectively block proteins that diminish memory may be developed. The analyses of the genome of people with senile dementia or Huntington disease may reveal the existence of gene products that compromise memory.

As the number of military members shrinks but the number of deployments increase, the force that can be maintained will depend to a great extent on how well individuals can be protected from the effects of sustained operations—most notably fatigue, sleep loss, and circadian desynchronosis. For example, replacement or additional personnel may not be available. Consequently, sleep will be sacrificed to extend duty times. When sleep time is available, it will often be fragmentary, and people will be expected to perform their missions immediately upon awakening. The detrimental impacts on mission performance are obvious. For example, 15 percent of mishaps in naval combat aircraft and 13 percent of the associated deaths can be attributed to sleep deprivation. fatigue, and/or circadian disruption. In a study of friendly fire casualties, most of the causal errors could be attributed to the negative consequences of sleep deprivation on operator performance.

Increasing knowledge about the hormonal and neurochemical basis of circadian rhythms and their role in memory, vigilance, decision-making, and other cognitive functions will lead to meaningful drug interventions. Under normal conditions, circadian rhythms are entrained to approximately 24 hours by the presence of a number of Zeitgebers (i.e., environmental events that provide the stimulus setting the biological clock), such as the occurrence of light and dark, that inhibit or permit the release of melatonin, a hormone produced by the pineal gland. Strategies that use melatonin to induce sleep or shift circadian rhythms to negate jet lag or block melatonin (e.g., bright light) or to sustain arousal will enhance performance during time zone transitions. Improvements (up to threefold increases) in the duration of effective duty times (up to 72 hours) and recovery from sustained operations will be made possible by fatigue management through pharmacological interventions (e.g., ergogenic compounds) that either induce sleep and circadian synchronization or produce controlled arousal through targeting neural center stimulation and inhibition (e.g., modafanil) without side effects. Administration of these pharmacological controls will be individually titrated to the needs of the warfighter through next-generation drug delivery systems.

TIMELINE FOR TECHNOLOGY AVAILABILITY

Predicting when specific human performance enhancement technologies will be routinely available to the warfighter is difficult because gradations of capability will be introduced progressively. This notwithstanding, projections for the near term, midterm, and far term can be made with some confidence. Near term implies 3 to 5 years, midterm implies greater than 5 years but less than 10 years, and far term implies greater than 10 years. The application of perception and cognition research to the design of intuitive interfaces, automation, and decision-making will begin in the near term; some real-time, rudimentary physiological monitoring that drives meaningful changes in task allocation will be available in the midterm; and true system fusion and, later, control of systems by thought will be available in the far term. Revolutionary changes in the human systems interface will appear at the end of the near term or in the midterm. Significant pharmacological interventions to manage fatigue will begin in the next few years; however drugs or gene products that will change memory, for example, will appear in the midterm.

ADDITIONAL DATA

Tables 3.1-1A and 3.1-1B present additional data on this developing critical technology.

WORLDWIDE TECHNOLOGY ASSESSMENT (See Figure 3.1-1)

Those nations that have advanced human genomics and strong human behavior research efforts are leaders in this area. The Human Genome Project is a multinational effort, with leading contributions from France, Japan, the United Kingdom, and United States. Since the genomic information is the result of a multi-national effort and the data are placed on the world wide web, most nations have ready access to the results. The methods used to transition these data to useful information resides in the laboratories in which the data are acquired. The human behavior research programs of Germany, the Netherlands, and the United States have been at the leading edge. Much of the equipment used to collect electrophysiological data related to human perception and response is used in the medical community. The information obtained from these studies has applicability for improvement of performance in the transportation, medical, manufacturing sectors and in the military. Germany, Japan, the Netherlands, and the United States have strong programs in information display and control. Pharmacological intervention expertise is reflected in development and production facilities in France, Switzerland, the United Kingdom, and the United States.

Table 3.1-1A. Human Performance Enhancement Militarily Critical Technologies

Technology	Developing Critical Technology Parameter	Rationale	Critical Materials	Technical Issues	Joint Vision 2010 DoD S&T Plan	Unique Test, Production, and Inspection Equipment
Miniaturized, non-obtrusive measures of eye movements; auditory- and visual-evoked potentials; EEG; cardiac interbeat interval; oxygen saturation	Drowsiness indicator; vigilance and fitness monitor; physiological indicators that predict decreased combat performance. Miniaturized auditory evolved potential in the 300-ms wave and visual in the 300- to 600-ms wave. Application ready in near term to midterm for evoked potential and eye movement.	Real-time assessment of operator workload, vigilance, and fatigue to drive adaptive automation and information formatting will maximize human system interface and safety. Provides human and automated intervention capability to assess and support vigilance and operational readiness.	Miniaturized sensors embedded in body cover or helmet; lightweight data fusion tools and energy-efficient signal transmissions.	Determination of reliable, useful correlations between physiological measures and measures of cognitive performance and combat readiness.	Information and systems integration; dominant battlespace awareness; dominant maneuver; agile organization training. See Table A-7.	None identified.
Functional MRI	Simulator training with fMRI correlates brain functional response to training to job criteria. Magnet level of 1.8 Tesla or higher.	fMRI enables evaluation of training by examining anticipated movement (intent before action).	Non-magnetic simulator system required to use MRI.	Long-term fMRI (time longitudinal map) will resolve baseline variations.	See Table A-7.	None identified.
Information display and control; robust voice control of military systems; 3-D audio/visual displays; brain activated control of machine interfaces; haptic interfaces; HMDs	Interface-evoked potentials with human-machine interfaces and virtual reality projectors. Response times in 10 to 100 ms.	"Hands-free" task execution and localization cues from multiple sensory modalities decrease reaction times and enhance control by operator.	Integrated audio/visual/haptic information presentation.	Development of multi-modal sensory presentation for all people. Development of interface to allow neural control of flight systems. Application ready in near term.	Information integration; dominant maneuver; agile organization training; focused logistics; precision engagement. See Table A-7.	None identified.

Table 3.1-1A. Human Performance Enhancement Militarily Critical Technologies (Continued)

Technology	Potential Developing Critical Technology Parameter	Rationale	Critical Materials	Technical Issues	Joint Vision 2010 DoD S&T Plan	Unique Test, Production, and Inspection Equipment
Pharmacological interventions, including drugs that affect object recognition and improve memory functions and drugs that control arousal or sleep and produce circadian synchronization	Permits selective intervention in time-limited and reversible manner. Intervention is effective within minutes, and duration is hours. A sustained high performance level, including sleep/work cycles synchronized to work demands.	Enhancement or degradation of memory will affect human performance of skilled tasks for C3 personnel, aircraft, or other complex systems.	Drugs that bind or otherwise affect CREB or offer AMPAkines functions.	Development of drugs that bind selectively to memory-related genes or to receptors that are involved in sleep control. Understanding the process underlying memory and maintenance and the differentiation of memory.	Dominant maneuver; precision engagement; full dimensional protection; full spectrum dominance. See Table A-7.	None identified.
Biomarkers for early detection of genetic susceptibility to toxicants and stress	Rapid identification (in hours) of human genes affecting susceptibility to toxicants, stress, and disease. Includes genes for cytochrome P450, GST, acute phase proteins. Application ready near term to midterm.	The human genome will reveal factors of susceptibility to toxicants and stress. Mutations affecting activity of toxicant metabolizing enzymes may result in increased susceptibility.	None identified.	Creation of a database relating human genome to toxicant and stress susceptibility.	Full dimensional protection. See Table A-7.	None identified.
Biomarkers for detection of environmental agents or conditions that adversely affect human performance	Detects body response to and insulate from toxicants that degrade human performance in either or both near term (minutes to hours) and long term (years). Glutathione transferases and cytochrome P450 are detector biomolecules. Application ready near term to midterm.	Rapid increases in the levels of these biomarkers may provide early warning of performance degradation.	None identified.	Determination of specific environmental factors that adversely affect human performance. Determination of specific physiological indicators that predict decreased human performance.	Full dimensional protection. See Table A-7.	None identified.

Table 3.1-1B. Human Performance Enhancement Militarily Critical Technologies

Technology	Military Applications	Unique Software	Center of Technology Development: Military or Commercial	Commercial Applications	Commercial Technology Requires Development for Military Use	Access To Technology	Other Important Data
Miniaturized, non-obtrusive measures of eye movements; auditory- and visual-evoked potentials; EEG; cardiac interbeat interval; oxygen saturation	Provides indicators of loss of vigilance and fitness and predicts decreased combat performance and readiness.	Signal-averaging miniaturization and data-sorting algorithms and programs.	Multiple commercial development centers include assessing production line assessments in factories; determining effectiveness of the medical care delivery (e.g., surgery); assessing human monitor of chemical process production (petroleum industry); measuring fitness of pilots and railway engineers; monitoring patients.	Provides equivalent indicators of loss of vigilance and fitness and predicts decreased performance and readiness in industry and commerce.	Commercial applications drive technology in biomedical arena. Leading-edge R&D is commercial.	Ready access for the military. Some proprietary issues apply.	None identified.
Functional MRI	Ensures performance in stress conditions, with correlation of brain and behavior functions.	None identified.	Training centers for airline pilots and railway engineers; medical monitors for patient status.	Diagnosis of disease in patients. Training for high-skilled jobs.	None identified. Still in development.	Ready access.	None identified.
Information display and control; robust voice control of military systems; 3-D audio/visual displays; brain activated control of machine interfaces; haptic interfaces; HMDs	Decreases response time under stress of battle; SUSOPS; informs operator of direction of target or incoming missile by multiple modalities; optimizes man-in-the-loop cognitive and motor workload and tasking.	Software for interfaces.	Biomedical applications in operating rooms for anesthesiology; precision machining of production processes involving toxic/radioactive materials.	Decrease response time under stress; motor workload and tasking; enhance performance in variable shift work.	Commercial applications drive the technology.	Ready access.	None identified.

Table 3.1-1B. Human Performance Enhancement Militarily Critical Technologies (Continued)

Technology	Military Applications	Unique Software	Center of Technology Development: Military or Commercial	Commercial Applications	Commercial Technology Requires Development for Military Use	Access To Technology	Other Important Data
Pharmacological interventions, including drugs that affect object recognition and improve memory functions and drugs that control arousal or sleep and produce circadian synchronization	Reduces training time; enhances situation awareness; improves reaction time; improves decision-making under stress; enhances fatigue management; SUSOPS.	None identified.	Military may drive selective use of drugs that enhance vigilance for 72 hours. Military will want to develop counter-measures for drugs that disrupt memory processes.	Training people for complex tasks; treating learning disabilities (e.g., ADD).	The commercial and military world have complementary needs.	Ready access to drugs for learning enhancement.	None identified.
Biomarkers for early detection of genetic susceptibility to toxicants and stress	Selection of soldiers for DECON and for TADMUS activities.	Database of human genome.	The medical and pharmaceutical industry and NIH fund 90 percent of this activity.	Assisting with liability assessment; enhancing biomedical diagnostics.	Commercial sector drives technology.	Ready access to human genome.	See Biomedical Section, Part III of MCTL. Ethic and privacy cultural issues involved.
Biomarkers for detection of environmental agents or conditions that adversely affect human performance	Enables proactive avoidance of hazards; creates opportunity to detoxify. Increases avoidance of threat situations; assists in implementation of strategies to improve mission outcome.	Database of correlative information.	Industrial hygiene; OSHA and industrial liability concerns have interest.	Improving safety of workplace and reduce liability.	Private sector and NIH are major supporters of technology development.	Ready access.	None identified.

Country	Physiological Monitoring	Information Display and Control	Pharmacological Intervention
Australia	●●●	●●	●●●
Canada	●●●	●●●	●●●
China	●	●	●●
Cuba	●	●	●●
Czech Republic	●●	●●●	●●
Egypt	●●	●●	●●
France	●●●	●●●	●●●●
Germany	●●●	●●●	●●●
Hungary	●●	●●●	●●
India	●●●	●●●	●●
Iran	●	●	●
Iraq	●	●	●
Israel	●●●	●●●	●●●●
Italy	●●●	●●	●●●
Japan	●●●	●●●●	●●●●
Malaysia	●●	●●	●●
Netherlands	●●●●	●●●●	●●●
North Korea	●	●	●
Norway	●●	●●	●●●
Poland	●●	●●	●●
Russia	●●●	●●●	●●●
Singapore	●●	●●●	●●
South Korea	●●●	●●●	●●
Sweden	●●	●●●	●●●●
Switzerland	●●●●	●●●●	●●●●
Syria	●	●	●●
Taiwan	●●●	●●●	●●●
Ukraine	●●	●	●
United Kingdom	●●●●	●●●●	●●●●
United States	●●●●	●●●●	●●●●

Legend: Capability in Technology Elements: ●●●● Most ●●● Many ●● Some ● At Least One

Figure 3.1-1. Human Performance Enhancement Technologies Area Worldwide Capability Assessment

SECTION 3.2 BIOLOGICAL SENSORS

Highlights

- Biological multi-array sensors will permit rapid and accurate acquisition, storage, and assessment of data for biological, chemical, and mechanical signals.
- Biological sensors are anticipated to be more compact, energy efficient, and sensitive than non-biological units.

OVERVIEW

A sensor provides the interface through which a system can detect the state of a changing environment in real time. It can also provide information about the interaction between the environment and individuals. Appropriate sensors and data fusion systems serve to integrate operational systems in much the same way as the sensory nervous system in the body. This analogy has given impetus to the construction of multi-spectral [shark-electromagnetic detection; pigeons and ultraviolet (UV) detection; swine and canine odor detection] and non-traditional (abiotic "sensored" insects) sensors for scanning the battlefield and for force deployment. The integration of advances in optics, electronics, microfabrication technology, and molecular biochemistry has made biosensor technology an area of rapid technological progress. Concepts that will result in improved field sensors by the year 2010 include gene probes, monoclonal and genetically engineered antibodies and other receptors, high-precision polymer molding, polymer liquid crystals, chemiluminescence, neuronal and protein DNA patterning, combinatorial chemistry, monolithic UV/visible/infrared (IR) laser light-emitting and light-detecting surfaces, Microelectromechanical Systems (MEMS), charge-coupled devices (CCDs), and neural networks.

At present and, in fact, in our entire history of chemical syntheses, it is estimated that less than 1 percent of all possible organic compounds have been synthesized. Combinatorial chemistry uses automated production and screening of thousands of compounds, typically for drug screening or for materials with novel electroconductive properties. The number of compounds that can be synthesized per day is now in the thousands, and the production size is growing. One method produces as many as 25,000 different substances on a 3-inch diameter substrate. The challenge of this technology is to develop methods that rapidly screen the newly synthesized materials for properties of interest. Technologies for large-scale screening remain to be developed.

Advances in synthesis and screening will lead to very small, lightweight, energy-efficient biosensors capable of simultaneous analysis for multiple analytes. All sensors would be reduced to a card format and integrated into a single case. Individual sensors can be formatted with both the biodetection components and the optoelectronic sensing elements in micron-scale arrays. Microfluidics, data processing, data transmission/display elements, and electropower sources will be incorporated with the optical or electrochemical detector and the molecular or cellular biodetection components in a fully automated, hand-held unit. For automated environmental monitoring, these units will be adaptable to a variety of platforms. Materials that have applications in display technology can be rapidly screened. Corporations in the United States with technology in this area include Affymetrix and Symyx.

In addition, biosensors will be incorporated into suites of sensors. An example of such a suite would be an environmental monitor including sensors that measure meteorological conditions, Global Positioning Systems (GPSs), determinants for viability, UV-laser detectors for organic particles, generic biosensors for detecting any toxin or pathogen, and biosensors for biological agent identification. Data fusion techniques could assess the threat and deliver an assessment based on multiple factors. Similarly, a sensor suite for medical diagnostics could integrate sensors for emperature, blood pressure, pH, CO_2, infectious agents, toxins, key enzymes, and markers of septic shock. Information acquired through the use of these sensors will markedly increase command and control (C2) awareness of the battlefield and other military environments.

In 10 to 30 years, these sensors and sensor suites will be "smart devices" that perform appropriate analyses, data fusion, and evaluation and then trigger an appropriate response, such as input to command. For instance, a sensor incorporated into a smart chemical and biological warfare (CBW) suit will close the suit's pores in response to a chemical agent, activate the mask only in response to a biological agent, regulate internal temperature, and trigger the injection of antidotes after agent identification. For medical applications, the on-line monitoring of blood chemistries, endotoxins, and drug levels might automatically trigger the introduction of appropriate levels of therapeutic agents.

At the close of the 20th century, the U.S. defense community is experiencing a reduction in people and is conscious of the cost of acquisitions. The following subareas are critical to increasing military superiority:

- Molecular recognition components
- Biological and biomimetic photon and electron transfer materials
- Self-assembly systems, data fusion arrays, and characterization of the sensor surface.

Array sensors for enzymes, analytes, antigens, antibodies, receptors, and nucleic acid targets are available on chips that contain the biocomponents and the integrated circuitry to record the results. These arrays are supplied in the form of small disposable cassettes into which a sample of whole blood or any other biological fluid can be added. The "chip" will process the sample as required and move the processed material—through either capillary action or a small micropump—to the sites of analysis. The results could be read by the corpsman or telemetered to an off-site location for analysis and decisions regarding treatment and combat response.

Detection and measurement of infectious and toxic threats and of hazard substances can be done before and during deployment to any global location. Application of this sensor information is profound for the protection and functional well-being and mission accomplishment of deploying forces. See the Biomedical section in Part III of the MCTL.

RATIONALE

Sensors provide data regarding materials and conditions in the surrounding environment. Converting these data to information can enable the performance of many functions, including threat awareness, functional awareness of individual groups, awareness of selected field conditions, and readiness of equipment. Military implications include well-being of the soldier if taken ill or wounded and determination of CBW attack. A soldier may even be able to carry one of these devices as a dog tag that would notify a command center or central processor that this soldier has been exposed to an agent or other toxicant.

The most recent approaches have focused on exploiting the high affinity binding of naturally occurring molecules to target materials of interest. The molecules include polynucleotides that recognize pathogenic genome sequences of biological agents or human genomic material of interest; monoclonal and polyclonal antibodies that recognize surface molecules of biological or chemical agents; and receptors that change conformation following exposure to biological or chemical agents and generate an electronic or photonic signal.

The major element in the U.S. biological defense program is early detection, identification, and warning to provide situational awareness and allow military forces to avoid or manage the threat. Detection and identification of biological agents and the prediction of future threat situations provide useful information to military commanders and individuals. Such information permits the pre-treatment of people before deployment and the management of medical care following exposure to infectious or toxic agents. The detection and identification of organisms in a deployed zone allow commanders to take steps to avoid contamination, determine the appropriate protection for continued operations, and initiate proper prophylaxis and therapy to minimize casualties and performance degradation. The appropriate care of the exposed people will increase the rate of their reinsertion into units and minimize long-term disability resulting from exposure to the agents. Biosensors based on genomics can rapidly distinguish pathogenic organisms and pathogen-producing organisms from non-pathogens.

TECHNOLOGY DESCRIPTIONS

Molecular Recognition Components (Technology Items I,L,N in Table 3.0-2)

Arrays based on antibodies, nucleic acid sequences, or receptors are in production at the close of the 20th century. Genomic sequencing of all known biological threat agents and infectious disease organisms is in progress. Some regions of the bacterial genome appear to be common among organisms pathogenic to humans (e.g., pathogenicity islands). Such regions could be used in detecting pathogens on nucleic acid arrays or other type of DNA/RNA detection systems. In addition, using anti-sense gene technology, stopping the pathogenic activity of an organism by tying up this region of the pathogens genome may be possible.

Nucleic acid oligomer arrays for the detection of specific gene sequences have been produced on patterned surfaces by Affymetrix, several other companies, and the Naval Research Laboratory (NRL). Similar arrays, for the sequencing of genes at very rapid rates to determine "viral load" and mutations, are being generated. Antibody and antigen arrays for detecting and quantifying biological agent antigens and antibodies that will permit the assay of several analytes simultaneously are in development. Sensors employing these components will allow the detection, characterization, and measurement of the concentration of biological agent and will predict the dispersal efficiency of the agent. Remote or standoff devices can be configured using this new technology.

This technology addresses real-world problems. Anthrax and other biological agents are normally present in particular environments. These agents vary in natural abundance as a function of geographical location, land use, season, and meteorological conditions. After a database has been established for natural abundance of each biological agent in a given setting, the multi-array sensor will inform the user when a threshold that requires action has been reached. This type of approach will be inexpensive, fast, and portable.

The application of receptors in analyses, as opposed to antibodies, is of great interest. Receptor- and antibody-based sensors are complementary and, in certain situations, may be useful together on an array. Antibodies provide information on mass but not on biological activity. Ligand binding to receptors provides information on biological activity but not on mass. Military interests include an array that detects bacteria and bacterial toxins. Since putting genes for toxins in common organisms is possible, "hiding" the presence of a dangerous toxin from detection is possible. An array containing antibodies to the organisms, receptors to the active toxin, and gene probes for the toxins may recognize threat conditions earlier than any of the single-component sensors. Using MEMS technology, placing these arrays on the soldier and sharing the information through telemetry will be possible.

A Human Vital Systems Card is an example of an integrated biosensor and data fusion component. The devices use different technologies and perform different functions by measurement of parameters, such as pO_2, pCO_2, pH, glucose, creatinine, blood urea nitrogen, blood electrolytes, and other components that would indicate the state of an individual. These devices can use MEMS and/or nanofabrication technologies and be small enough to be used either by being implanted under the skin or attached to the skin surface. Advances in non-invasive technologies will be required for this type of sensor.

Biological and Biomimetic Photon and Electron Transfer Materials (Technology Items K and Q in Table 3.0-2)

Bacterial Rhodopsin is a transmembrane protein of approximately 27 kilodaltons and is light-sensitive. When exposed to yellow light, it pumps a proton from the interior of the cell-membrane to the exterior through a photocycle. During this photocycle, it passes through several intermediate states with unique characteristics and life times. This protein could be faster than a Josephson Junction for switching and could operate at room temperature. It has potential for holography, information storage and retrieval at the molecular level, and molecular computing. An artificial retina using a light-sensitive material, such as bacterial Rhodopsin or one of the porphyrin containing dyes on a chip, has been produced in prototype form. Such a device could be implanted in the eye and electrically connected to the optical nerve. Although the signals sent to the brain will not be the same as those in normal eyes, the brain can learn the meaning of the signals and, in effect, see. Using this material as an artificial eye for both human and robot application will be possible.

Phthalocyanine is a biomimetic molecule that has properties similar in some respects to bacterial Rhodopsin. It is a photosensitive porphyrin-like dye. When placed in a lipid membrane together with other compounds, it shows photo effects. As with other photosensitive materials that are reversible, it could be useful for molecular electronic applications.

Self-Assembly Systems, Data Fusion Arrays, and Characterization of the Sensor Surface (Technology Items H,O,R in Table 3.0-2)

Most recent approaches have focused on exploiting the interaction of binding molecules with the surface of the sensor to construct sensors and provide quality control in the production of sensors. Techniques with sensitivities that can be applied to this problem include, in part, Atomic Force Microscopy (AFM), optical tweezers, membrane micropipette manipulations, and manipulation of functionalized magnetic beads by magnetic force fields. All these approaches have evolved toward methods used in analyzing molecular interactions that are most suitable for interrogating the interactions of a single molecule. AFM has been used to measure individual molecular interactions by separately modifying the surfaces of the AFM probe and the target surface with complementary binding agents. Atomic force microscopes and Near Field Scanning Optical Microscopy (NSOM) will make it possible to provide quality control for microscale and nanoscale fabrication. As an example, avidin passively adsorbed to the surface of the AFM probe was used to scan the surface that was derivatized with biotin. The rapid association of avidin with biotin was observed as the two surfaces came into closer contact and eventually touched. On retracting the AFM probe from the biotin-coated surface, the force fluctuations between the avidin and the biotin could be measured. The AFM system allows structural defects in thin films to be recognized at Angstrom-level resolution.

The speed at which the AFM probe pulls away from the substrate significantly affects the force that is needed to rupture the receptor-ligand complex. Direct measurements of the forces between complementary strands of DNA were measured by AFM, where sample DNA was bound to the AFM probe and substrate surfaces by complementary capture of oligonucleotides. For nucleic acid detection of DNA or RNA sequences in samples where these analytes are in the range of 100 to 1,000,000 copies, some form of amplification is required.

ADDITIONAL DATA

Tables 3.2-1A and 3.2-1B present additional data on this developing critical technology.

*WORLDWIDE TECHNOLOGY ASSESSMENT **(See Figure 3.2-1)***

Several nations, including Canada, France, Germany, Japan, the Netherlands, Sweden, and Switzerland, the United Kingdom, and the United States, have advanced capabilities in sensor technology. Russia and Israel are also advanced in these technologies. Much of foreign investment in biosensors is in the pharmaceutical and commercial areas. Most of the current, field-hardened detection, warning, and identification systems are for chemical agents. Several multi-array sensor systems are in development and should be available for use in mobile systems in the 5-year period. These multi-array systems will be able to detect and identify biological agents based on genomics, immunogenics, combinatorial chemistry, and proteomics. The leading edge technologies are present in Germany, France, Israel, Japan, the United Kingdom, and the United States.

Table 3.2-1A. Biological Sensors Militarily Critical Technologies

Technology	Potential Developing Critical Technology Parameter	Rationale	Critical Materials	Technical Issues	Joint Vision 2010 DoD S&T Plan	Unique Test, Production, and Inspection Equipment
Multi-array sensors for detecting biological agents (bacterial, viral, fungal), genome sequences, or antigenic epitopes (see Biomedical Technologies, Part III MCTL)	The gene sequence codes of PAIs; determine epitopes specific for AG BW agents.	Genomic elements of biological agents determine pathogenicity; epitope can identify biological agents; multi-array sensors allow rapid unambiguous detection and identification of biological agents.	Complimentary DNA sequence or epitope/ Gke surface.	Placement of complimentary gene sequences or antibodies on micron-level pmod elements in addressable manner.	Full dimensional protection. See Table A-7.	None identified.
Biosensors for odors/light/sound/pressure	Chemical to electron/ phonon transduction; pressure to electron/ phonon transduction. Incorporate sensor in thin film; MEMS. Design sensor surface to accommodate attachment of molecular recognition elements (10^4 per pixel) in a functionally active state.	Specific chemical or sound patterns that provide unambiguous identification of target (midterm).	Host transducing materials.	Production of sensor surface containing biomolecules in an active form; data fusion correlating known patterns with real-time threat conditions.	Full dimensional protection; precision engagement; dominant maneuver. See Table A-7.	None identified.

Table 3.2-1B. Biological Sensors Militarily Critical Technologies

Technology	Military Applications	Unique Software	Center of Technology Development: Military or Commercial	Commercial Applications	Commercial Technology Requires Development for Military Use	Access To Technology	Other Important Data
Multi-array sensors for detecting biological agents (bacterial, viral, fungal), genome sequences, or antigenic epitopes (see Biomedical Technologies, Part III MCTL)	Multiple applications, including protecting soldiers against biological agent and disease in deployed areas.	Database of PAIs and cell surface markers.	Medical/ pharmaceutical food industries lead.	Clinical detection of infectious agents; food processing.	Biomedical and food industry lead military applications.	Ready access.	None identified.
Biosensors for odors/light/sound/pressure	Identifies movement of personnel, armor, and target location; stores high-density information.	Algorithm of chemical light/ sound or pattern unique to targets.	Food industry; chemical processing; military.	Food spoilage; toxicant release.	Military and food industries can have similar impact on technology.	Ready access.	None identified.

SECTION 3.3 BIOMATERIALS AND NANOFABRICATION

Highlights

- Miniaturization at the nanometer-scale level will facilitate high-density information storage, retrieval, and processing.
- Biomaterial-based circuit and switching devices will be available on the nanometer scale to enable rapid and accurate responses to changing requirements for applications of military force.
- Biopolymer-based artificial retinas have the potential to advance night vision, missile guidance, and reconnaissance systems.

OVERVIEW

The rapid progress in biochemistry and molecular biology has provided industry with many new classes of materials. These include structural materials (silks and bioceramics, such as chitins); electron/photon conductive polymers, such as cytochrome and polyporphyrins; and ion gating molecules, including bioreceptors (adhesives from barnacles that can function on wet surfaces); and biocompatible lubricants. The military uses of these materials include miniaturization of electron circuits; lightweight, energy-efficient piezoelectric sensors for detection of delamination; exoskeletal supports for personnel; and camouflage/stealth. Because many biomaterials can self-assemble, these materials have been used to pattern sensor surfaces having thicknesses in the nanometer range. The biomaterials are produced in living organisms. The cost of production is modest, requiring only fermentation-like facilities. Because the living cells replicate, small starter cultures can be used to produce large numbers of organisms. The living organisms require aqueous systems for growth and, therefore, are environmentally advantageous. Organic solvent requirements are minimized, thereby reducing remediation, treatment, and pollution costs. The technology of genetic engineering permits proteins synthesized in one order or species of animal to be synthesized in other organisms. By this method, proteins normally made only in mammals, including humans, can be produced on a large scale in bacteria, yeast, or plants. One form of such engineering involves the use of cassette mutation, most studied in yeast. Cassette mutation allows entire genes of interest to be inserted into a microorganism for large-scale manufacture of the gene product.

The following subareas are critical to increased military superiority and advanced economic competitiveness:

- **Biomaterials.** Production of high-tensile-strength, low-weight materials.
- **Nanofabrication.** Fabrication of miniaturized electronic and photonic switches, circuits, and molecular motors

RATIONALE

Biopolymer-based systems have specificity and selectivity with regard to the chemical processes they facilitate. A major advantage of biosystems is their self-assembly properties. Such self-assembly occurs at the nanometer scale. The materials provide opportunities in the manufacture of artificial retinas, microsensor chips, and electron/phonon switching devices. These technologies provide large numbers of chips and electron/phonon switching devices. These technologies provide large numbers of lightweight sensors that operate in ambient temperatures. Military applications include protective clothing, stealth coatings, missile guidance systems, and underwater acoustic sensors. Production and maintenance quality control at the molecular level have profound implications for reliable operation of military systems in high-stress environments.

TECHNOLOGY DESCRIPTIONS

Biomaterials (Technology Items K and Q in Table 3.0-2)

Within a decade, scalable, artificial imagers with optical response functions that more closely resemble human retinal response will be developed. These artificial imagers will be scalable to large sizes, will be low cost, and will have on-chip image processing functions. The materials include amorphous Silicon (a-Si) thin-film transistor (TFT) imaging arrays in conjunction with photosensitive layers comprised of materials such as bacteriorhodopsin (bR). Bacteriorhodopsin has the unique attribute that the generated photo response of a bR film yields the integral characteristic of the incident optical image. This means that the bR layer generates a photovoltage whose magnitude is proportional to the time rate of change of the incident optical signal. Thus, an imaging system with bR as the photosensitive layer performs optical processing functions that lead to systems that respond as edge or motion detectors. Combined with more conventional imager response functions, these bR-based imagers can yield a better picture of remote battlefield conditions. In addition. fabricating imagers with pixels that use the orientation properties of bR films is possible. By fabricating individual pixels with regions that exhibit opposite orientation, generating optical response functions that closely mimic the response of retinal ganglion cells is possible. Such cells have particular sensitivity to moving or stationary target tracking. These are required elements of smart targeting weapons. These imagers could replace traditional sensor elements for strike, night vision, or reconnaissance missions.

Lightweight, chemical-resistant, shrapnel-resistant, self-cooling materials are under development, using biological systems as models. The ceramic coatings of insects and shellfish are metal-protein-carbohydrate complexes. Spider silk, plant tassle silk, and silkworm silk all have high strength-to-mass ratios. These materials may be readily modified because proteins and carbohydrates contain many functional groups for grafting other materials. This ease of grafting has advantages in attachment of sensors to the clothing fabric.

Signature reduction of a soldier may be achieved using a uniform made from a polymer material with side chains of a chiral material that reduces his signature to radar. A material that will reduce the infrared (IR) signature of a soldier at night is also very likely within the next 5 to 10 years—if it is not already available.

Nanofabrication (Technology Items H,O,R in Table 3.0-2)

Biomaterials, such as proteins, lipids, and nucleic acids, can self-assemble. Self-assembly is the formation of organized, patterned structures without external direction. The fidelity of the self-assembly is extraordinarily high. These self-assembled components can be complex and perform unique functions. An example of self-assembly is the formation of pores by the addition of hemolysin from Staph. aureus to a solution containing lipid membranes. Without any external manipulation, the proteins will join together and form a pore in the membrane. Other examples include the appropriate folding of proteins as they are synthesized from messenger RNA and the assembly of a biomolecular motor that consists entirely of proteins (dynein and kinesin) combined in such a fashion that the product is an engine. Many polymers can self-assemble into membrane-like structures. Examples are the formation of monolayer of alkylthiols on gold surfaces, silanes on glass and metal oxides, and organic molecules into polymers on conducting surfaces by electropolymerization.

Within the next 5 to 15 years, using self-assembling techniques to produce nano-size gears, motors, and other mechanical and electronic components in the nanometer size range will be possible. Such devices will find application in monitoring human health, behavior, and the environment. They may even be capable of assembling within the body by simple addition in "monomeric form." These devices will perform functions presently requiring complex, expensive, and proportionally large integrated circuitry. They can also perform tasks not related to humans.

Using Microelectromechanical Systems (MEMS) technology and possibly self-assembly and nanofabrication, it will be possible to produce small devices capable of switching. These devices may be used to change wavelength in optical systems, turn electronic devices or circuits on and off, or monitor specific phenomenon in military systems, the environment, or the human body. The ability to change wavelength can lead to low-weight iconographic presentation formats on helmet mounted displays (HMDs).

Atomic Force Microscopy (AFM) and Near Field Scanning Optical Microscopy (NSOM) use a technique that involves maintaining close contact between a cantilevered probe and a surface by maintaining either constant potential, resistance or other measurable quality. The device produces a "picture" of surfaces at the molecular level. This picture can be the physical surface, the electrical potential, the resistance, the magnetic differences, or several other qualities that this surface may possess. In all these cases, the device produces a diagram of that surface or that surface property at the molecular level.

In the next decade, the devices will be available in a hardened and portable form. With portable devices, examining surfaces in our external environment for changes, contaminants, or other characteristics that may have commercial or military application will be possible. One possible example of a commercial application would be the production of silicon or diamond chips for integrated circuits. Using atomic force microscopes and NSOM, the quality control of microscale and nanoscale fabrication can be realized. Such a device on a production line would be useful in looking for defects if it could be designed to scan surfaces at a relatively rapid rate.

ADDITIONAL DATA

Tables 3.3-1A and 3.3-1B present additional data on this developing critical technology.

WORLDWIDE TECHNOLOGY ASSESSMENT (See Figure 3.3-1)

Technologically advanced nations have recognized the civilian and the military opportunities provided in biomaterials. France, Germany, Japan, the Netherlands, Sweden, Switzerland, the United Kingdom, and the United States are overall world leaders in this area. Other nations, including Israel, also conduct leading-edge research and development (R&D). Russia had a leading edge in the development of artificial retinas, but this advantage has diminished in recent years. In Germany, Japan, and the United States, the chemical industry has been a dominant driver of advances in biomaterials and bioelectronics.

Table 3.3-1A. Biomaterials and Nanofabrication Militarily Critical Technologies

Technology	Potential Developing Critical Technology Parameter	Rationale	Critical Materials	Technical Issues	Joint Vision 2010 DoD S&T Plan	Unique Test, Production, and Inspection Equipment
Smart biomaterials	Optimizes match or contrast color of cloth to environment (e.g., electroactive polymers, bioceramics, HMD); indicates condition of structure by changing color.	Biopolymers can change wavelength absorbed and emitted or change function as piezoelectrically active materials.	None identified.	Develop biopolymers that have tunable absorbing/emitting properties in the IR, visible and UV range (800- to 260-nm wave-length); embed piezoelectrically active biopolymers in matrix while retaining structural integrity (midterm).	Dominant maneuver; full dimensional protection; full spectral dominance. See Table A-7.	None identified.
Self assembly biomaterials	Produces nanometer-scale materials (midterm).	Biopolymers self assemble and can form contractile, vectorially oriented filaments, which allow for production of nanometer-scale motors, transducers, and guidance systems.	Contractile proteins that self assemble.	Insertion of contractile proteins into stable lipid-fluid monotypes with vector movement control.	Dominant maneuver; full dimensional protection; full spectral dominance. See Table A-7.	None identified.

Table 3.3-1B. Biomaterials and Nanofabrication Militarily Critical Technologies

Technology	Military Applications	Unique Software	Center of Technology Development: Military or Commercial	Commercial Applications	Commercial Technology Requires Development for Military Use	Access To Technology	Other Important Data
Smart biomaterials	Uses in camouflage; determines structural integrity of aircraft.	Algorithm for matching tuned materials to environment.	Primarily military.	Structural integrity of materials that become deformed.	None identified.	Limited.	None identified.
Self assembly biomaterials	Micromotors for guidance systems.	None identified.	Medicine; manufacturing.	Nanometer-scale motors; microfluidics.	None identified.	Ready access.	None identified.

Country	Biomaterials	Nanofabrication
Australia	●●	●●
Canada	●●●	●●●
China	●●	●●
Cuba	●●	●
Czech Republic	●●	●●
Egypt	●	●
France	●●●●	●●●●
Germany	●●●●	●●●●
Hungary	●●	●
India	●●	●●
Iran	●	●
Iraq	●	●
Israel	●●●	●●●
Italy	●●●	●●
Japan	●●●●	●●●●
Malaysia	●●	●
Netherlands	●●●●	●●●●
North Korea	●	●
Norway	●●	●●
Poland	●●	●●
Russia	●●●	●●●
Singapore	●●	●●
South Korea	●●	●●
Sweden	●●●●	●●●●
Switzerland	●●●●	●●●●
Syria	●	●
Taiwan	●●●	●●
Ukraine	●	●
United Kingdom	●●●●	●●●●
United States	●●●●	●●●●

Legend: Capability in Technology Elements: ●●●● Most ●●● Many ●● Some ● At Least One

Figure 3.3-1. Biomaterials and Nanofabrication Technologies Area Worldwide Capability Assessment

SECTION 3.4 INDIVIDUAL AND GROUP PROTECTION

Highlights

- Novel biomaterials will enable reinsertion of military persons into active service at a rate several times faster than current techniques.
- Performance indicators will enable timely and appropriate intervention and maintain readiness and fitness assessment.
- Pharmaceuticals and performance enhancers will protect the combat force from endemic disease and extend mission performance capability.
- Biological tests will decrease combat morbidity and mortality by a factor several times faster than existing techniques.
- Prompt protection of human systems through analysis and assessment of food, water, and environmental factors will reduce the occurrence of disabling human disease and increase capability in a fighting force with fewer combatants.

OVERVIEW

Technological developments in novel materials, containment capability, and telecommunications have improved survivability and effectiveness of military forces. New insights regarding the molecular mechanisms involved in pathogenicity have given rise to countermeasures that can be employed if people are exposed to infectious diseases or agents. During the next two decades, novel material developments will provide filters that bind and inactivate biological agents, with minimal reduction of air flow, thereby protecting soldiers in the field or in contained vehicles. Telecommunications and the development of haptic devices are key to the introduction of remote surgical procedures that medical corpsmen can perform under the guidance of surgeons located at a distance from the battle arena. These methods of telemedicine have applications in training physicians and in delivering medicine to people residing in less densely populated areas. This innovation is important because early effective treatment of trauma provides higher survival rates and because injury of forces requires extensive deployment of support forces. It is also important because the military forces are being downsized and because physicians are more frequently deployed in sites distant from battle. The Biomedical Science section of Part III of the Militarily Critical Technologies List (MCTL) has an extensive consideration of telemedicine.

Biotechnology advances have also provided insights into the molecular basis of immune protection, vascular fluid loss, and neural regeneration. Enhanced immune competency permits forces to function at maximal strength after deployment. Sick calls are normally a problem during the first few weeks after deployment. Enhancement of the immune response by vaccination with super-antigens, by treatment with biological response modifiers (BRMs), or by food additives may ameliorate the problem. Extensive fluid loss is a primary cause of loss of life from traumatic injury. Applications of newly developed compatible blood clotting biopolymers, administered intra-abdominally to stanch blood loss, will help to ensure the survivability of ground forces. Spinal cord injuries constitute a major cause of long-term disability because current techniques do not permit regeneration of severed Central Nervous

System (CNS) connections. New techniques, using growth factors and biocompatible guidance materials, are being developed to facilitate regrowth and repair of the CNS.

The following subareas are critical to increasing individual and group protection and survivability:

- Biomedical imaging and transporting pods
- Encapsulation
- Pathogen Genome Project and pathogenicity islands (PAIs)
- Blood substitutes and biocompatible clotting matrices
- Remote locator of persons to permit rescue of captured personnel
- Water purification
- Biomarkers for toxicant/stress exposure.

Enhanced resistance of people to disease organisms (biological agents and endemic organisms) before and after deployment/exposure to impact trauma and to burn injury and the subsequent reduction in sick calls are goals to be achieved by the newly emerging biotechnologies and medical advances. The Biomedical Science section of Part III of the MCTL discusses the role of environmental toxicants on performance and the protection of people from such toxicants.

Enhanced prophylaxis can be achieved by using active vaccination against biological agents or infectious agents endemic in deployed areas. Vaccination with super-antigens or advanced adjuvants (i.e., improved antigen presentation) can upregulate the immune system before deployment and provide protection in a 4-day period rather than the usual 10-day post-vaccination period. The enabling technologies include genomic sequencing of all known threat agents and infectious disease organisms; multi-component, multi-valent vaccination systems to upregulate the immune system; and the development of immune response modifiers, including interferons and interleukins, having the potential to enhance immune response under crisis situations. The sequencing of PAIs can reduce the total number of vaccines needed.

New technologies that protect people before deployment include DNA vaccines. These vaccines have become available as an additional technology to the current vaccine systems that have eliminated small pox and controlled childhood diseases.

Recent advances have been realized in technology and molecular biology to treat personnel in the 12- to 72-hour period after exposure to B agents. Some of these advances followed research on the treatment of AIDS. People can be protected from the clinical signs of viral infection after exposure to agents because the virus in the lung or intestinal system must attack cells at secondary sites, such as the kidney, nervous system, blood cell, or other organ system. For the virus to attack successfully, it must enter the cell and must often travel to the nucleus. Then, the newly synthesized nucleic acid must be coated by the protein capsid. Cysteine protease inhibitors can block entry of the virus to cells, inhibitors of cytoplasmic nuclear transport can inhibit the second step, and inhibitors of molecular chaperones can inhibit the nucleic acid-viral protein coat interaction.

New sets of antibiotics that inactivate PAIs may have utility in individual and group protection. These antibiotics can be anti-sense materials or chemicals similar to traditional antibiotics. Developing antibiotics that have not been used to eliminate the development of drug resistance in the general population is important.

RATIONALE

A defensive capability against biological threat agents provides a nation with the technology to continue operations in a contaminated environment, thereby diminishing the likelihood that such agents will compromise a well-protected military force. New understandings of the human organism have led to the development of pharmaceuticals that sustain human performance. Information about the human genome also provides a new capability to determine peoples' susceptibility to naturally occurring toxicants and infectious agents. Biotechnology has the capability to fashion materials that bind toxicants with a high affinity and, thereby, inactivate such material. An awareness of the peoples' predisposition to environmental toxicants and of the nature of materials that bind such chemicals will allow military capability to be maintained in a symmetric manner and provide maximum deterrence. These technologies have broad applications in the civilian chemical industry, hazardous industrial operations, and the detection of contraband. The interface of medicine and space satellite communication has given rise to a new field: telemedicine. This new area will markedly improve the surgical care provided to those in the military who experience traumatic injury, blood loss, and shock. It will permit placement of support physicians at a distance from the front line of combat while providing continuous visual, auditory, and haptic input to medics in the field.

TECHNOLOGY DESCRIPTIONS

Biological Imaging and Transporting Pods (Technology Item J in Table 3.0-2)

Pod-like units that permit patients who have been exposed to biological agents to be transferred to remote sites have been built. This protects troop cohorts from infection. The pods have a controlled environment, can be incorporated into a battle dress uniform (BDU), are lightweight, and can be decontaminated and reused. The pod contains a dosimeter of chemical and biological agent exposure (similar to a film badge for radiation).

The pod is a bi-directional system that reports the personal information and also receives information about the environment, allowing the individual to take the appropriate action. This does not necessarily require individual-level monitoring if the device can be electronically linked to a nearby area monitor or sensor. The pod can encapsulate a patient in a small, automated decontamination (DECON) pod that can be managed by a crew smaller than that needed to encapsulate a large DECON team.

Encapsulation (Technology Item L in Table 3.0-2)

Many biologically active materials lose activity when exposed to oxidizing environments or to dilution in aqueous environments. Encapsulation of the biologically active material in a variety of matrices (e.g., liposomes, lactides, hydrogels) stabilizes the biomaterials and can provide a delivery vehicle for slow release. The slow release can be activated by enzymes normally found in the body or by pulsed electromagnetic signals. The release is a function of the chemical nature of the encapsulating material. Encapsulation can also provide targeted delivery of the biologically active material to specific organs if the capsules are coated with antibodies, lectins, or polycations. Encapsulation technology has potential applications in delivering taggants for a variety of military applications.

Pathogen Genome Project and Pathogenicity Islands (PAIs) (Technology Item N in Table 3.0-2)

The entire genome of many pathogenic organisms and of identified biological agents has been determined. As a result, the genomic sequences responsible for increased virulence and PAIs of these organisms have been characterized. It is apparent that the number of PAIs is smaller than the number of biological agents because the same PAIs are shared by several organisms. This observation can lead to the development of new antibiotics and anti-sense technologies to reduce virulence.

Blood Substitutes and Biocompatible Clotting Matrices (Technology Item S in Table 3.0-2)

Polymers having oxygen-carrying capabilities are available. Many of these are fluorinated compounds but have a disadvantage of showing hepatotoxicity. Biopolymeric oxygen-carrying compounds and clot-inducing compounds are being developed with anticipated product availability in the 2005 time period. These new products should have universal acceptor compatibility and should not require human donors.

Remote Locator of Persons To Permit Rescue of Captured Personnel (Technology Item T in Table 3.0-2)

Compounds may be developed that, when metabolized, will result in the excretion of compounds that are strong chromophores. Operators must be able to turn on or off for covert missions. The signal generator can be placed internally so it cannot "easily" be removed by the enemy for counter use. The device will be low cost and contain a physiological "honesty" checker so if "scared" Johnny cuts his finger, he will not activate this device unnecessarily and draw needed assets. The system may be used to locate a sailor overboard.

Water Purification (Technology Item U in Table 3.0-2)

Current technology requires use of reverse osmosis (RO). Desalinization of sea water and purification of contaminated water can be achieved using a combination of filtration and RO. New technologies that are effective in defouling and regenerating the capillary tubes in the RO system are being developed.

Transgenic flora and fauna are being developed to produce renewable resources (water, food, therapeutics). Microbes capable of bioremediation and terraforming terrain to shape the environment are being examined.

Biomarkers for Toxicant/Stress Exposure (Technology Items M,P,V in Table 3.0-2)

Humans and other animals synthesize a variety of biological compounds when challenged with fever, infection, or emotional stress. The synthesized compounds may be of low molecular weight (e.g., adrenalin) or high molecular weight (e.g., acute phase proteins, glutathione transferases, cytochrome P 450). Exposure of individuals to stress, exhaustion, or other adverse conditions also results in changes in electrophysiological and behavioral characteristics of the affected people. The changes in the profiles of body chemicals, electrophysiological patterns, and behavior may be used as signatures to evaluate fitness and readiness.

ADDITIONAL DATA

Tables 3.4-1A and 3.4-1B present additional data on this developing critical technology.

WORLDWIDE TECHNOLOGY ASSESSMENT (WTA) (See Figure 3.4-1)

Nations with advanced capability in individual and group protection include France, Germany, Israel, the Netherlands, Sweden, Switzerland, the United Kingdom, and the United States. Other nations have advanced technologies in pharmaceuticals or biopolymer grafting onto fibers. This area also requires leading-edge communications and space satellite industries for the telemedicine component. France, Japan, Russia, the United Kingdom, and the United States have strength in this area although the economic realities in Russia have diminished their strength. China is rapidly increasing capability in the space satellite area and, as a consequence, will have capability in telemedicine. Many leaders in these technologies are multi-national (e.g., Celanese, Hoechist, Dupont), and the rapid advances are not nation specific.

Table 3.4-1A. Individual and Group Protection Militarily Critical Technologies

Technology	Potential Developing Critical Technology Parameter	Rationale	Critical Materials	Technical Issues	Joint Vision 2010 DoD S&T Plan	Unique Test, Production, and Inspection Equipment
Encapsulation	Releases chemical or biological response modifiers on demand (near term).	Biologically active materials, such as vaccines, antitoxins and performance sustainers, can be encapsulated, inserted in soldier, and released on demand.	Encapsulation vehicles that stabilize biochemicals and release high percentage of active biological (greater than 80 percent).	Develop matrices that bind biological materials to delay release.	Full dimensional protection.	None identified.
Novel antivirals and antibiotics (see Biomedicine section, Part III MCTL)	Intervenes in infectious entry into target and infection of secondary organ systems; replicates in cells and releases infectious particles (near term to midterm).	Antivirals and novel antibiotics can protect person exposed to infectious agents from clinical signs of disease.	Antivirals and novel antibiotics.	Inhibit viral uptake; intracellular transport and maturation; create new antibacterials and reduce development of antibiotic resistance.	Full dimensional protection.	None identified.
Organ culture; tissue growth (see Biomedicine section, Part III MCTL	Replaces organs (midterm).	Traumatic injury results in long-term damage to internal organs and neural function. Replacing heart valves and bridging of transected spinal cord can lead to restored function.	Production of organ stem cells and organs that are accepted in immune-competent people.	Grow organs that will not be subjected to tissue rejection; grow organs in large enough mass to assume function of destroyed abdominal organs.	Full dimensional protection.	None identified.
Temperature and biotic system controlled transport pods	Develops pod that can stanch bleedings, have sterile environment, and contain B agents (midterm).	Protect troop cohort from B-agent-contaminated individuals.	Materials suitable for filtration, body covering, weight bearing, shock absorbing, and humidity control.	Maximize survivability of exposed/wounded soldier while eliminating exposure of cohort.	Full dimensional protection.	None identified.

Table 3.4-1B. Individual and Group Protection Militarily Critical Technologies

Technology	Military Applications	Unique Software	Center of Technology Development: Military or Commercial	Commercial Applications	Commercial Technology Requires Development for Military Use	Access To Technology	Other Important Data
Encapsulation	Provides protection against B agent; SUSOPS; sustains performance.	None identified.	Biomedical, pharmaceutical; cosmetics; drug delivery.	Biomedical, pharmaceutical; cosmetics; drug delivery.	Drug delivery; vaccination.	Ready access.	None identified.
Novel antivirals and antibiotics (see Biomedicine section, Part III MCTL)	Protects soldier from infectious agents.	None identified.	Biomedical and pharmaceutical industry and food industry.	Biomedical and pharmaceutical industry and food industry.	Health, food quality.	Ready access.	None identified.
Organ culture; tissue growth (see Biomedicine section, Part III MCTL	Stanches abdominal bleeding; repairs injuries to spinal cord, brainstem, and brain.	None identified.	Health care dominates.	Health care; pharmaceutical production.	Medical organ replacement.	Ready access.	None identified.
Temperature and biotic system controlled transport pods	Allows reinsertion of soldier; protects group.	Vital sign monitor.	Primarily military; emergency preparedness teams in isolated locations.	Biomedical.	Rural medical emergencies and preparedness tasks.	Ready access.	None identified.

Country	Individual Protection (Encapsulation, PAIs, Blood Substitutes, Biomarkers)	Group Protection (Biological Imaging and Transporting Pods, Remote Locator, Water Purification)
Australia	●●	●●
Canada	●●●	●●●
China	●●	●
Cuba	●●	●
Czech Republic	●●●	●●
Egypt	●	●
France	●●●●	●●●
Germany	●●●●	●●●
Hungary	●●	●
India	●●	●
Iran	●	●
Iraq	●	●
Israel	●●●	●●●●
Italy	●●	●●
Japan	●●●	●●●
Malaysia	●	●
Netherlands	●●●●	●●●●
North Korea	●	●
Norway	●●	●●
Poland	●●	●
Russia	●●●	●●●
Singapore	●●	●
South Korea	●●	●
Sweden	●●●●	●●●
Switzerland	●●●●	●●●●
Syria	●	●
Taiwan	●●	●
Ukraine	●●	●
United Kingdom	●●●●	●●●●
United States	●●●●	●●●●

Legend:

Capability in Technology Elements: ●●●● Most ●●● Many ●● Some ● At Least One

Figure 3.4-1. Individual and Group Protection Technologies Area Worldwide Capability Assessment